T0189131

Signals and Communication Technology

More information about this series at http://www.springer.com/series/4748

Jose Maria Giron-Sierra

Digital Signal Processing with Matlab Examples, Volume 3

Model-Based Actions and Sparse Representation

Jose Maria Giron-Sierra
Systems Engineering and Automatic Control
Universidad Complutense de Madrid
Madrid
Spain

ISSN 1860-4862 ISSN 1860-4870 (electronic)
Signals and Communication Technology
ISBN 978-981-10-9644-0 ISBN 978-981-10-2540-2 (eBook)
DOI 10.1007/978-981-10-2540-2

Softcover reprint of the hardcover 1st edition 2016

Printed on acid-free paper

This Springer imprint is published by Springer Nature
The registered company is Springer Nature Singapore Pte Ltd.
The registered company address is: 152 Beach Road, #22-06/08 Gateway East, Singapore 189721, Singapore

Preface

This is the third book of a trilogy. As in the other books, a series of MATLAB programs are embedded in the chapters for several purposes: to illustrate the techniques, to provide implementation examples, to encourage for personal exploration departing from a successful start.

The book has two parts, each having just one chapter. These chapters are long and have a considerable number of bibliographic references.

When using a GPS on a car, sometimes it is not possible to keep contact with satellites, like for instance inside tunnels. In this case, a model of the car motion—a dynamic model—can be used for data substitution. The adequate combination of measurements and models is the key idea of the Kalman filter, which is the central topic of the first part of the book. This filter was formulated for linear conditions. There are modifications for nonlinear conditions, like the extended Kalman filter, or the unscented Kalman filter. A new idea is to use particle filters. These topics are covered in the chapter under an important perspective: Bayesian filtering.

Compressed sensing has emerged as a promising idea. One of the intended applications is networked devices or sensors, which are becoming a reality. This topic is considered in the second part of the book. Some experiments that demonstrate image denoising applications were included.

For easier reading of the book, the longer programs have been put in an appendix. And a second appendix on optimization has been added to support some contents of the last chapter.

The reader is invited to discover the profound interconnections and commonalities that exist behind the variety of topics in this book. This common ground would become surely the humus for the next signal processing future.

As said in the preface of the other books, our particular expertise on signal processing has two main roots: research and teaching. I belong to the Faculty of Physics, University Complutense of Madrid, Spain. During our experimental research on autonomous vehicles, maritime drones, satellite control, etc., we practiced main methods of digital signal processing, for the use of a variety of sensors and for prediction of vehicle motions. From years ago, I teach Signal Processing in a Master of Biomedical Physics, and a Master on New technologies.

The style of the programs included in the book is purposively simple enough. The reader is invited to typeset the programs included in the book, for it would help for catching coding details. Anyway, all programs are available from the book web page: www.dacya.ucm.es/giron/SPBook3/Programs.

A lot of different materials have been used to erect this book: articles, blogs, codes, experimentation. I tried to cite with adequate references all the pieces that have been useful. If someone has been forgotten, please contact me. Most of the references cited in the text are available from Internet. We have to express our gratitude to the public information available in this way.

Please, send feedback and suggestions for further improvement and support.

Acknowledgments

Thanks to my University, my colleagues and my students. Since this and the other book required a lot of time taken from nights, weekends and holidays, I have to sincerely express my gratitude to my family.

Madrid, Spain Jose Maria Giron-Sierra

Contents

List of Figures

Listings

Part I
Model-Based Actions: Filtering, Prediction, Smoothing

Chapter 1
Kalman Filter, Particle Filter and Other Bayesian Filters

1.1 Introduction

Consider the case of satellite tracking. You must determine where is your satellite, using a large antenna. Signals from the satellite are noisy. Measurements of antenna angles have some uncertainty margins. But you have something that may help you: the satellite follows a known orbit, so at time T must be at position P. However this help should be taken with caution, since there are orbit perturbations.

Satellite tracking is an example of a more general scenario. The target is to estimate the state of a dynamic system, a state that is changing along time. The means you have are measurements and a mathematical model of the system dynamics. These two means should be combined as best as possible.

Some more examples could be useful to capture the nature of the problem to be considered in this chapter.

According with the description given in a scientific conference, a research team was developing a small *UAV* (*unmanned aerial vehicle*) to fly on their university campus. They used distance measurement with an ultrasonic sensor, for flight altitude control. Sometimes the altitude measurements were lost or completely wrong. In such moments, these measurements were substituted by a reasonable value. At least a simplistic (or even implicit) model should be used here for two reasons: to determine that there is a wrong measurement, and to obtain a reasonable value.

Nowadays a similar problem is found with vehicular GPS. In certain circumstances of the travel—for instance, a tunnel—the connection with satellites is lost, but the information to the driver should continue.

Another example is the case of biosignals. When measuring electroencephalograms (EEG) or electrocardiograms (ECG), etc., sometimes artifacts appear due to bad electrodes contact, eye blinking, interferences, etc. These bad measurements should be correctly identified as outliers, and some correction procedure should be applied.

This chapter deals with optimal state estimation for dynamic systems. The Bayesian methodology provides a general framework for this problem. Along time,

J.M. Giron-Sierra, *Digital Signal Processing with Matlab Examples, Volume 3*,
Signals and Communication Technology, DOI 10.1007/978-981-10-2540-2_1

a set of important practical methods, which can be seen as Bayesian instances, have been developed, such as the Kalman filter and the particle filter.

Given a dynamic system, the **Bayesian approach** to state estimation attempts to obtain the posterior PDF of the sate vector using all the available information. Part of this information is provided by a model of the system, and part is based on measurements. The state estimation could be done with a recursive filter, which repeats a prediction and an update operation.

Denote the posterior PDF at time step k as $p(\mathbf{x}_k | \mathbf{Y}_k)$, where $\mathbf{Y}_k$ is the set of all previous measurements $\mathbf{Y}_k = \{\mathbf{y}_j,\ j = 1, 2, \ldots, k\}$.

The prediction operation propagates the posterior PDF from time step $k - 1$ to k, as follows:

$$\underbrace{p(\mathbf{x}_k | \mathbf{Y}_{k-1})}_{A} = \int \underbrace{p(\mathbf{x}_k | \mathbf{x}_{k-1})}_{B}\ \underbrace{p(\mathbf{x}_{k-1} | \mathbf{Y}_{k-1})}_{C}\, d\mathbf{x}_{k-1} \tag{1.1}$$

where A is the prior at k, B is given by the system model, and C is the posterior at $k - 1$.

The update operation takes into account the new measurement $\mathbf{y}_k$ at k:

$$\underbrace{p(\mathbf{x}_k | \mathbf{Y}_k)}_{D} = \left(\underbrace{p(\mathbf{y}_k | \mathbf{x}_k)}_{L}\ \underbrace{p(\mathbf{x}_k | \mathbf{Y}_{k-1})}_{A} \right) \Big/ \underbrace{p(\mathbf{y}_k | \mathbf{Y}_{k-1})}_{N} \tag{1.2}$$

where D is the posterior, L is the measurement likelihood, and N is a normalizing denominator.

The system model could be of the form:

$$\mathbf{x}(k+1) = \mathbf{f}(\mathbf{x}(k),\ \mathbf{u}(k),\ \mathbf{w}(k)) \tag{1.3}$$

$$\mathbf{y}(k) = \mathbf{h}(\mathbf{x}(k),\ \mathbf{v}(k)) \tag{1.4}$$

where $\mathbf{w}(k)$ is the process noise, and $\mathbf{v}(k)$ is the observation noise.

The first equation of the system model can be used for the term B, and the second equation for the term L.

If some cases, with linear dynamics and measurement, the system model can be a Gauss-Markov model:

$$\mathbf{x}(n+1) = A\,\mathbf{x}(n) + B\,\mathbf{u}(n) + \mathbf{w}(n) \tag{1.5}$$

$$\mathbf{y}(n) = C\,\mathbf{x}(n) + \mathbf{v}(n) \tag{1.6}$$

Notice that, in order to align with the notation commonly employed in the Bayesian filters literature, we denote vectors in boldface letters (in previous chapters we used bars over letters).

The Kalman filter [55] offers an optimal solution for state estimation, provided the system is linear, and noises are Gaussian. It is a particular case of the Bayesian recursive filter.

In more general cases with non-linear dynamics, the system model is based on the two functions $\mathbf{f}(\cdot)$ and $\mathbf{h}(\cdot)$. A linearization strategy could be applied to still use the Kalman filter algorithm, and this is called *'Extended Kalman Filter'* (EKF). An alternative is to propagate a few special points (*'sigma points'*) through the system equations; with these points it is possible to approximate the prior and posterior PDFs. An example of this alternative is the *'Unscented Kalman Filter'* (UKF). For non-linear/non-Gaussian cases that do not tolerate approximations, the *'Particle Filter'* could be used; the filter is based on the propagation of many points.

The chapter starts with the standard Kalman filter, after some preliminaries. Then nonlinearities are considered, and there are sections on EKF, UKF, and particle filters. All these methods are naturally linked to the use of computers or digital processors, and so the world of numerical computation must be visited in another section. Smoothing is another important field that is treated in Sect. 1.10. The last sections intend to at least introduce important extensions and applications of optimal state estimation.

Some limits should be admitted for this chapter, and therefore there is no space for many related topics, like H-infinity, game theory, exact filters, etc. The last section offer some links for the interested reader.

The chapter is mainly based on [8, 15, 19, 40, 72, 96].

In general, many problems related with state estimation remain to be adequately solved. The field is open for more research.

1.2 Preliminaries

The main stream of this chapter is estimation. Therefore, it is convenient to harvest relevant concepts and results about it.

In particular, the Kalman filter is concerned with joint distributions of states $\mathbf{x}$ and observations $\mathbf{y}$ (both are stochastic variables).

The basic mathematical treatment of stochastic variables uses the following magnitudes:

- Scalar case:

$$mean: \ \mu_x = E(x(n))$$

$$variance: \ \sigma_x^2 = E((x(n) - \mu_x)^2)$$

- Vector case (vectors in column format):

$$covariance\ matrix: \ \Sigma_x = E((\mathbf{x}(n) - \boldsymbol{\mu}_x)(\mathbf{x}(n) - \boldsymbol{\mu}_x)^T)$$

A typical element σ_{ij} of a covariance matrix is:

$$\sigma_{ij} = E((x_i(n) - \mu_{xi})(x_j(n) - \mu_{xj})^T)$$

- Two processes:

$$cross - covariance\ matrix:\ \Sigma_{xy} = E((\mathbf{x}(n) - \boldsymbol{\mu}_x)(\mathbf{y}(n) - \boldsymbol{\mu}_y)^T)$$

Suppose there is a function $\mathbf{f}(\mathbf{y})$ that gives us an estimation of $\mathbf{x}$, the variance of the estimation (a scalar value) would be:

$$E(||\,\mathbf{f}(\mathbf{y}) - \mathbf{x}||^2)$$

It can be shown that this variance is minimized when:

$$\mathbf{f}(\mathbf{y}) = E(\mathbf{x}|\mathbf{y})$$

Therefore, the minimum variance estimate (MVE) of $\mathbf{x}$, given $\mathbf{y}$ is the conditional expectation:

$$\hat{\mathbf{x}} = E(\mathbf{x}|\mathbf{y})$$

In other words, the conditional mean gives the MVE.

The Kalman filter is a linear minimum variance estimator (LMVE). If the process and observation noises are Gaussian, the filter gives the MVE.

Note that the MAP estimate is the **mode** of the *a posteriori* PDF, while the MVE is the **mean** of the *a posteriori* PDF. In some applications there are several modes, and specific filters–modal filters-should be employed. In other applications, it may happen that the mean differs from the mode. Bayesian estimation obtains the MAP estimation. In the case of linear systems with process and observation Gaussian noises, the mode and the mean are coincident.

Starting from an initial state, the state of dynamic systems evolves along time. The estimation of this state is linked to propagation of states and noises. It would be a good exercise for the reader to simulate this propagation using a linear state model with noises.

When taking data for humanoid robot research, it is common to record from top the motion of a walking man. It is not purely straight, even trying to follow a line. There are state oscillations. The same happen with satellites, it was supposed to follow a precise orbit, but there are oscillations. Notice that these are real oscillations. Process noise corresponds to real perturbations.

Measurement noise tends to confound you, not giving the real state. A main target of state estimation is to filter out this problem. And speaking of problems, when possible it is better to use good measurements, avoiding sources of noise, and trying to obtain linear measurements. As it would be demonstrated in this chapter, nonlinearities usually cause estimation difficulties. Sometimes change of variables could help to obtain linear conditions.

1.2.1 A Basic Example

This is an example taken from [95]. It is the case of driving a car, keeping a distance x from the border of the road (Fig. 1.1).

In this example is useful to take into account that:

$$N(\mu_1, \Sigma_1) \times N(\mu_2, \Sigma_2) = const \times N(\mu, \Sigma) \tag{1.7}$$

with:

$$\begin{aligned} \mu &= \Sigma\, \Sigma_1^{-1}\, \mu_1 + \Sigma\, \Sigma_2^{-1}\, \mu_2 \\ \Sigma^{-1} &= \Sigma_1^{-1} + \Sigma_2^{-1} \end{aligned} \tag{1.8}$$

(it is usual in the literature to denote the Gaussian PDF as $N(\mu, \Sigma)$).

At time 0, a passenger guess that the distance to the border is $y(0)$. It could be considered as an inexact measurement with a variance σ_{y0}^2. One could assume a Gaussian conditional density $f(x|y(0))$ with variance σ_{y0}^2. In this moment, the best estimate is $\hat{x}(1|0) = y(0)$.

At time 1, the driver also takes a measurement $y(1)$, which is more accurate. Assume again a Gaussian conditional density $f(x|y(1))$, with $\sigma_{y1}^2 < \sigma_{y0}^2$. The joint distribution that combines both measurements (product of Gaussians) would have:

$$\mu_x = \frac{\sigma_{y1}^2}{\sigma_{y0}^2 + \sigma_{y1}^2}\, y(0) + \frac{\sigma_{y0}^2}{\sigma_{y0}^2 + \sigma_{y1}^2}\, y(1) \tag{1.9}$$

$$\frac{1}{\sigma_x^2} = \frac{1}{\sigma_{y0}^2} + \frac{1}{\sigma_{y1}^2} \tag{1.10}$$

The new estimate of the distance would be μ_x. The corresponding variance, σ_x^2, is smaller than either σ_{y0}^2 or σ_{y1}^2. The combination of two estimates is better.

Fig. 1.1 Keeping the car at a distance from the road border

By the way, the so-called '*Fisher information*' of a Gaussian distribution is the inverse of the variance: $1/\sigma^2$. Large uncertainty means poor information.

Now, write the new estimate as follows:

$$\hat{x}(1|1) = \mu_x = y(0) + \frac{\sigma_{y0}^2}{\sigma_{y0}^2 + \sigma_{y1}^2}\,(y(1) - y(0)) = \hat{x}(1|0) + K(1)(y1 - \hat{x}(1|0)) \tag{1.11}$$

where:

$$K(1) = \frac{\sigma_{y0}^2}{\sigma_{y0}^2 + \sigma_{y1}^2} \tag{1.12}$$

Equation (1.11) has the form of a *Kalman filter*; $K()$ is the *Kalman gain*. The equation says that the best estimate can be obtained using the previous best estimate and a correction term. This term compares the latest measurement with the previous best estimate.

Likewise, the variance could be written as follows:

$$\sigma_x^2(1|1) = \sigma_x^2(1|0) - K(1)\,\sigma_x^2(1|0) \tag{1.13}$$

so the variance decreases.

The car moves. A simple model of the lateral motion, with respect to the border of the road, could be:

$$\frac{dx}{dt} = u + w \tag{1.14}$$

The term w is random perturbation with variance σ_w^2. The lateral velocity is u, set equal to zero. After some time T, the best estimate (prediction) would be:

$$\hat{x}(2|1) = \hat{x}(1|1) + Tu \tag{1.15}$$

The variance of the estimate increases:

$$\sigma_x^2(2|1) = \sigma_x^2(1|1) + T\sigma_w^2 \tag{1.16}$$

The increase of variance is not good. A new measurement is welcome. Suppose that at time 2 a new measurement is taken. Again the product of Gaussians appears, as we combine the prediction $\hat{x}(2|1)$ and the measurement $y(2)$. Therefore:

$$\hat{x}(2|2) = \hat{x}(2|1) + K(2)(y2 - \hat{x}(2|1)) \tag{1.17}$$

$$\sigma_x^2(2|2) = \sigma_x^2(2|1) - K(2)\,\sigma_x^2(2|1) \tag{1.18}$$

where:

$$K(2) = \frac{\sigma_x^2(2|1)}{\sigma_x^2(2|1) + \sigma_{y2}^2} \tag{1.19}$$

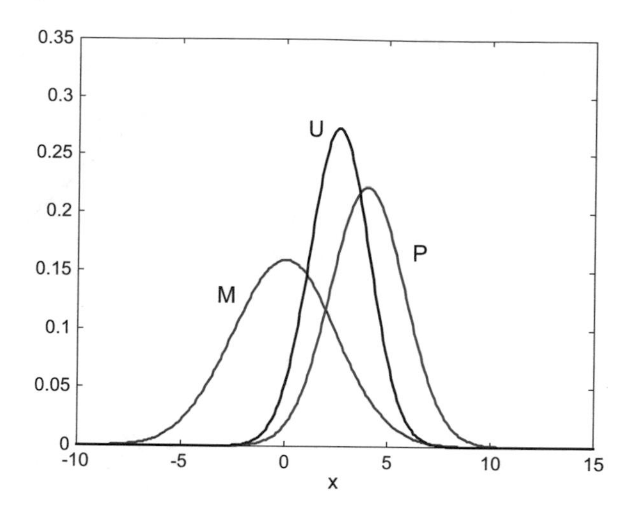

Fig. 1.2 Prediction (*P*), measurement (*M*) and update (*U*) PDFs

The philosophy of the Kalman filter is to obtain better estimates by combining prediction and measurement (the combination of two estimates is better). The practical procedure is to update the prediction with the measurement.

Figure 1.2 shows what happens with the PDFs of the prediction (P), the measurement (M) and the update (U). The update PDF has the smallest variance, so it is a better estimation.

Program 1.1 Combining prediction and measurement

```
%Combining prediction and measurement
%
xiv=0.1;
x=-10:xiv:15;
%prediction pdf
pmu=4; psig=1.8;
ppdf=normpdf(x,pmu,psig);
%measurement pdf
mmu=0; msig=2.5;
mpdf=normpdf(x,mmu,msig);
%combine pdfs
cpdf=ppdf.*mpdf;
%area=1
KA=1/(sum(cpdf)*xiv);
cpdf=KA*cpdf;
figure(1)
plot(x,ppdf,'b'); hold on;
plot(x,mpdf,'r'); plot(x,cpdf,'k');
title('combine prediction and measurement: update');
xlabel('x');
```

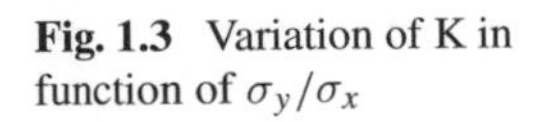

Fig. 1.3 Variation of K in function of σ_y/σ_x

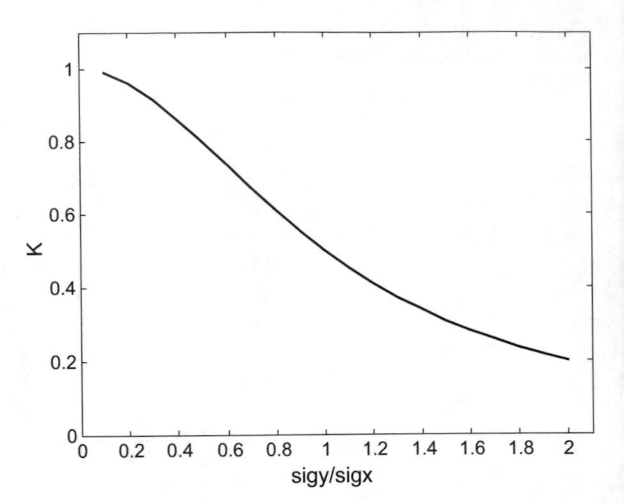

Another aspect of the Kalman filter philosophy is that, in order to estimate the present system state, the value of $K()$ is modulated according with the confidence offered by measurements or by prediction. Suppose that the prediction variance is constant, then, if measurement uncertainty increases, K decreases: it could be said that the correction exerted by K becomes more 'prudent'.

Figure 1.3 shows how K depends on σ_y/σ_x. The figure has been generated with the Program 1.17.

Program 1.2 K in function of sigmay/sigmax

```
%K in function of sigmay/sigmax
%
%values of sigmay/sigmax
r=0.1:0.1:2;
%reserve space
K=zeros(1,length(r));
figure(1)
K=1./(1+r.^2); %vectorized code
plot(r,K,'k');
title('K vs. sigy/sigx');
xlabel('sigy/sigx'); ylabel('K');
axis([0 2.1 0 1.1]);
```

1.2.2 Prediction with the Gauss-Markov Model

This subsection focuses on the propagation of mean and covariance. The basis of the next study is an important lemma, which applies to a partition of a set of Gaussian variables:

$$\mathbf{x} = \begin{pmatrix} \mathbf{x}_1 \\ \mathbf{x}_2 \end{pmatrix}$$

With:

$$\mu_x = \begin{pmatrix} \mu_{x1} \\ \mu_{x2} \end{pmatrix} \; ; \quad S_x = \begin{pmatrix} S_{11} S_{12} \\ S_{21} S_{22} \end{pmatrix}$$

Lemma: the conditional distribution of $\mathbf{x}_1(n)$, given a $\mathbf{x}_2(n) = \mathbf{x}_2^*(n)$ is Gaussian with:

$$mean = \mu_{x1} + S_{12}\, S_{22}^{-1}\, (\mathbf{x}_2 - \mu_{x2})$$

$$cov = \Sigma_{11} - \Sigma_{12}\, \Sigma_{22}^{-1}\, \Sigma_{21}$$

Now, let us focus on dynamic systems described with Gauss-Markov models.

The Gauss-Markov model deals with states $\mathbf{x}(n)$ and observations $\mathbf{y}(n)$ (also denoted as outputs or measurements). Here is the model:

$$\mathbf{x}(n+1) = A\,\mathbf{x}(n) + B\,\mathbf{u}(n) + \mathbf{w}(n) \tag{1.20}$$

$$\mathbf{y}(n) = C\,\mathbf{x}(n) + \mathbf{v}(n) \tag{1.21}$$

Let us use the following notation:

- $\hat{\mathbf{x}}(n+1)$ is the conditional mean of $\mathbf{x}(n+1)$ given a set of observations $\{\mathbf{y}(0), \mathbf{y}(1), \ldots \mathbf{y}(n)\}$.
- The state error is:

$$\mathbf{er}(n) = \mathbf{x}(n) - \hat{x}(n)$$

- The state error covariance is:

$$\Sigma(n) = E(\mathbf{er}(n) \cdot \mathbf{er}(n)^T)$$

Now, consider the following partition:

$$\begin{pmatrix} \mathbf{x}(n+1) \\ \mathbf{y}(n) \end{pmatrix} \tag{1.22}$$

Taking into account the model, the four covariance components are:

$$\Sigma_{xx} = A\,\Sigma(n)\,A^T + \Sigma_w \tag{1.23}$$

$$\Sigma_{xy} = A\,\Sigma(n)\,C^T + \Sigma_{wv} \tag{1.24}$$

$$\Sigma_{yx} = C\,\Sigma(n)\,A^T + \Sigma_{wv}^T \tag{1.25}$$

$$\Sigma_{yy} = C\,\Sigma(n)\,C^T + \Sigma_v \tag{1.26}$$

Thus, the mean and the variance of the partitioned process are the following:

$$\boldsymbol{\mu}_p = \begin{pmatrix} A\,\hat{\mathbf{x}}(n) \\ C\,\hat{\mathbf{x}}(n) \end{pmatrix} + \begin{pmatrix} B \\ 0 \end{pmatrix} \mathbf{u}(n) \tag{1.27}$$

$$\Sigma_p = \begin{pmatrix} A\,\Sigma(n)\,A^T + \Sigma_w & A\,\Sigma(n)\,C^T + \Sigma_{wv} \\ C\,\Sigma(n)\,A^T + \Sigma_{wv}^T & C\,\Sigma(n)\,C^T + \Sigma_v \end{pmatrix} \tag{1.28}$$

Now, according to the lemma, the conditional mean is:

$$\hat{\mathbf{x}}(n+1) = A\,\hat{\mathbf{x}}(n) + B\,\mathbf{u}(n) + K(n)\,(\mathbf{y}(n) - C\,\hat{\mathbf{x}}(n)) \tag{1.29}$$

And the conditional covariance is:

$$\Sigma(n+1) = [A\,\Sigma(n)\,A^T + \Sigma_w] - K(n)\,[C\,\Sigma(n)\,C^T + \Sigma_v]\,K^T(n) \tag{1.30}$$

where:

$$K(n) = \Sigma_{xy} \cdot \Sigma_{yy}^{-1} = [A\,\Sigma(n)\,C^T + \Sigma_{wv}] \cdot [C\,\Sigma(n)\,C^T + \Sigma_v]^{-1} \tag{1.31}$$

The last three equations provide a one-step prediction. The term $K(n)$ is called the *Kalman gain*.

And the important point is that since we know the conditional mean, then we have the minimum variance estimate (MVE) of the system state: this is the Kalman filter.

1.2.3 Continuation of a Simple Example of Recursive Wiener Filter

This subsection links with subsection (5.2.5) where a simple example of Wiener filter was described. In that example, no model of $x(n)$ was considered. A recursive estimation was obtained, with the following expression:

$$\hat{x}(n+1) = \hat{x}(n) + g(n+1)\,(y(n+1) - \hat{x}(n)) \tag{1.32}$$

As it was said in the second book, the above expression has the typical form of recursive estimators, where the estimation is improved in function of the estimation error.

In the preceding subsection the Kalman filter was derived by studying the propagation of means and variances. Notice that $\hat{\mathbf{x}}(n+1)$ was obtained using $\mathbf{y}(n)$. This is *prediction*.

Now, let us establish a second version of the filter where $\hat{\mathbf{x}}(n+1)$ is obtained using $\mathbf{y}(n+1)$. This is *filtering*. A scalar Gauss-Markov example will be considered. The derivation of the Kalman filter will be done based on minimization of estimation variance. This is a rather long derivation borrowed from [8]. Although long, it is an interesting deduction exercise that exploits orthogonality relations.

Let us proceed along three steps:

1. **Problem statement**

The scalar system is the following:

$$x(n+1) = A\,x(n) + w(n) \tag{1.33}$$

$$y(n) = C\,x(n) + v(n) \tag{1.34}$$

where A and C are constants (they are not matrices, however we prefer to use capital letters).
Assumptions:

$$x(0) = 0; \;\; w(0) = 0$$

$x(n)$ and $w(n)$ are uncorrelated

$v(n)$ and $w(n)$ are uncorrelated, and zero-mean
The following recursive estimation is proposed:

$$\hat{x}(n+1) = f(n+1)\,\hat{x}(n) + k(n+1)\,y(n+1) \tag{1.35}$$

(it will reach the typical form, after the coming development)
The estimation variance is:

$$\Sigma(n+1) = E\,\{(\hat{x}(n+1) - x(n+1))^2\} \tag{1.36}$$

To obtain the MVE:

$$\frac{\partial\,\Sigma(n+1)}{\partial\,f(n+1)} = 2\,E\,\{(\hat{x}(n+1) - x(n+1)) \cdot \hat{x}(n)\} = 0 \tag{1.37}$$

$$\frac{\partial\,\Sigma(n+1)}{\partial\,k(n+1)} = 2\,E\,\{(\hat{x}(n+1) - x(n+1)) \cdot y(n+1)\} = 0 \tag{1.38}$$

From these equations, one obtains the following orthogonality relations:

$$E\{e(n+1)\,\hat{x}(n)\} = 0 \tag{1.39}$$

$$E\{e(n+1)\,y(n+1)\} = 0 \tag{1.40}$$

where we introduced the estimation error as $e(n) = \hat{x}(n) - x(n)$.

2. **The $f(n+1)$ term**
Let us focus on the $f(n+1)$ term. Taking into account the estimator expression, the variance derivatives would be:

$$\frac{\partial\,\Sigma(n+1)}{\partial\,f(n+1)} = 2\,E\,\{(f(n+1)\,\hat{x}(n)\ + k(n+1)y(n+1) - x(n+1))\cdot \hat{x}(n)\,\} = 0 \tag{1.41}$$

$$\frac{\partial\,\Sigma(n+1)}{\partial\,k(n+1)} = 2\,E\,\{(f(n+1)\hat{x}(n)\ + k(n+1)y(n+1) - x(n+1))\cdot y(n+1)\,\} = 0 \tag{1.42}$$

Take Eq. (1.41), add and subtract $f(n+1)\,x(n)$. After reordering:

$$\begin{aligned} &E\{[f(n+1)\,(\hat{x}(n) - x(n)) + f(n+1)\,x(n)]\cdot \hat{x}(n)\} = \\ &= E\{[x(n+1) - k(n+1)\,y(n+1)]\cdot \hat{x}(n)\} \end{aligned} \tag{1.43}$$

Since $y(n+1) = C\,x(n+1) + v(n+1)$:

$$\begin{aligned} &f(n+1)\,E\{[e(n) + x(n))]\cdot \hat{x}(n)\} - \\ &= E\{[x(n+1)(1 - C\,k(n+1)) - k(n+1)\,v(n+1)]\cdot \hat{x}(n)\} \end{aligned} \tag{1.44}$$

Due to $E\{v(n+1)\,\hat{x}(n)\} = 0$, and to the rest of orthogonality relations, the previous equation reduces to:

$$f(n+1)\,E\{x(n)\,\hat{x}(n)\} = (1 - C\,k(n+1))E\{x(n+1)\hat{x}(n)\} \tag{1.45}$$

Since $x(n+1) = A\,x(n) + w(n)$:

$$E\{x(n+1)\hat{x}(n)\} = E\{\,(A\,x(n) + w(n))\,\hat{x}(n)\,\} = A\,E\{\,x(n)\,\hat{x}(n)\} \tag{1.46}$$

Therefore, Eq. (1.43) becomes:

$$f(n+1) = A\cdot(1 - C\,k(n+1)) \tag{1.47}$$

3. **The $k(n+1)$ term** Using the expression of $f(n+1)$ in (1.47), the estimator could be written as follows:

$$\hat{x}(n+1) = A\cdot\hat{x}(n) + k(n+1)\,(y(n+1) - A\,C\,\hat{x}(n)) \tag{1.48}$$

Now, let us find a convenient expression of the variance. From definitions, one obtains:

$$\Sigma(n+1) = E\{e^2(n+1)\} = E\{e(n+1)\,[\hat{x}(n+1) - x(n+1)]\} \quad (1.49)$$

Since:

$$\hat{x}(n+1) = f(n+1)\,\hat{x}(n) + k(n+1)\,y(n+1)$$

and using orthogonality relations:

$$\Sigma(n+1) = -E\{e(n+1)\,x(n+1)\} \quad (1.50)$$

Considering that:

$E\{e(n+1)\,y(n+1)\} = 0$ and $y(n+1) = C\,x(n+1) + v(n+1)$

one could write:

$$C\,E\{e(n+1)\,x(n+1)\} = -\,E\{e(n+1)\,v(n+1)\} \quad (1.51)$$

Then, coming back to the variance:

$$\begin{aligned}\Sigma(n+1) &= \tfrac{1}{C} E\{e(n+1)\,v(n+1)\} = \\ &= \tfrac{1}{C}\,E\{[f(n+1)\hat{x}(n) + k(n+1)\,y(n+1) - x(n+1)]\cdot v(n+1)\} = \\ &= \tfrac{1}{C}\,k(n+1)\,E\{y(n+1)\,v(n+1)\} = \tfrac{1}{C}\,k(n+1)\,\sigma_v^2\end{aligned} \quad (1.52)$$

Therefore:

$$k(n+1) = \frac{1}{\sigma_v^2}\,\Sigma(n+1) \quad (1.53)$$

Let us continue a bit with our development effort, in order to put $\Sigma(n+1)$ in function of $\Sigma(n)$. Start from definition:

$$\begin{aligned}\Sigma(n+1) &= E\{[\hat{x}(n+1) - x(n+1)]^2\} = \\ &= E\{[\,(A\,\hat{x}(n) + k(n+1)\cdot\,[y(n+1) - A\,C\,\hat{x}(n)]\,) - x(n+1)]^2\,\} = \\ &= E\{[\,[-A\cdot(1 - C\,k(n+1))\cdot e(n)] - [1 - C\,k(n+1)\,w(n)] + \\ &\quad + [k(n+1)\,v(n+1)]\,]^2\,\}\end{aligned}$$

After squaring most terms do cancel, because error and noises are uncorrelated. The result is:

$$\Sigma(n+1) = A^2\,[1 - C\,k(n+1)]^2\,\Sigma(n) + [1 - C\,k(n+1)]^2\,\sigma_w^2 + k^2(n+1)\,\sigma_v^2 \quad (1.54)$$

Finally, using (1.53):

$$k(n+1) = \frac{C\,[A^2\,\Sigma(n) + \sigma_w^2]}{\sigma_v^2 + C^2\,\sigma_w^2 + C^2 A^2\,\Sigma(n)} \tag{1.55}$$

To conclude this subsection, let us write a summary. The Kalman filter is given by the next three equations:

$$\hat{x}(n+1) = A \cdot \hat{x}(n) + k(n+1)\,(y(n+1) - C\,A\,\hat{x}(n)) \tag{1.56}$$

$$k(n+1) = \frac{C\,[A^2\,\Sigma(n) + \sigma_w^2]}{\sigma_v^2 + C^2\,\sigma_w^2 + C^2 A^2\,\Sigma(n)} \tag{1.57}$$

$$\Sigma(n+1) = [A^2\,\Sigma(n) + \sigma_w^2]\,(1 - C\,k(n+1)) \tag{1.58}$$

These equations can be easily generalized for multivariable systems.

1.3 Kalman Filter

The Kalman filter was introduced in 1960 [55]. Now, there are thousands of related publications. This fact reflects the eminent importance of the Kalman filter, which is a fundamental tool for many real life applications.

An overview of the academic literature shows that the Kalman filter can be derived in several ways. Likewise, there are several different notations that could be used.

This section establishes the Kalman filter equations, and then introduces a typical algorithm for its use. Then, there is a series of subsections devoted to important aspects related to the Kalman filter.

Let us recapitulate what we have seen, and write now a summary of the Kalman filter equations.

The system (or signal) model is:

$$\mathbf{x}(n+1) = A\,\mathbf{x}(n) + B\,\mathbf{u}(n) + \mathbf{w}(n) \tag{1.59}$$

$$\mathbf{y}(n) = C\,\mathbf{x}(n) + \mathbf{v}(n) \tag{1.60}$$

In the pair of equations above, the first equation is frequently called the *'transition equation'* (transition from one state to the next one), the second equation corresponds to measurement.

In order to simplify the expressions, let us introduce the following matrices:

$$M(n+1) = A\,\Sigma(n)\,A^T + \Sigma_w = \Sigma_{xx} \tag{1.61}$$

$$N(n+1) = C\,\Sigma(n)\,C^T + \Sigma_v = \Sigma_{yy} \tag{1.62}$$

The most used versions of the Kalman filter are the (at present) filter, and the one-step prediction filter.

(a) At present filter (AKF):

$$\hat{\mathbf{x}}(n+1) = A\,\hat{\mathbf{x}}(n) + K(n+1)\,[\mathbf{y}(n+1) - C\,A\,\hat{\mathbf{x}}(n)] \tag{1.63}$$

$$K(n+1) = M(n+1)\,C^T\,[C\,M(n+1)\,C^T + \Sigma_v]^{-1} \tag{1.64}$$

$$\Sigma(n+1) = M(n+1) - K(n+1)\,C\,M(n+1) \tag{1.65}$$

(b) One-step prediction filter (OPKF):

$$\hat{\mathbf{x}}(n+1) = A\,\hat{\mathbf{x}}(n) + B\,\mathbf{u}(n) + K(n)\,[\mathbf{y}(n) - C\,\hat{\mathbf{x}}(n)] \tag{1.66}$$

$$K(n) = [A\,\Sigma(n)\,C^T + \Sigma_{wv}]\,N(n+1)^{-1} \tag{1.67}$$

$$\Sigma(n+1) = M(n+1) - K(n)\,N(n+1)\,K(n)^T \tag{1.68}$$

Usually the literature does not includes Σ_{wv}, either because noises are uncorrelated, or because, if noises are correlated, the problem could be re-written to make this term disappear.

It is also common in the literature to replace names as follows:

$$P(n) = \Sigma(n)\,;\quad Q = \Sigma_w\,;\quad R = \Sigma_v \tag{1.69}$$

The noise covariance matrix could be written as:

$$\begin{pmatrix} Q & S \\ T & R \end{pmatrix} \tag{1.70}$$

Note that the AKF filter corresponds to the development in subsection (8.2.3), based on [8]; while the OPKF filter corresponds to subsection (8.2.2), based on [40].

1.3.1 *The Algorithm*

The standard way for the application of the Kalman filter is by repeating a two-step algorithm. This is represented with the diagram shown in Fig. 1.4.

The equations to be applied in each step are written below. They correspond to the AKF filter with input. Note slight changes of notation, which are introduced for easier translation to code.

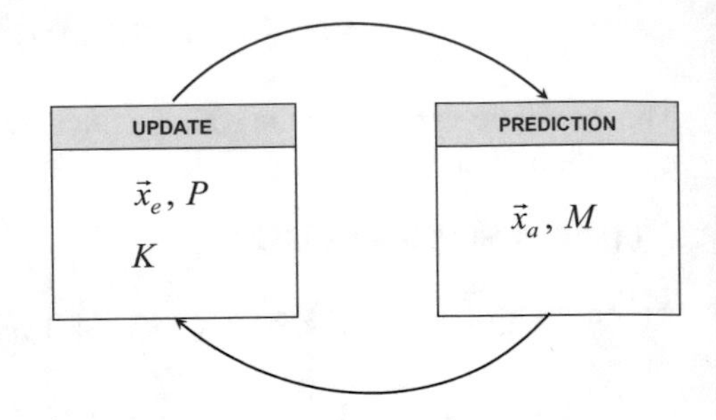

Fig. 1.4 The algorithm is a cycle

(a) Prediction

$$\mathbf{x}_a(n+1) = A\,\mathbf{x}_e(n) + B\,\mathbf{u}(n) \tag{1.71}$$

$$M(n+1) = A\,P(n)\,A^T + \Sigma_w \tag{1.72}$$

(b) Update

$$K(n+1) = M(n+1)\;C^T\;[C\,M(n+1)\,C^T + \Sigma_v]^{-1} \tag{1.73}$$

$$P(n+1) = M(n+1) - K(n+1)\;C\,M(n+1) \tag{1.74}$$

$$\mathbf{x}_e(n+1) = \mathbf{x}_a(n+1) + K(n+1)\,[\mathbf{y}(n+1) - C\,\mathbf{x}_a(n+1)] \tag{1.75}$$

(the subscript a refers to 'a priori').

It is interesting, also for comparison purposes, to continue with the example employed in the second book, in the section devoted to observers. It is again the two tank system. A change to the matrix C has been made, to allow for the measurement of both tanks height (Fig. 1.5).

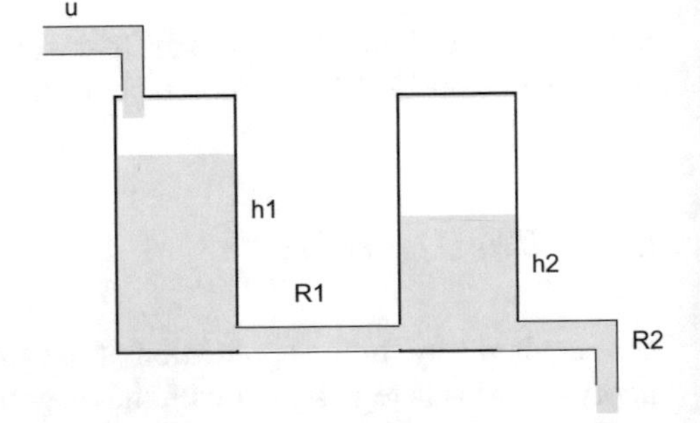

Fig. 1.5 A two-tank system example

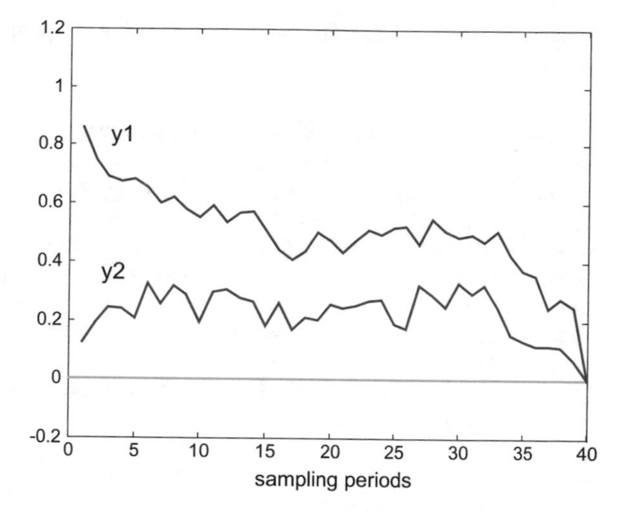

Fig. 1.6 System outputs (measurements)

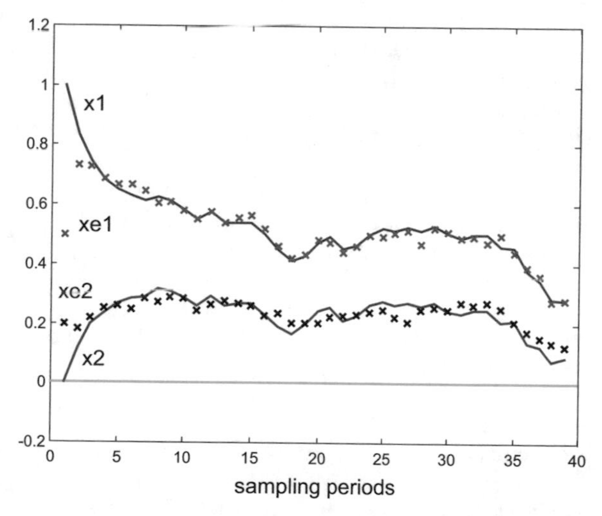

Fig. 1.7 System states, and states estimated by the Kalman filter

Program 1.3 includes an implementation of the AKF Kalman filter. The program is somewhat long because it has to prepare for the algorithm and reserve space for matrices and vectors.

The Program 1.3 includes a simulation of the noisy process. Figure 1.6 depicts the outputs of the system, which are the measurements of tank heights. A great virtue of the Kalman filter is that it is able to get good estimates of the states from severely corrupted measurements.

Figure 1.7 compares the evolution of the 2-variable system, in continuous curves, and the state estimation yield by the Kalman filter, depicted by x marks. Since perhaps the initial system state is not known, the program considers different initial states for the system and for the Kalman filter. The initial values of covariance matrices are set to zero, but it could be initialized to other values depending on previous knowledge about the process.

Program 1.3 Kalman filter example, in noisy conditions

```
%Kalman filter example, in noisy conditions
%state space system model (2 tank system):
A1=1; A2=1; R1=0.5; R2=0.4;
cA=[-1/(R1*A1) 1/(R1*A1); 1/(R1*A2) -(1/A2)*((1/R1)+(1/R2))];
cB=[1/A1; 0]; cC=[1 0; 0 1]; cD=0;
Ts=0.1; %sampling period
csys=ss(cA,cB,cC,cD); %setting the continuous time model
dsys=c2d(csys,Ts,'zoh'); %getting the discrete-time model
[A,B,C,D]=ssdata(dsys); %retrieves discrete-time model matrices
%simulation horizon
Nf=40;
%process noise
Sw=[12e-4 0; 0 6e-4]; %cov
sn=zeros(2,Nf);
sn(1,:)=sqrt(Sw(1,1))*randn(1,Nf);
sn(2,:)=sqrt(Sw(2,2))*randn(1,Nf);
%observation noise
Sv=[6e-4 0; 0 15e-4]; %cov.
on=zeros(2,Nf);
on(1,:)=sqrt(Sv(1,1))*randn(1,Nf);
on(2,:)=sqrt(Sv(2,2))*randn(1,Nf);
% system simulation preparation
%space for recording x1(n), x2(n)
x1=zeros(1,Nf-1); x2=zeros(1,Nf-1);
x=[1;0]; % state vector with initial tank levels
u=0.4; %constant input
% Kalman filter simulation preparation
%space for matrices
K=zeros(2,2,Nf); M=zeros(2,2,Nf); P=zeros(2,2,Nf);
%space for recording er(n), xe1(n), xe2(n),ym(n)
er=zeros(2,Nf-1); xe1=zeros(1,Nf-1);
xe2=zeros(1,Nf-1);rym=zeros(2,Nf);
xe=[0.5; 0.2]; % filter state vector with initial values
%behaviour of the system and the Kalman filter
% after initial state
% with constant input u
for nn=1:Nf-1,
x1(nn)=x(1); x2(nn)=x(2); %recording the system state
xe1(nn)=xe(1); xe2(nn)=xe(2); %recording the observer state
er(:,nn)=x-xe; %recording the error
%
%system simulation
xn=(A*x)+(B*u)+sn(nn); %next system state
x=xn; %system state actualization
ym=(C*x)+on(:,nn); %output measurement
```

```
rym(:,nn)=ym;
%
%Prediction
xa=(A*xe)+(B*u); %a priori state
M(:,:,nn+1)=(A*P(:,:,nn)*A')+ Sw;
%Update
K(:,:,nn+1)=(M(:,:,nn+1)*C')*inv((C*M(:,:,nn+1)*C')+Sv);
P(:,:,nn+1)=M(:,:,nn+1)-(K(:,:,nn+1)*C*M(:,:,nn+1));
%estimated (a posteriori) state:
xe=xa+(K(:,:,nn+1)*(ym-(C*xa)));
end;
%------------------------------------------
% display of system outputs
figure(1)
plot([0 Nf],[0 0],'g'); hold on; %horizontal axis
plot([0 0],[-0.2 1.2],'k'); %vertical axis
plot(rym(1,:),'r'); %plots y1
plot(rym(2,:),'b'); %plots y2
xlabel('sampling periods');
title('system outputs');
% display of state evolution
figure(2)
plot([0 Nf],[0 0],'g'); hold on; %horizontal axis
plot([0 0],[-0.2 1.2],'k'); %vertical axis
plot(x1,'r'); %plots x1
plot(x2,'b'); %plots x2
plot(xe1,'mx'); %plots xe1
plot(xe2,'kx'); %plots xe2
xlabel('sampling periods');
title('system and Kalman filter states');
```

1.3.2 Evolution of Filter Variables

1.3.2.1 Evolution of the State Error

The state error $\mathbf{er} = \mathbf{x} - \hat{\mathbf{x}}$ could also be considered the estimation error.

From OPKF:

$$\hat{\mathbf{x}}(n+1) = A\,\hat{\mathbf{x}}(n) + B\,\mathbf{u}(n) + K(n)\,[\mathbf{y}(n) - C\,\hat{\mathbf{x}}(n)]$$

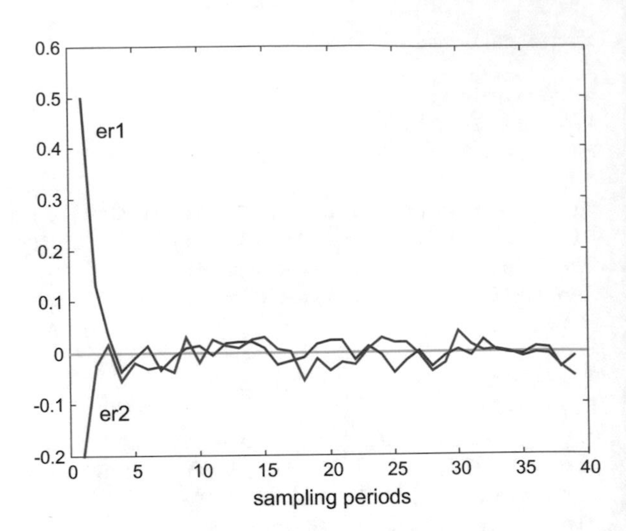

Fig. 1.8 Error evolution

It is easily deduced that:

$$\begin{aligned} \mathbf{er}(n+1) &= A\,\mathbf{er}(n) + \mathbf{w}(n) - K(n)\,[\mathbf{y}(n) - C\,\hat{\mathbf{x}}(n)] = \\ &= [A - K(n)\,C]\,\mathbf{er}(n) + \mathbf{w}(n) - K(n)\,\mathbf{v}(n) \end{aligned} \tag{1.76}$$

Then, the error mean evolves as follows:

$$\boldsymbol{\mu}_e\,(n+1) = [A - K(n)\,C] \cdot \boldsymbol{\mu}_e\,(n) \tag{1.77}$$

(let us denote $\Psi = [A - K(n)\,C]$)

Assuming that at the beginning the mean of the error is zero, then (depending on Ψ), the mean would stay at $\boldsymbol{\mu}_e(n) = 0$.

Figure 1.8 shows the evolution of the error for the two-tank system example. It has been computed by program an extension of the Program 1.3 that has been included in Appendix B.

1.3.2.2 Stationary Kalman Filter

If the error dynamics is stable, the values of $K(n)$ converge to constants, and so:

$$\Sigma(n+1) = \Sigma(n) \tag{1.78}$$

This equation translates to an algebraic Riccati equation (*ARE*), which for OPKF [40] is:

$$\Sigma - A\,\Sigma A^T + [A\,\Sigma C^T\,[C\,\Sigma C^T + \Sigma_v]^{-1}\,C\,\Sigma A^T] - \Sigma_w = 0 \tag{1.79}$$

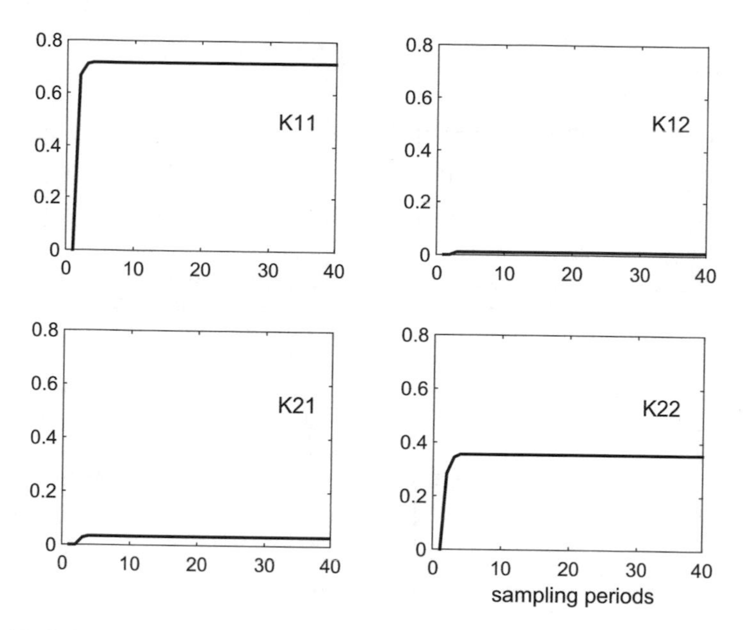

Fig. 1.9 Evolution of the Kalman gains

The stationary Kalman filter would be:

$$\hat{\mathbf{x}}(n+1) = \Psi\,\hat{\mathbf{x}}(n) + K\,\mathbf{y}(n) + B\,\mathbf{u}(n) \tag{1.80}$$

with:

$$\begin{aligned} \Psi &= A - K\,C \\ K &= A\,\Sigma\,C^T\,[A\,\Sigma\,C^T + R]^{-1} \end{aligned} \tag{1.81}$$

In certain cases, it would be convenient to use from the beginning this constant K, which could be pre-computed.

Figure 1.9 shows the evolution of the Kalman filter gains in the two-tank system example. Like before, the figure has been generated using the extended version of Program 1.3. Clearly, the Kalman gains rapidly evolve to constant values.

Continuing with the example of the two-tank system, Fig. 1.10 shows the evolution of the state covariance (the four matrix components). It has been computed with the extended version of Program 1.3. It is clear that all components tend to constant values.

1.3.2.3 The Steps of the Kalman Filter in Terms of Uncertainties

As said before, a main idea of the Kalman filter is to combine prediction and measurement to get good state estimation (the best in terms of minimum variance). An

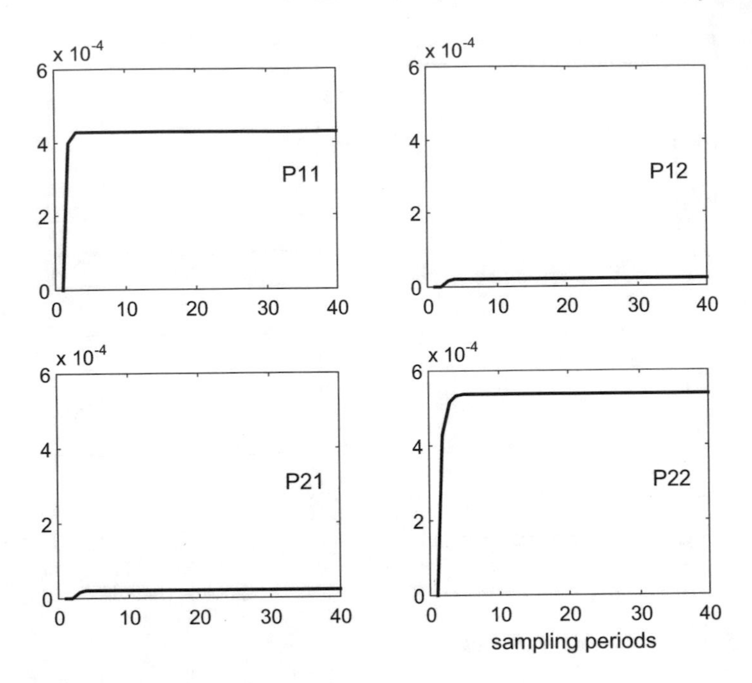

Fig. 1.10 Evolution of the state covariance

interesting feature of the two-tank example being used in this section, is that it is a two-dimensional system so PDFs of the variables could be plotted as ellipsoids. This yields a valuable opportunity to graphically illustrate the Kalman filter steps.

The next three figures have been generated with a program included in Appendix B. This program focuses on the transition from state number 3 to state number 4.

Figure 1.11 corresponds to the prediction step. The transition equation is used to predict the next state $\mathbf{x}_a(n+1)$ using the present estimated state $\mathbf{x}_e(n)$. There is process noise, with covariance Sw, that increases the uncertainty of the predicted state.

Figure 1.12 corresponds to the output measurement equation. There is measurement noise, with covariance Sv, that increases the uncertainty of the result. The actual measurement, provided by the system simulation, has been marked with a star on the right-hand plot. Recall that the update step considers the error between the actual measurement and $C\,\mathbf{x}_a(n+1)$ (this term has been called 'estimated **y**' in Figs. 1.12 and 1.13).

Finally, Fig. 1.13 contains, from left to right, the two things to combine in the update step, prediction and measurement, and the final result. Notice that the result, which is the estimated state, has smaller variance.

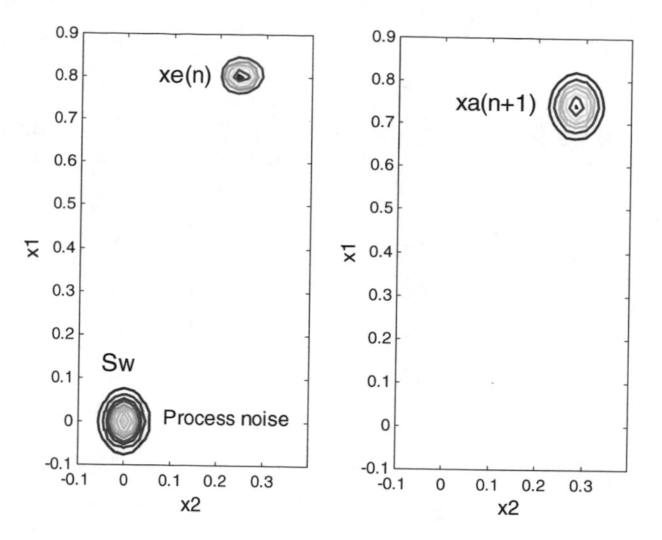

Fig. 1.11 The prediction step, from *left* to *right*

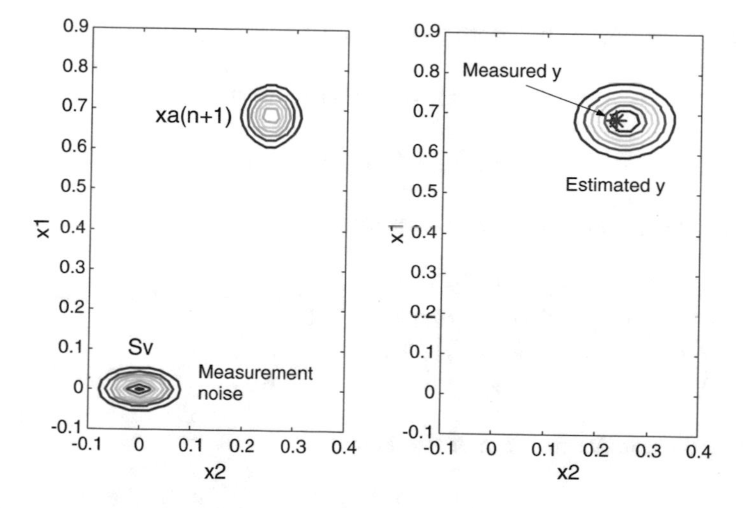

Fig. 1.12 The measurement

1.3.2.4 Observation Noise and the Stationary Kalman Filter

Let be the model of an autonomous system (no input):

$$\mathbf{x}(n+1) = A\,\mathbf{x}(n) + \mathbf{w}(n) \tag{1.82}$$

$$\mathbf{y}(n) = C\,\mathbf{x}(n) + \mathbf{v}(n) \tag{1.83}$$

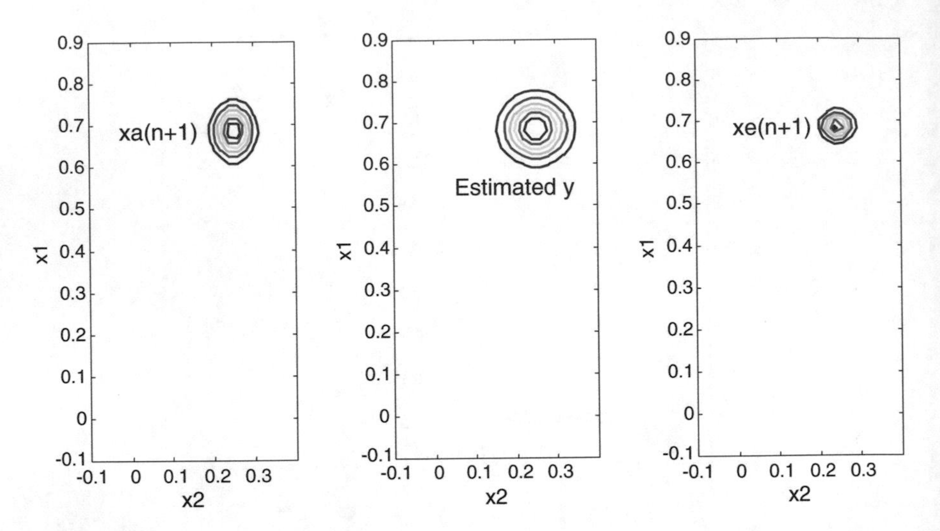

Fig. 1.13 Estimation of the next state

Suppose that the observation noise is coloured noise:

$$\mathbf{z}(n+1) = U\,\mathbf{z}(n) + \varphi(n) \tag{1.84}$$

$$\mathbf{v}(n) = F\,\mathbf{z}(n) + \rho(n) \tag{1.85}$$

In this case, the stationary Kalman filter would be:

$$\begin{aligned}\begin{pmatrix}\hat{\mathbf{x}}(n+1)\\ \hat{\mathbf{z}}(n+1)\end{pmatrix} &= \begin{pmatrix}A & 0\\ 0 & U\end{pmatrix}\begin{pmatrix}\hat{\mathbf{x}}(n)\\ \hat{\mathbf{z}}(n)\end{pmatrix} + \begin{pmatrix}K_1\\ K_2\end{pmatrix}(\mathbf{y}(n) - F\,\hat{\mathbf{z}}(n)) = \\ &= \begin{pmatrix}A - K_1 C & -K_1 F\\ -K_2 C & U - K_2 F\end{pmatrix}\begin{pmatrix}\hat{\mathbf{x}}(n)\\ \hat{\mathbf{z}}(n)\end{pmatrix} + \begin{pmatrix}K_1\\ K_2\end{pmatrix}\mathbf{y}(n)\end{aligned} \tag{1.86}$$

Therefore, the transfer function of the filter is:

$$H(z) = (I\ 0)\begin{pmatrix}zI - A + K_1 C & K_1 F\\ K_2 & zI - U + K_2 F\end{pmatrix}\begin{pmatrix}K_1\\ K_2\end{pmatrix} \tag{1.87}$$

It is found that the zeros of this transfer function are given by:

$$\det(zI - U) = 0 \tag{1.88}$$

Consequently, the zeros of the Kalman filter are placed on the poles of the observation noise.

When it is desired that the Kalman filter rejects certain frequencies, it is opportune to raise these frequencies in the observation noise model.

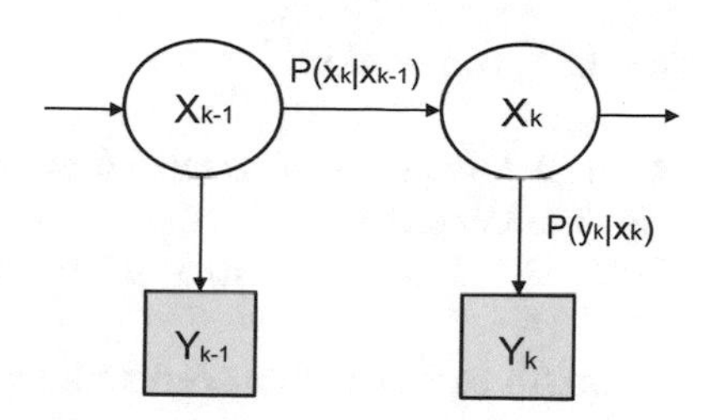

Fig. 1.14 Bayes net corresponding to Kalman filter

1.3.3 Several Perspectives

Looking at the Kalman filter scheme, it is possible to emphasize several interpretations. Some of these are introduced in this subsection.

1.3.3.1 With Respect to Bayes Nets

Consider the Bayes net depicted in Fig. 1.14. It corresponds to the Gauss-Markov model used in the Kalman filter.

Notice that the diagram can be regarded as a representation of a Hidden Markov Model (HMM), where the observed variables have been represented as shaded squares.

The filtering problem is to estimate, sequentially, the states of a system as measurements are obtained. The Kalman filter gives a procedure to get $p(\mathbf{x}_k | \mathbf{Y}_k)$.

1.3.3.2 As State Observer

Recall from Chap. 5 that the Luenberger observer was:

$$\mathbf{x}_e(n+1) = A\,\mathbf{x}_e(n) + B\,\mathbf{u}(n) + K\,(\mathbf{y}(n) - C\,\mathbf{x}_e(n)) \tag{1.89}$$

where $\mathbf{x}_e$ is the observer's estimation of the system state.

If you compare with the OPKF filter:

$$\hat{\mathbf{x}}(n+1) = A\,\hat{\mathbf{x}}(n) + B\,\mathbf{u}(n) + K(n)\,[\mathbf{y}(n) - C\,\hat{\mathbf{x}}(n)] \tag{1.90}$$

The only difference between both equations is evident: to use K or $K(n)$. Clearly, the Kalman filter can be considered an optimal state observer. Likewise, differences are even lighter if the stationary Kalman filter is chosen; in this case, Kalman can be regarded as an algorithm to compute a suboptimal K for the observer.

1.3.3.3 Innovations

Along the treatment of Kalman filter aspects, some authors [72] consider the following difference:

$$\eta(n+1) = \mathbf{y}(n+1) - \hat{\mathbf{y}}(n+1|n) \tag{1.91}$$

This difference, which we denoted as η, is called *'innovations process'*, or *'prediction error process'*, or *'measurement residual process'*.

The last term in (1.39) is the prediction of the output, that is:

$$\hat{\mathbf{y}}(n+1|n) = C\,\hat{\mathbf{x}}(n+1|n) = C\,A\,\hat{\mathbf{x}}(n) \tag{1.92}$$

Therefore, the first equation of the AKF Kalman filter could be written as follows:

$$\hat{\mathbf{x}}(n+1) = A\,\hat{\mathbf{x}}(n) + B\,\mathbf{u}(n) + K(n+1)\,\eta(n+1) \tag{1.93}$$

Notice that the innovations correspond to the error term $[\mathbf{y}(n+1) - CA\,\hat{\mathbf{x}}(n)]$.

It can be shown that the innovations process is a zero-mean Gaussian white noise. The covariance matrix of the innovations process is:

$$\begin{aligned} &\Sigma_{in}(n+1) = cov\,(\eta(n+1)) = E(\eta(n+1)\cdot\eta(n+1)^T) = \\ &= C\,\Sigma(n+1)\,C^T + \Sigma_v = N(n+1) \end{aligned} \tag{1.94}$$

Looking now at the OPKF filter, the error term is $[\mathbf{y}(n) - C\hat{\mathbf{x}}(n)]$. It is the case that authors concerned with this type of filter [40], define innovations as follows:

$$\eta(n) = \mathbf{y}(n) - C\,\hat{\mathbf{x}}(n) \tag{1.95}$$

Then, the OPKF filter could be formulated as:

$$\hat{\mathbf{x}}(n+1) = A\,\hat{\mathbf{x}}(n) + B\,\mathbf{u}(n) + K(n)\,\eta(n) \tag{1.96}$$

$$\mathbf{y}(n) = C\,\hat{\mathbf{x}}(n) + \eta(n) \tag{1.97}$$

In this form, the filter could be interpreted as a model that generates the sequence $\mathbf{y}(n)$; it is denoted then as innovations model.

Another formulation could be:

$$\hat{\mathbf{x}}(n+1) = [A - K(n)\,C]\,\hat{\mathbf{x}}(n) + B\,\mathbf{u}(n) + K(n)\,\mathbf{y}(n) \tag{1.98}$$

$$\eta(n) = \mathbf{y}(n) - C\,\hat{\mathbf{x}}(n) \tag{1.99}$$

This last expression could be interpreted as a white noise generator, departing from the sequence $\mathbf{y}(n)$. In this case, it is a whitening filter.

1.3.4 Some Connections

The Kalman filter is connected with other topics, mainly because it is a recursive procedure to minimize a quadratic criterion.

1.3.4.1 Using Kalman Algorithm Por Parameter Identification

Given a vector of data $\mathbf{d}(n)$ and a set of parameters $\boldsymbol{\theta}(n)$ to be estimated, the parameter identification problem could be written as follows [40]:

$$\boldsymbol{\theta}(n+1) = \boldsymbol{\theta}(n) \tag{1.100}$$

$$\mathbf{y}(n) = \mathbf{d}(n)^T \boldsymbol{\theta}(n) + \mathbf{v}(n) \tag{1.101}$$

The equations above are a particular case of the Gauss-Markov model for $A = I$, $B = 0$, $C = \mathbf{d}(n)^T$, $\mathbf{w}(n) = 0$. In this case, the Kalman filter that gets the optimal estimation of $\boldsymbol{\theta}(n)$ is:

$$\hat{\boldsymbol{\theta}}(n+1) = \hat{\boldsymbol{\theta}}(n) + K(n+1)\,[\mathbf{y}(n+1) - \mathbf{d}(n+1)^T\,\hat{\boldsymbol{\theta}}(n)] \tag{1.102}$$

$$K(n+1) = \frac{P(n)\mathbf{d}(n+1)}{\Sigma_v + \mathbf{d}(n+1)^T\,P(n)\,\mathbf{d}(n)} \tag{1.103}$$

$$P(n+1) = P(n) - K(n+1)\,\mathbf{d}(n+1)^T P(n) \tag{1.104}$$

Notice that the equations above are the same already obtained for least squares recursive parameter identification (see the section on parameter identification in Chap. 6). Of course, this is not surprising.

1.3.4.2 The Kalman Filter as an ARMAX Model

Let us use the OPKF innovations model for a single-input single output system:

$$\hat{\mathbf{x}}(n+1) = A\,\hat{\mathbf{x}}(n) + B\,u(n) + K(n)\,\eta(n) \tag{1.105}$$

$$y(n) = C\,\hat{\mathbf{x}}(n) + \eta(n) \tag{1.106}$$

Suppose that the model comes in observer form [40]. Then:

$$A = \begin{pmatrix} -a_1 \ 1 \ 0 \ldots 0 \\ -a_2 \ 0 \ 1 \ldots 0 \\ \ldots \ \ldots \ \ldots \ \ldots \\ -a_m \ 0 \ 0 \ldots 1 \end{pmatrix} \ ; \ B = \begin{pmatrix} b_1 \\ b_2 \\ \vdots \\ b_m \end{pmatrix} \ ; \ K(n) = \begin{pmatrix} K_1(n) \\ K_2(n) \\ \vdots \\ K_m(n) \end{pmatrix} \tag{1.107}$$

$$C = [1 \ 0 \ldots 0] \tag{1.108}$$

By means of successive substitutions and using these matrices, the innovations model can be written as an ARMAX model:

$$A(q^{-1})\, y(n) \ = \ B(q^{-1})\, u(n) \ + \ C(n, q^{-1})\, \eta(n) \tag{1.109}$$

with:

$$A(q^{-1}) \ = \ 1 \ + \ a_1 q^{-1} \ + \ a_2 q^{-2} + \ \ldots + a_m q^{-m} \tag{1.110}$$

$$B(q^{-1}) \ = \ b_1 q^{-1} \ + \ b_2 q^{-2} + \ \ldots + b_m q^{-m} \tag{1.111}$$

$$\begin{aligned} C(n,\, q^{-1}) \ &= 1 + (K_1(n-1) \ + a_1)\, q^{-1} \ + \ (K_2(n-1) \ + a_2) q^{-2} + \ \ldots \\ &+ (K_m(n-1) \ + a_m) q^{-m} \end{aligned} \tag{1.112}$$

In the case of a stationary Kalman filter, $C(n,\, q^{-1})$ would be $C(q^{-1})$.

The Kalman filter could be seen as a model that can be given in ARMAX format. Likewise, the ARMAX model could be interpreted as a Kalman filter.

1.3.5 Numerical Issues

It was soon noticed, in the first aerospace applications of the Kalman filter, that the programs may suffer from numerical instability. Numerical errors may yield non-symmetric or non positive definite covariance matrix $P(n)$.

One of the difficulties on first spacecrafts was the limited digital processing precision (15 bits arithmetics). Square root algorithms were introduced (Potter 1963) to get more precision with the same wordlength. The Cholesky factorization (upper and lower triangular factors) was used.

An alternative factorization was introduced (Bierman 1977) that do not require explicit square root computations. It is the U-D factorization.

MATLAB provides the *chol()* function for Cholesky factorization, and the *ldl()* function for U-D factorization.

There are some techniques to avoid the loss of $P(n)$ symmetry. One is to symmetrise the matrix at each algorithm step by averaging with its transpose. Another is to use and propagate just the upper or the lower triangular part of the matrix. And another is to use the *Joseph's form*:

$$P(n+1) = (I - K(n+1)\,C)\,M(n+1)(I - K(n+1)\,C)^T \; + K(n+1)\,R\,K(n+1)^T \tag{1.113}$$

If the measurement noise vector components are uncorrelated, the state update can be done one measurement at a time. Then there are not matrix inversions, only scalar operations. This is called *sequential processing*.

See [112] for details of numerical aspects of Kalman filtering. Several chapters of [96] are devoted to different mechanizations of the Kalman filter.

1.3.6 Information Filter

Some difficulties could appear with the initialization of variables in the Kalman filter. For instance, having no knowledge of state error covariance could lead to suppose large or infinite values as starting point. The consequence could be an undetermined value of K, (∞/∞). This problem could be avoided by using the inverse of the state error covariance.

The idea is to use the following equations:

$$(\Sigma(n))^{-1} \; = \; (M(n))^{-1} \; + \; C^T\,\Sigma_v^{-1} C \tag{1.114}$$

$$K(n) = \; \Sigma(n) C^T\,\Sigma_v^{-1} \tag{1.115}$$

This is an equivalent alternative to the AKF Kalman filter. The filtering algorithm would be:

$$L = C^T\,\Sigma_v^{-1} \tag{1.116}$$

-Iterate:

$$\mathbf{x}_a(n) \; = \; A\,\mathbf{x}_e\,(n-1) \; + \; B\,\mathbf{u}(n-1) \tag{1.117}$$

$$(\Sigma(n))^{-1} \; = \; (M(n))^{-1} \; + \; LC \tag{1.118}$$

$$K(n) = \; \Sigma(n)\,L \tag{1.119}$$

$$\mathbf{x}_e(n) \; = \; \mathbf{x}_a(n) \; + \; K(n)\,(\mathbf{y}(n) \; - \; C\,\mathbf{x}_a(n)) \tag{1.120}$$

$$M(n+1) \; = \; A\,\Sigma(n)\,A^T \; + \; \Sigma_w \tag{1.121}$$

$$n \leftarrow n+1 \tag{1.122}$$

The inverse Σ^{-1} of the covariance matrix is called information matrix (also called Fisher information matrix). This is the reason for the name 'information filter'.

Recently, the research on mobile sensors and robots has shown preference for the information filter, since it leads to sparse covariance representations [105, 114].

1.4 Nonlinear Conditions

It is usual in real life applications to face nonlinearities. Then, instead of the linear model that has been used up to now in this chapter, it would be more appropriate to represent the system with the following equations:

$$\mathbf{x}(n+1) = \mathbf{f}(\mathbf{x}(n),\ \mathbf{u}(n),\ \mathbf{w}(n)) \tag{1.123}$$

$$\mathbf{y}(n) = \mathbf{h}(\mathbf{x}(n),\ \mathbf{v}(n)) \tag{1.124}$$

The nonlinearities could appear in the transition equation and/or in the measurement equation.

Examples of nonlinear phenomena are saturations, thresholds, or nonlinear laws with trigonometric functions or exponentials or squares, etc. In many cases sensors, friction, logarithmic variables, modulation, etc. are the cause of nonlinearity.

There are versions of the Kalman filter that are able to cope with certain nonlinear conditions. Several sections of this chapter cover this aspect. Now, it is opportune to focus on a preliminary consideration of nonlinear conditions under the optics of Kalman filtering..

1.4.1 Propagation and Nonlinearity

An example frequently used to illustrate problems caused by nonlinearity is the case of tracking a satellite from Earth. Distance and angle are measured, each with an amount of uncertainty.

Indeed, one could take many measurements, and obtain the average of radius measurements by one side, and the average of angle measurements by the other side. Figure 1.15 depicts the scenario, with a cloud of measurements. The point marked with a plus sign (denoted S), is the expected value of the satellite location, correctly obtained with the averages.

Now, suppose you apply a change of coordinates and obtain the Cartesian coordinates of each point obtained by the measurements. If you compute the average of

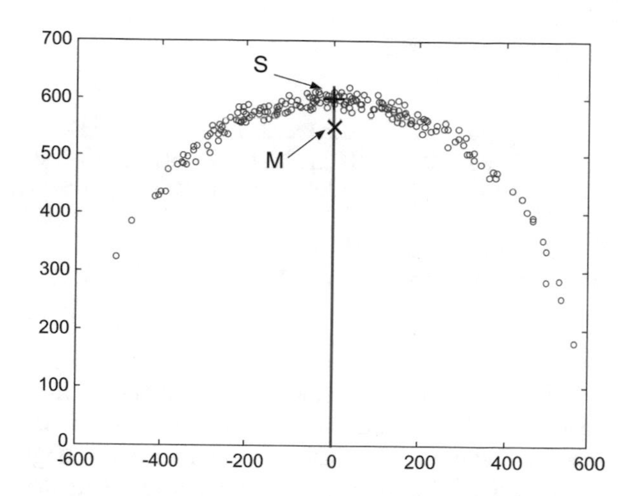

Fig. 1.15 Satellite position under disturbances

the vertical position, and the average of the horizontal position, you get the point marked with a cross in the figure (denoted M). This point is deviated from the correct expected value.

The change of coordinates, from polar to Cartesian, is nonlinear. The propagation of statistical distributions through nonlinear transformations should be handled with care.

Program 1.4 Example of satellite position at T0

```
%Example of satellite position at T0
%
%reference position at T0
r0=600;
alpha0=pi/2;
Np=200; %number of points to display
%reserve space for points
px=zeros(1,Np);
py=zeros(1,Np);
%influence of perturbations and noise
%variances
sigr=10;
siga=0.45; %radians
nr=sigr*randn(1,Np);
na=siga*randn(1,Np);
for nn=1:Np,
r=r0+nr(nn);
a=alpha0+na(nn);
px(nn)=r*cos(a);
py(nn)=r*sin(a);
end;
xmean=sum(px/Np);
```

```
ymean=sum(py/Np);
%display
figure(1)
plot(px,py,'r.'); hold on; %the points
plot([0 0],[0 r0+20],'b'); %vertical line
plot(0,r0,'k+','MarkerSize',12); %+ for reference satellite
%
%X for mean cartesian position:
plot(xmean,ymean,'kx','MarkerSize',12);
%
title('Satellite position data');
```

Now, let us study in more detail a nonlinear measurement case. The original data have a Gaussian distribution. The nonlinear function, due to the measurement method, is *arctan()*. For instance, you measure an angle using x-y measurements.

Figure 1.16 depicts the nonlinear behaviour of the *arctan()* function (which is *atan()* in MATLAB).

Figure 1.17 shows how a Gaussian PDF, with $\sigma = 0.1$, propagates through the nonlinearity. It can be noticed that the original and the propagated PDFs are almost equal. Figures 1.16 and 1.17 have been generated with the Program 1.5, which uses a simple code vectorization for speed.

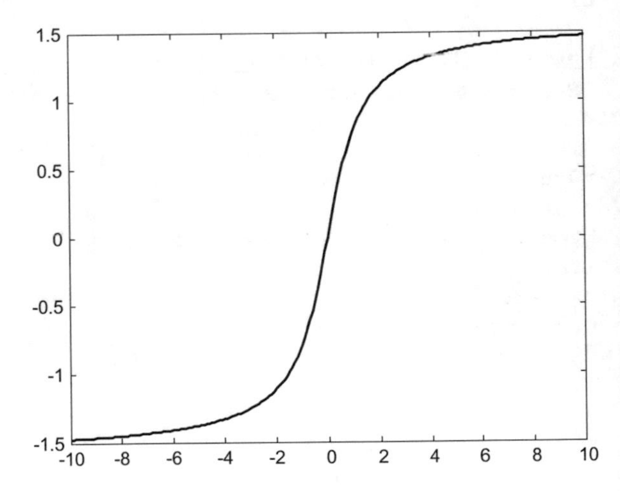

Fig. 1.16 Example of nonlinear function: arctan()

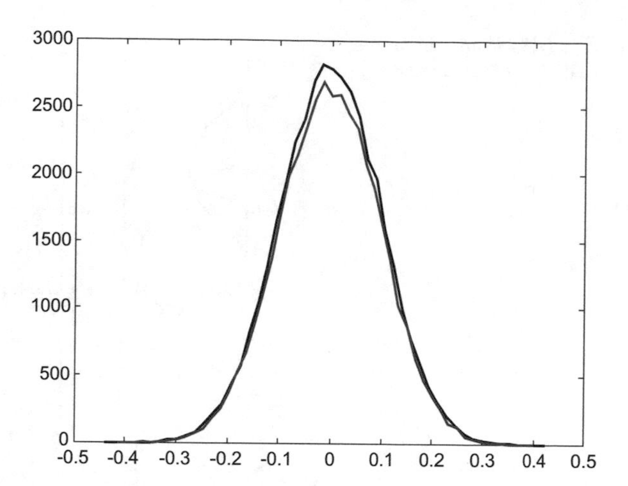

Fig. 1.17 Original and propagated PDFs

Program 1.5 Propagation through nonlinearity

```
%Propagation through nonlinearity
%
%nonlinear function
vx=-10:0.1:10;
vy=atan(vx);
%gaussian random data
Nd=40000;
sig=0.1;
adat=sig*randn(1,Nd);
%histogram of a priori data
Nbins=50;
[ha,hax]=hist(adat,Nbins);
%propagated random data
pdat=atan(adat);
%histogram of posterior data
[hp,hpx]=hist(pdat,Nbins);
figure(1)
plot(vx,vy);
title('arctan function');
figure(2)
plot(hax,ha,'k'); hold on;
plot(hpx,hp,'r');
title('propagation through arctan: PDFs');
```

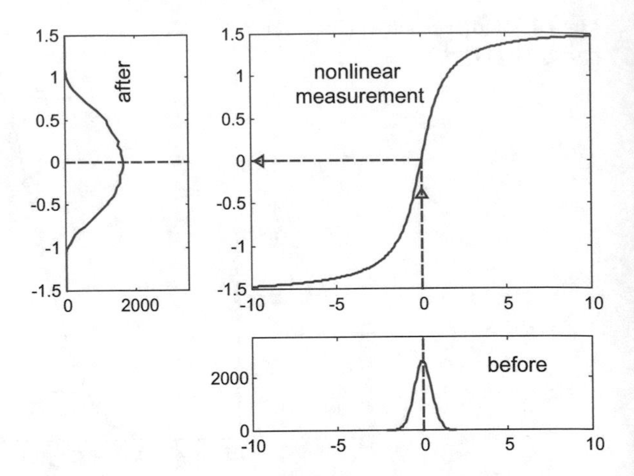

Fig. 1.18 Propagation of a PDF through nonlinearity

Figure 1.18 shows a practical perspective of the PDF propagation through the nonlinear measurement function. The standard deviation of the Gaussian PDF has been set to 0.5. This kind of representation will be used again in next sections.

Program 1.6 Propagation through nonlinearity

```
%Propagation through nonlinearity
%
%nonlinear function
vx=-10:0.1:10;
vy=atan(vx);
%gaussian random data
Nd=40000;
bx=-1.5:0.05:1.5; %location of histogram bins
sig=0.5;
adat=sig*randn(1,Nd);
%histogram of a priori data
Nbins=50;
[ha,hax]=hist(adat,Nbins);
%propagated random data
pdat=atan(adat);
%histogram of posterior data
hpt=hist(pdat,bx);
figure(1)
pl1=0.1; pb1=0.35; pw1=0.2; ph1=0.55;
pl2=0.4; pb2=0.05; pw2=0.55; ph2=0.2;
subplot('position',[pl1 pb1 pw1 ph1]) %left plot
plot(hpt,bx); hold on;
plot([0 3500],[0 0],'r--');
axis([0 3500 -1.5 1.5]);
ylabel('after');
subplot('position',[pl2 pb1 pw2 ph1]) %central plot
```

```
plot(vx,vy); hold on;
plot([-10 0],[0 0],'r--',-9.5,0,'r<');
plot([0 0],[-1.5 0],'r--',0,-0.4,'r^');
title('nonlinear measurement');
subplot('position',[pl2 pb2 pw2 ph2]) %bottom plot
plot(hax,ha); hold on;
plot([0 0],[0 3500],'r--');
axis([-10 10 0 3500]);
title('before');
```

It is clear that the *arctan(x)* function is almost linear for small values of x. A curvature appears when x becomes larger, and then saturation becomes predominant as x is further increased. A program has been made to investigate how the Gaussian PDF propagates when the standard deviation is large enough to enter into nonlinear effects. This is Program 1.7. The propagation has been applied to three Gaussian PDFs, the narrower with $\sigma = 0.7$, the intermediate with $\sigma = 1$, the wider with $\sigma = 2$.

Figure 1.19 shows the results. When the standard deviation becomes larger, the propagated PDF can exhibit two peaks.

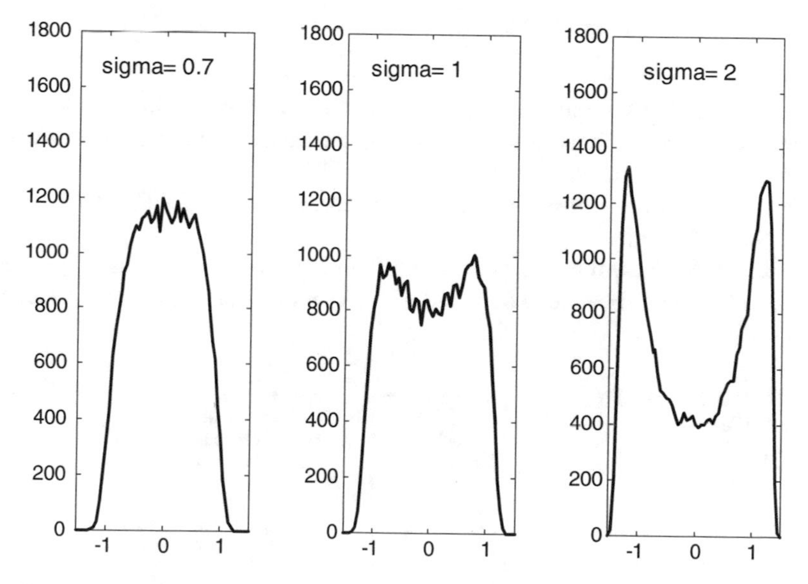

Fig. 1.19 Propagated PDFs for sigma = 0.7, 1, 2

Program 1.7 Propagation through nonlinearity

```
%Propagation through nonlinearity
%Several values of variance
%gaussian random data
Nd=40000;
rdat=randn(1,Nd);
bx=-1.5:0.05:1.5; %location of histogram bins
figure(1)
for nn=1:3,
switch nn,
case 1, sig=0.7;
case 2, sig=1;
case 3, sig=2;
end;
adat=sig*rdat;
pdat=atan(adat);
hpt=hist(pdat,bx);
subplot(1,3,nn)
plot(bx,hpt,'k');
tit=['sigma= ',num2str(sig)];
title(tit);
axis([-1.5 1.5 0 1800]);
end
```

It may happen that experimental data have a non-zero mean. In other words, the corresponding PDF is horizontally shifted. If the data are propagated through the *arctan()* function the result is a non-symmetric PDF.

Figure 1.20 shows a possible situation, when the original PDF, with $\sigma = 2$, is shifted 0.8 to the right.

Notice in Fig. 1.20 that the propagated PDF is asymmetrical. The mean of the propagated PDF has been also depicted with a cross on the vertical axis. Following

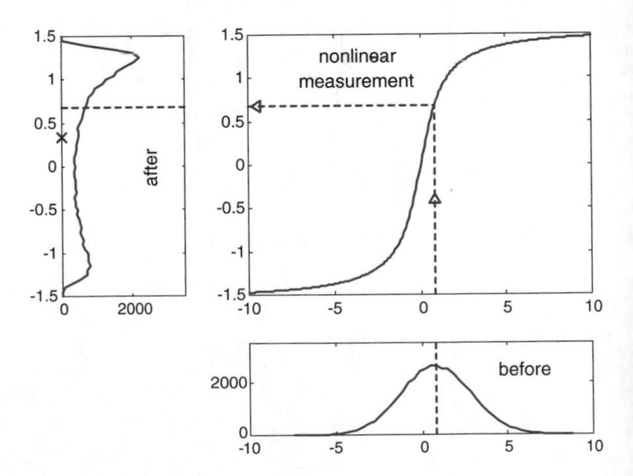

Fig. 1.20 Propagation of a shifted PDF through nonlinearity

the arrows, it is clear that the mean of the original Gaussian PDF propagates to a different point. Figure 1.20 has been generated with an extension of the Program 1.6 that has been included in Appendix B.

1.4.2 Jacobian. Hessian. Change of Coordinates.

Changing of coordinates, like from polar coordinates to Cartesian, can be the cause of nonlinearity. This has been shown in the previous subsection. Actually, many applications involving tracking or positioning of vehicles or objects incur in this situation.

In two dimensions, the tangent to a curve $f(x)$ can be obtained using the first derivative $d\,f(x)/dx$. A generalization of the first derivative is the gradient of a scalar function of several variables. A larger generalization is the 'Jacobian', which can be described as the first derivative of a vector function of several variables. Here is the expression of a Jacobian:

$$J_f = \frac{\partial \mathbf{f}}{\partial \mathbf{x}} = \begin{pmatrix} \frac{\partial f_1}{\partial x_1} & \frac{\partial f_1}{\partial x_2} & \frac{\partial f_1}{\partial x_3} & \cdots & & \frac{\partial f_1}{\partial x_N} \\ \frac{\partial f_2}{\partial x_1} & \frac{\partial f_2}{\partial x_2} & \cdots & & \cdots & \frac{\partial f_2}{\partial x_N} \\ \cdots & & \cdots\cdots & & & \cdots \\ \cdots & & \cdots\cdots & & & \cdots \\ \frac{\partial f_M}{\partial x_1} & \frac{\partial f_M}{\partial x_2} & \cdots & & \cdots & \frac{\partial f_M}{\partial x_N} \end{pmatrix} \tag{1.125}$$

The use of Jacobians appears quite naturally when dealing with changes of coordinates. For instance, if you change from variables x, y to variables u, v, the differentials of surface are related as follows:

$$dx\,dy = |J|\,du\,dv \tag{1.126}$$

where the Jacobian of the transformation of coordinates is:

$$J = \begin{bmatrix} \frac{\partial x}{\partial u} & \frac{\partial x}{\partial v} \\ \frac{\partial y}{\partial u} & \frac{\partial y}{\partial v} \end{bmatrix} \tag{1.127}$$

The Jacobian tells us how a surface expands or shrinks when coordinates are transformed. For instance, when changing from polar to Cartesian coordinates:

$$\begin{aligned} x &= r\cos\varphi \\ y &= r\sin\varphi \end{aligned} \tag{1.128}$$

The Jacobian is:

$$J = \begin{bmatrix} \frac{\partial x}{\partial r} & \frac{\partial x}{\partial \varphi} \\ \frac{\partial y}{\partial r} & \frac{\partial y}{\partial \varphi} \end{bmatrix} = \begin{bmatrix} \cos\varphi & -r\sin\varphi \\ \sin\varphi & r\cos\varphi \end{bmatrix} \tag{1.129}$$

the determinant is: $|J| = r$. Therefore:

$$dx\, dy = r\, dr\, d\varphi \tag{1.130}$$

In the case of satellites or aerial vehicles it is pertinent to consider the relation between Cartesian and spherical coordinates. The Jacobian in this case is:

$$J = \begin{bmatrix} \frac{\partial x}{\partial r} & \frac{\partial x}{\partial \theta} & \frac{\partial x}{\partial \phi} \\ \frac{\partial y}{\partial r} & \frac{\partial y}{\partial \theta} & \frac{\partial y}{\partial \phi} \\ \frac{\partial z}{\partial r} & \frac{\partial z}{\partial \theta} & \frac{\partial z}{\partial \phi} \end{bmatrix} = \begin{bmatrix} \cos\theta\sin\phi & -r\sin\theta\sin\phi & r\cos\theta\cos\phi \\ \sin\theta\sin\phi & -r\cos\theta\sin\phi & r\sin\theta\cos\phi \\ \cos\phi & 0 & -r\sin\phi \end{bmatrix} \tag{1.131}$$

the determinant is: $|J| = r^2 \sin\phi$

Given the Jacobian J_f of a function $\mathbf{f}$, and supposing that the function is invertible, the Jacobian of the inverse function is the inverse of J_f.

Consider a random variable with PDF p_x, and an invertible function $f(x)$. The PDF of $y = f(x)$ is:

$$p_y = p_x \left| \frac{dx}{dy} \right| \tag{1.132}$$

This relation can be generalized using the inverse of the Jacobian and joint distributions. In particular, for a change of coordinates,

$$p_{\mathbf{y}} = p_{\mathbf{x}} \left| J_f^{-1} \right| \tag{1.133}$$

For example, in the transformation from polar to Cartesian coordinates:

$$p_{x,y} = p_{r,\varphi} \cdot \frac{1}{r} \tag{1.134}$$

Jacobians are also used for the analysis of nonlinear systems behaviour around certain states of interest. For instance, given a pendulum:

$$\begin{aligned} \frac{d\varphi}{dt} &= f_1(\varphi, \omega) = \omega \\ \frac{d\omega}{dt} &= f_2(\varphi, \omega) = -\frac{g}{L}\sin\varphi \end{aligned} \tag{1.135}$$

The Jacobian is:

$$J = \begin{bmatrix} \frac{\partial f_1}{\partial \varphi} & \frac{\partial f_1}{\partial \omega} \\ \frac{\partial f_2}{\partial \varphi} & \frac{\partial f_2}{\partial \omega} \end{bmatrix} = \begin{bmatrix} 0 & 1 \\ -\frac{g}{L}\cos\varphi & 0 \end{bmatrix} \tag{1.136}$$

If the Jacobian is evaluated for $\varphi = 0$, the eigenvalues of the matrix are $\pm\, j\, \sqrt{g/L}$. If it is evaluated for $\varphi = \pi$, the eigenvalues are $\pm\, \sqrt{g/L}$. For real eigenvalues, it can be shown that if any eigenvalue is positive the state is unstable. For complex eigenvalues, if any eigenvalue has positive real part the state is unstable. Therefore, when $\varphi = \pi$ the pendulum is unstable.

A generalization of the second derivative is the '*Hessian*'. Given a scalar function of several variables, the Hessian has the following expression:

$$H_f = \frac{\partial^2 f}{\partial \mathbf{x}^2} = \begin{pmatrix} \frac{\partial^2 f}{\partial x_1^2} & \frac{\partial^2 f}{\partial x_1 \partial x_2} & \frac{\partial^2 f}{\partial x_1 \partial x_3} & \cdots & \frac{\partial^2 f}{\partial x_1 \partial x_N} \\ \frac{\partial^2 f}{\partial x_2 \partial x_1} & \frac{\partial^2 f}{\partial x_2^2} & \cdots & \cdots & \frac{\partial^2 f}{\partial x_2 \partial x_N} \\ \cdots & \cdots\cdots & & \cdots & \\ \cdots & \cdots\cdots & & \cdots & \\ \frac{\partial^2 f}{\partial x_N \partial x_1} & \frac{\partial^2 f}{\partial x_N \partial x_2} & \cdots & \cdots & \frac{\partial^2 f}{\partial x_N^2} \end{pmatrix} \tag{1.137}$$

In the case of a vector function, the expression would become a tensor.

The Hessian is used for the study of curvatures and for discrimination of maxima, minima, etc.

Scalar functions of several variables can be approximated with a Taylor expansion as follows:

$$y = f(\mathbf{x} + \Delta\mathbf{x}) \approx f(\mathbf{x}) + J_f(\mathbf{x})\Delta\mathbf{x} + \frac{1}{2}\Delta\mathbf{x}^T H_f(\mathbf{x})\,\Delta\mathbf{x} + \ldots \tag{1.138}$$

where $J_f(\mathbf{x})$ is the Jacobian (in this case a vector: the gradient) evaluated at $\mathbf{x}$, and $H_f(\mathbf{x})$ is the Hessian evaluated at $\mathbf{x}$.

1.4.3 Local Linearization

When there are nonlinearities, it is very common to use linear approximations. A basic case is represented in Fig. 1.21 where a tangent is used to approximate the nonlinear curve $f(x)$ near the point $(x_0,\ f(x_0))$. Obviously, the quality of the approximation depends on the shape of the curve.

In mathematical terms, there is a curve:

$$y = f(x) \tag{1.139}$$

and a linear approximation:

$$f(x) \approx f(x_0) + \left.\frac{df(x)}{dx}\right|_{x_0} (x - x_0) \tag{1.140}$$

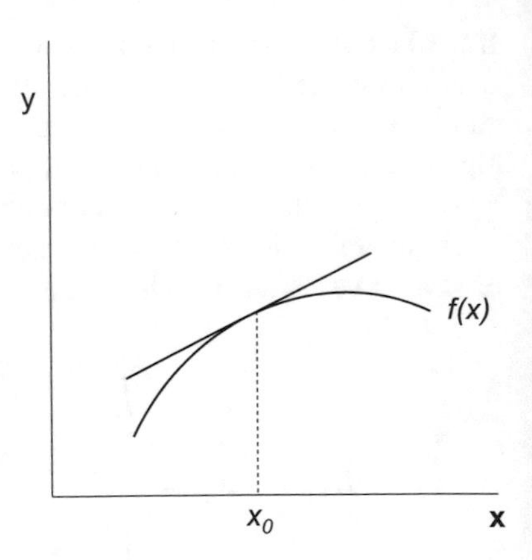

Fig. 1.21 Basic linear approximation using tangent

The approximation could be improved using a Taylor series, with higher order derivatives.

Also, this approach can be generalized for n dimensions. So it is possible to write:

$$\mathbf{f}(\mathbf{x}) \approx \mathbf{f}(\mathbf{x}_0) + \left.\frac{\partial \mathbf{f}(\mathbf{x})}{\partial \mathbf{x}}\right|_{\mathbf{x}_0} \delta\,\mathbf{x} + \frac{1}{2}\left.\frac{\partial^2 \mathbf{f}(\mathbf{x})}{\partial \mathbf{x}^2}\right|_{\mathbf{x}_0} \delta\,\mathbf{x}^2 + \frac{1}{3!}\left.\frac{\partial^3 \mathbf{f}(\mathbf{x})}{\partial \mathbf{x}^3}\right|_{\mathbf{x}_0} \delta\,\mathbf{x}^3 + \ldots \tag{1.141}$$

Suppose that $\delta\,\mathbf{x}$ is a Gaussian variable with covariance P. It can be shown that the mean and covariance of $\mathbf{f}(\mathbf{x})$ are:

$$\mu_{\mathbf{y}} \approx \mathbf{f}(\mathbf{x}_0) + \frac{1}{2}\left.\frac{d^2 \mathbf{f}(\mathbf{x})}{d\mathbf{x}^2}\right|_{\mathbf{x}_0} P + \frac{1}{2}\left.\frac{d^4 \mathbf{f}(\mathbf{x})}{d\mathbf{x}^4}\right|_{\mathbf{x}_0} E(\delta\,\mathbf{x}^4) + \ldots \tag{1.142}$$

$$P_{\mathbf{y}} \approx F\,P\,F^T + \frac{1}{2\times 4!}\left.\frac{d^2 \mathbf{f}(\mathbf{x})}{d\mathbf{x}^2}\right|_{\mathbf{x}_0} (E(\delta\,\mathbf{x}^4) + \ldots \tag{1.143}$$

where $E()$ is expected value, and F is the Jacobian:

$$F = \left.\frac{d\mathbf{f}(\mathbf{x})}{d\mathbf{x}}\right|_{\mathbf{x}_0}.$$

The approximation by Taylor series can be applied to the type of models being used in this chapter. For instance, take a model of nonlinear state transition:

$$\dot{\mathbf{x}} = \mathbf{f}(\mathbf{x},\ \mathbf{w}) \tag{1.144}$$

A first order approximation of the nonlinear function $\mathbf{f}(\,)$can be:

$$\mathbf{f}(\mathbf{x},\, \mathbf{w}) \approx \mathbf{f}(\mathbf{x}_0,\, \mathbf{w}_0) + \left.\frac{\partial \mathbf{f}(\mathbf{x},\, \mathbf{w})}{\partial \mathbf{x}}\right|_{\mathbf{x}_0, \mathbf{w}_0} \delta\, \mathbf{x} + \left.\frac{\partial \mathbf{f}(\mathbf{x},\, \mathbf{w})}{\partial \mathbf{w}}\right|_{\mathbf{x}_0, \mathbf{w}_0} \delta\, \mathbf{w} + \ldots \tag{1.145}$$

A similar approximation can be applied to nonlinear measurement model:

$$\mathbf{y} = \mathbf{h}(\mathbf{x},\, \mathbf{v}) \tag{1.146}$$

(we shall denote the Jacobian of **y** with respect to **x** as H)

1.4.4 Example of a Body Falling Towards Earth

The next sections of this chapter are devoted to filtering in nonlinear situations. A common example will be used along these sections. It is the case of a falling body being tracked by radar [61]. The body falls from high altitude, where atmospheric drag is small. As altitude decreases the density of air increases and the drag changes.

Figure 1.22 pictures the example. The body falls vertically. The radar is placed a distance L from the body vertical. The radar measures the distance y from the radar to the body.

Body state variables are chosen such that x_1 is altitude, x_2 is velocity, and x_3 is the ballistic coefficient. The falling body is subject to air drag, which could be approximated with:

$$drag = d = \frac{\rho\, x_2^2}{2\, x_3} \tag{1.147}$$

$$\rho = \rho_0 \,\exp(-\,\frac{x_1}{k}) \tag{1.148}$$

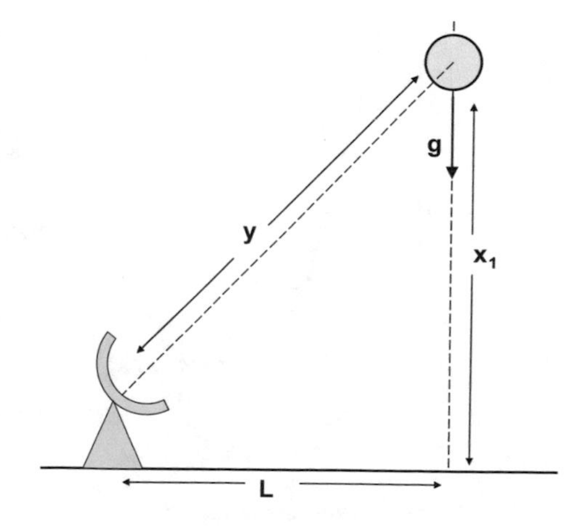

Fig. 1.22 Falling body example

where ρ is air density and k is a constant.

The motion equations of the body are:

$$\begin{aligned} \dot{x}_1 &= x_2 \\ \dot{x}_2 &= d + g \\ \dot{x}_3 &= 0 \end{aligned} \tag{1.149}$$

where $g = -9.81\ \mathrm{m/s^2}$ is gravity acceleration.

The measurement taken by the radar is:

$$y = \sqrt{L^2 + x_1^2} \tag{1.150}$$

Notice the two nonlinearities of the example: the drag, and the square root.

A simple Euler discretization is chosen for the motion equations. Hence:

$$\begin{aligned} x_1(n+1) &= x_1(n) + x_2(n) \cdot T \\ x_2(n+1) &= x_2(n) + (d+g) \cdot T \\ x_3(n+1) &= x_3(n) \end{aligned} \tag{1.151}$$

Program 1.8 simulates the evolution of body variables. Figure 1.23 shows the evolution of altitude and velocity.

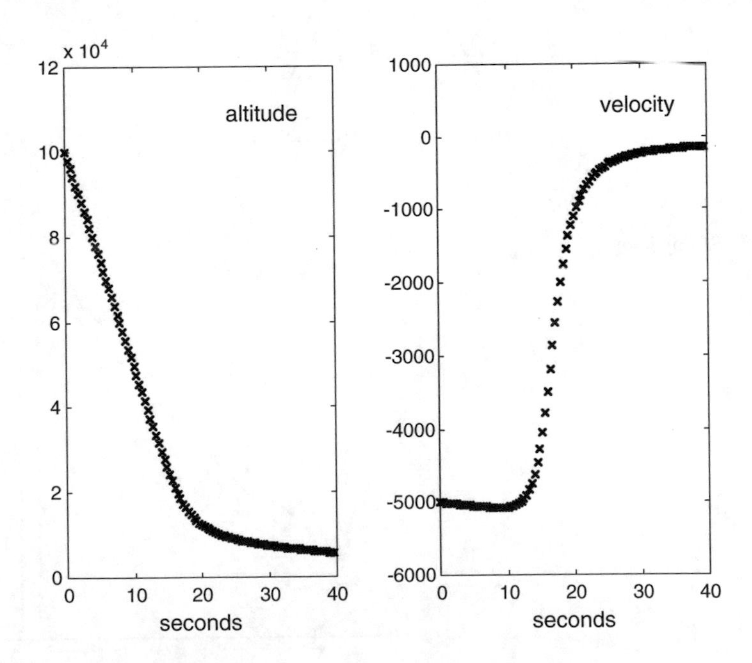

Fig. 1.23 System states (cross marks)

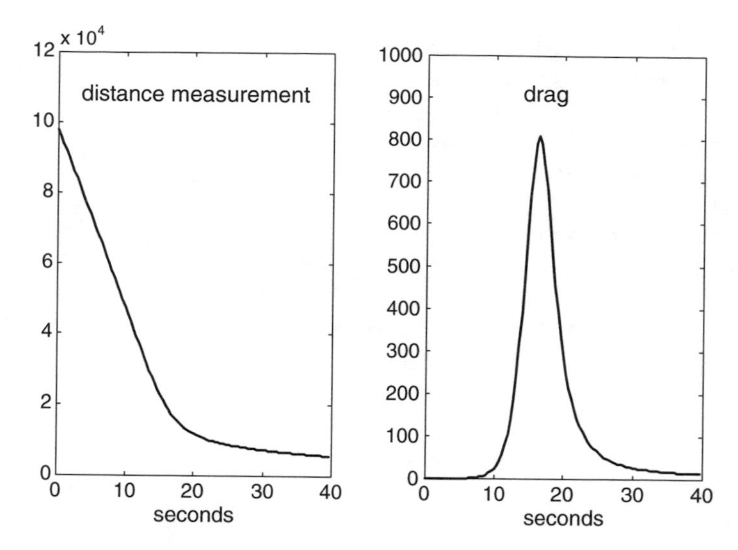

Fig. 1.24 Distance measurement and drag

Figure 1.24 shows the evolution of distance measurements obtained by the radar, and the evolution of air drag.

Program 1.8 Radar monitoring of falling body

```
%Example of nonlinear dynamics and measurement
%Radar monitoring of falling body
%----------------------------------------
%Prepare for the simulation of the falling body
T=0.4; %sampling period
g=-9.81;
rho0=1.225; %air density, sea level
k=6705.6; %density vs. altitude constant
L=100; %horizontal distance radar<->object
L2=L^2;
Nf=100; %maximum number of samples
rx=zeros(3,Nf); %space for state record
rd=zeros(1,Nf); %space for drag record
ry=zeros(1,Nf); %space for measurement record
tim=0:T:(Nf-1)*T; %time
x=[10^5; -5000; 400]; %initial state
%----------------------------------------
%simulation
nn=1;
while nn<Nf+1,
%system
```

```
rx(:,nn)=x; %state recording
rho=rho0*exp(-x(1)/k); %air density
d=(rho*(x(2)^2))/(2*x(3)); %drag
rd(nn)=d; %drag recording
%next system state
x(1)=x(1)+(x(2)*T);
x(2)=x(2)+((g+d)*T);
x(3)=x(3);
%system output
ym=sqrt(L2+(x(1)^2)); %measurement
ry(nn)=ym; %measurement recording
nn=nn+1;
end;
%----------------------------------------
%display
figure(1)
subplot(1,2,1)
plot(tim,rx(1,1:Nf),'kx');
title('altitude'); xlabel('seconds')
axis([0 Nf*T 0 12*10^4]);
subplot(1,2,2)
plot(tim,rx(2,1:Nf),'kx');
title('velocity'); xlabel('seconds');
axis([0 Nf*T -6000 1000]);
figure(2)
subplot(1,2,1)
plot(tim,ry(1:Nf),'k');
title('distance measurement');
xlabel('seconds');
axis([0 Nf*T 0 12*10^4]);
subplot(1,2,2)
plot(tim,rd(1:Nf),'k');
title('drag');
xlabel('seconds');
axis([0 Nf*T 0 1000]);
```

The Jacobians corresponding to this example are:

$$\frac{\partial \mathbf{f}}{\partial \mathbf{x}} = \begin{pmatrix} 0 & 0 & 0 \\ f_{21} & f_{22} & f_{23} \\ 0 & 0 & 0 \end{pmatrix} \tag{1.152}$$

$$\frac{\partial \mathbf{f}}{\partial \omega} = \begin{pmatrix} 1 & 0 & 0 \\ 0 & 1 & 0 \\ 0 & 0 & 1 \end{pmatrix} \tag{1.153}$$

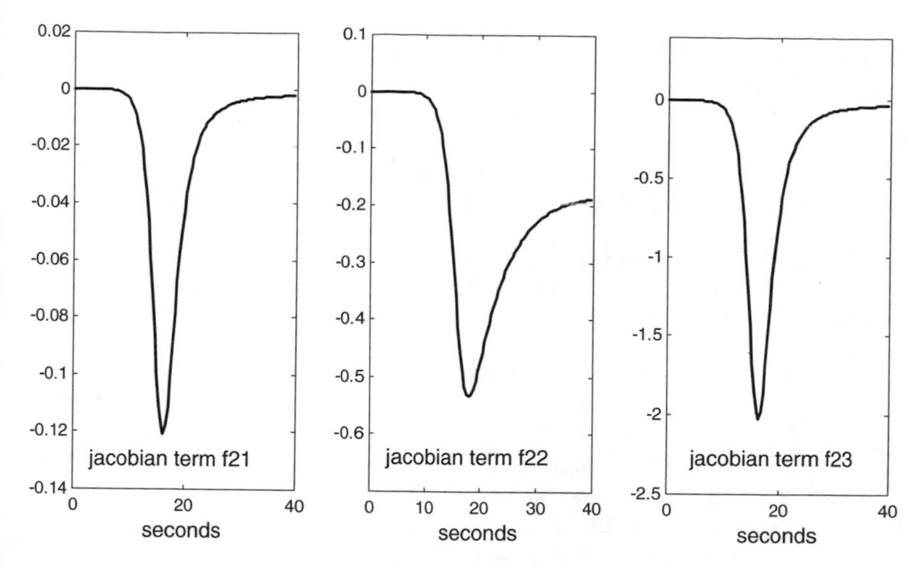

Fig. 1.25 The three non-zero components of the $\partial \mathbf{f} / \partial \mathbf{x}$ Jacobian

where:

$$f_{21} = -\frac{\rho\, x_2^2}{2\, k\, x_3} \tag{1.154}$$

$$f_{22} = \frac{\rho\, x_2}{x_3} \tag{1.155}$$

$$f_{23} = -\frac{\rho\, x_2^2}{2\, x_3^2} \tag{1.156}$$

and:

$$\frac{\partial \mathbf{h}}{\partial \mathbf{x}} = [\frac{x_1}{\sqrt{L^2 + x_1^2}} \quad 0 \quad 0] \tag{1.157}$$

$$\frac{\partial \mathbf{h}}{\partial v} = 1 \tag{1.158}$$

Program 1.9 computes the value of the three non-zero components of the $\partial \mathbf{f} \,/\, \partial \mathbf{x}$ Jacobian along the body fall. The program includes the same simulations as in the Program 1.8. Figure 1.25 shows the evolution of the three non-zero Jacobian components. They all have a minimum peak corresponding to maximum drag.

Program 1.9 Radar monitoring of falling body: Jacobians

```
%Jacobians
%Radar monitoring of falling body
%-----------------------------------------
%Prepare for the simulation of the falling body
T=0.4; %sampling period
g=-9.81;
rho0=1.225; %air density, sea level
k=6705.6; %density vs. altitude constant
L=100; %horizontal distance radar<->object
L2=L^2;
Nf=100; %maximum number of samples
tim=0:T:(Nf-1)*T; %time
%space for recording jacobians
rf2=zeros(3,Nf); rh1=zeros(1,Nf);
W=eye(3,3); %process noise jacobian
V=1; %observation noise jacobian
x=[10^5; -5000; 400]; %initial state
%-----------------------------------------
%Simulation
nn=1;
while nn<Nf+1,
%system
rx(:,nn)=x; %state recording
rho=rho0*exp(-x(1)/k); %air density
d=(rho*(x(2)^2))/(2*x(3)); %drag
rd(nn)=d; %drag recording
%next system state
x(1)=x(1)+(x(2)*T);
x(2)=x(2)+((g+d)*T);
x(3)=x(3);
%system output
ym=sqrt(L2+(x(1)^2)); %measurement
ry(nn)=ym; %measurement recording
%jacobians
f21=-d/k; f22=(rho*x(2)/x(3)); f23=-(d/x(3)); %state jacob.
ya=sqrt(L2+x(1)^2);
h1=xa(1)/ya; %measurement jacob.
rf2(:,nn)=[f21; f22; f23]; rh1(nn)=h1; %jacob. recording
nn=nn+1;
end;
%-----------------------------------------
%display
figure(1)
subplot(1,3,1)
plot(tim,rf2(1,1:Nf),'k');
title('jacobian term f21');
xlabel('seconds');
```

```
axis([0 Nf*T -0.14 0.02]);
subplot(1,3,2)
plot(tim,rf2(2,1:Nf),'k');
title('jacobian term f22');
xlabel('seconds');
axis([0 Nf*T -0.7 0.1]);
subplot(1,3,3)
plot(tim,rf2(3,1:Nf),'k');
title('jacobian term f23');
xlabel('seconds');
axis([0 Nf*T -2.5 0.4]);
```

To complete the initial analysis of the example, it is interesting to portray how the nonlinearities influence the state transitions.

In this case, for a fast and simple computation, the option was to create a set of perturbed states forming circles around a selected central state. The state number 43 (near the drag peak) was selected as the central state. Then, the perturbed states were propagated to the next state, state number 44, according with the transition equation. The result was plotted in Fig. 1.26. This all was the work of Program 1.10.

The propagated perturbed states form a series of closed curves showing certain asymmetries. Notice the scaling of the vertical axis.

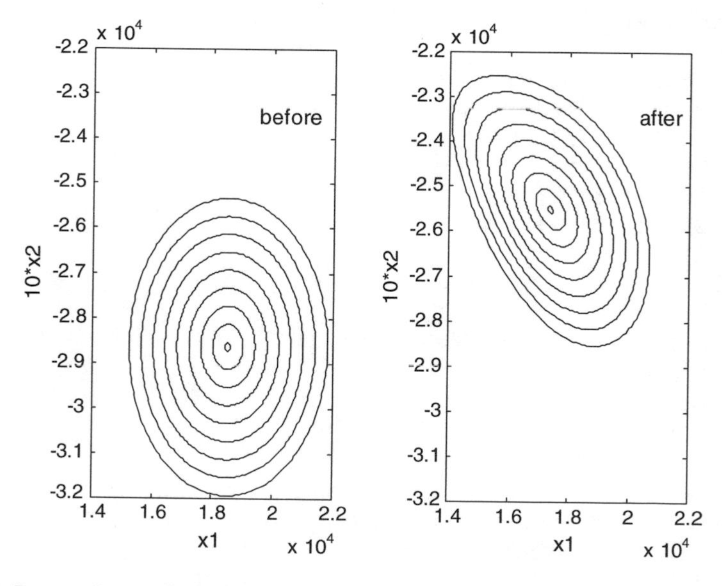

Fig. 1.26 Propagation of ellipsoids (state N=43 -> 44)s

Program 1.10 Radar monitoring of falling body: ellipsoids

```
%Propagation of ellipsoids
%Radar monitoring of falling body
%------------------------------------------
%Prepare for the simulation of the falling body
T=0.4; %sampling period
g=-9.81;
rho0=1.225; %air density, sea level
k=6705.6; %density vs. altitude constant
L=100; %horizontal distance radar<->object
L2=L^2;
Nf=42; %maximum number of samples
%space for ellipsoids
Ne=200;
ax1=zeros(1,Ne);
ax2=zeros(1,Ne);
px1=zeros(1,Ne);
px2=zeros(1,Ne);
tim=0:T:(Nf-1)*T; %time
x=[10^5; -5000; 400]; %initial state
%------------------------------------------
%simulation till a desired moment
nn=1;
while nn<Nf+1,
%system
rx(:,nn)=x; %state recording
rho=rho0*exp(-x(1)/k); %air density
d=(rho*(x(2)^2))/(2*x(3)); %drag
%next system state
x(1)=x(1)+(x(2)*T);
x(2)=x(2)+((g+d)*T);
x(3)=x(3);
nn=nn+1;
end;
%propagation of ellipsoids
% with display
for R=10:40:330,
for m=1:Ne+1,
%a priori ellipsoid
phi=(2*pi*(m-1))/Ne;
ax1(m)=x(1)+10*R*cos(phi);
ax2(m)=x(2)+R*sin(phi);
%posterior ellipsoid
rho=rho0*exp(-ax1(m)/k); %air density
d=(rho*(ax2(m)^2))/(2*x(3)); %drag
px1(m)=ax1(m)+(ax2(m)*T);
px2(m)=ax2(m)+((g+d)*T);
end;
```

```
%display
m=1:Ne+1;
figure(1)
subplot(1,2,1)
plot(ax1(m),10*ax2(m),'k'); hold on;
title('before');xlabel('x1'); ylabel('10*x2');
axis([1.4e4 2.2e4 -3.2e4 -2.2e4]);
subplot(1,2,2);
plot(px1(m),10*px2(m),'k'); hold on;
title('after');xlabel('x1'); ylabel('10*x2');
axis([1.4e4 2.2e4 -3.2e4 -2.2e4]);
end;
```

In the next sections, additive process and observation noises are considered. That means simple changes to the equations of the example:

$$\begin{aligned} x_1(n+1) &= x_1(n) + x_2(n) \cdot T + w_1(n) \\ x_2(n+1) &= x_2(n) + (d+g) \cdot T + w_2(n) \\ x_3(n+1) &= x_3(n) + w_3(n) \end{aligned} \tag{1.159}$$

$$y(n) = \sqrt{L^2 + x_1^2(n)} + v(n) \tag{1.160}$$

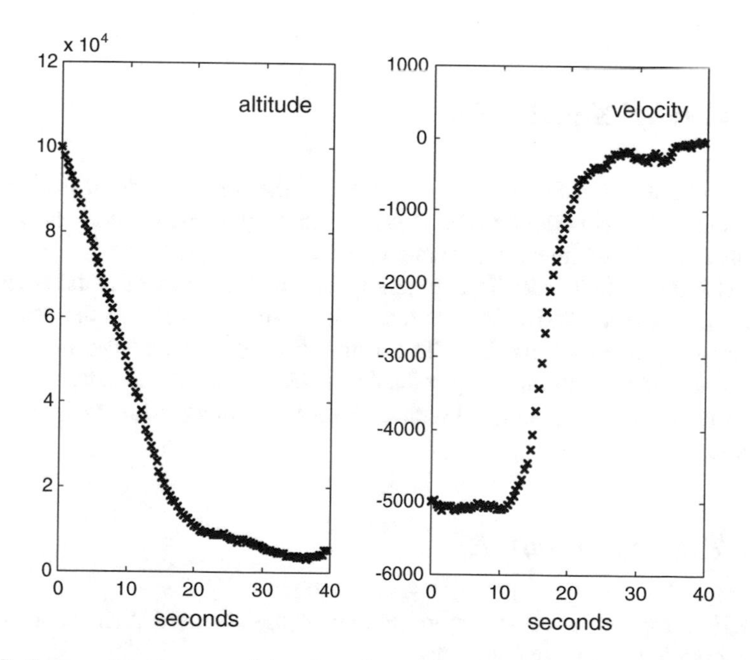

Fig. 1.27 System states (cross marks) under noisy conditions

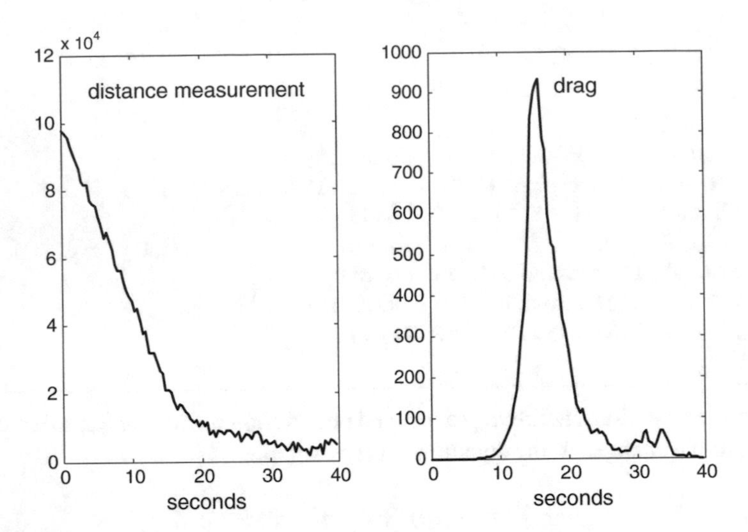

Fig. 1.28 Distance measurement. Drag

Next Figs. 1.27 and 1.28 show the evolution of state variables, measurements, and air drag during the fall under noisy conditions. These figures have been generated with a program that has been included in Appendix B and that is similar to the programs listed in this subsection.

1.5 Extended Kalman Filter (EKF)

The Extended Kalman Filter (EKF) is based on the linearization of the transition and the measurement equations [96, 119]. In this way, it can be used for nonlinear situations as far as the linear approximations are good enough.

The EKF uses a first order Taylor approximation of nonlinear functions via Jacobians, as it was described in subsection (8.4.3). The process and the observation noises are modelled as Gaussians. The distribution of propagated states, obtained with the transition equation, is approximated as Gaussian PDF. Likewise, the distribution of measurements, obtained with the measurement equation, is approximated as Gaussian PDF.

1.5.1 The EKF Algorithm

The EKF is applied as an algorithm similar to the standard AKF Kalman filter, repeating prediction and update steps.

Given the (nonlinear) system:

$$\mathbf{x}(n+1) = \mathbf{f}(\mathbf{x}(n),\ \mathbf{u}(n),\ \mathbf{w}(n)) \tag{1.161}$$

$$\mathbf{y}(n) = \mathbf{h}(\mathbf{x}(n),\ \mathbf{v}(n)) \tag{1.162}$$

The two steps of the EKF algorithm are the following

(a) Prediction

$$\mathbf{x}_a(n+1) = f(\mathbf{x}_e(n),\ \mathbf{u}(n)) \tag{1.163}$$

$$M(n+1) = F(n)\, P(n)\, F(n)^T + W(n)\, \Sigma_w\, W(n)^T \tag{1.164}$$

(b) Update

$$K(n+1) = M(n+1)\ H(n)^T \cdot \cdot [H(n)\, M(n+1)\, H(n)^T + V(n)\, \Sigma_v\, V(n)^T]^{-1} \tag{1.165}$$

$$P(n+1) = M(n+1) - K(n+1)\ H(n)\ M(n+1) \tag{1.166}$$

$$\mathbf{x}_e(n+1) = \mathbf{x}_a(n+1) + K(n+1)\,[\mathbf{y}(n+1) - h(\mathbf{x}_a(n+1, 0)] \tag{1.167}$$

The transition equation is directly used to predict the next state. According with subsection (8.4.3.), the associated covariance matrix M is computed using the following Jacobians, evaluated at $\mathbf{x}_e(n)$, $\mathbf{w}(n)$:

$$F(n) = \frac{d\mathbf{f}(\mathbf{x}, \mathbf{w})}{d\mathbf{x}}.$$

$$W(n) = \frac{d\mathbf{f}(\mathbf{x}, \mathbf{w})}{d\mathbf{w}}.$$

The update step computes the covariance matrix P using the following Jacobians, evaluated at $\mathbf{x}_a(n+1)$, $\mathbf{v}(n)$:

$$H(n) = \frac{d\mathbf{h}(\mathbf{x}, \mathbf{v})}{d\mathbf{x}}.$$

$$V(n) = \frac{d\mathbf{h}(\mathbf{x}, \mathbf{v})}{d\mathbf{v}}.$$

Notice that the measurement equation is directly used to compute the estimation error (see the last equation in the update step).

Comparing the equations of the prediction and update steps, with these steps in the standard Kalman filter, it could be said that the Jacobian $F(n)$ plays the role of the A matrix, and the Jacobian $H(n)$ plays the role of the C matrix.

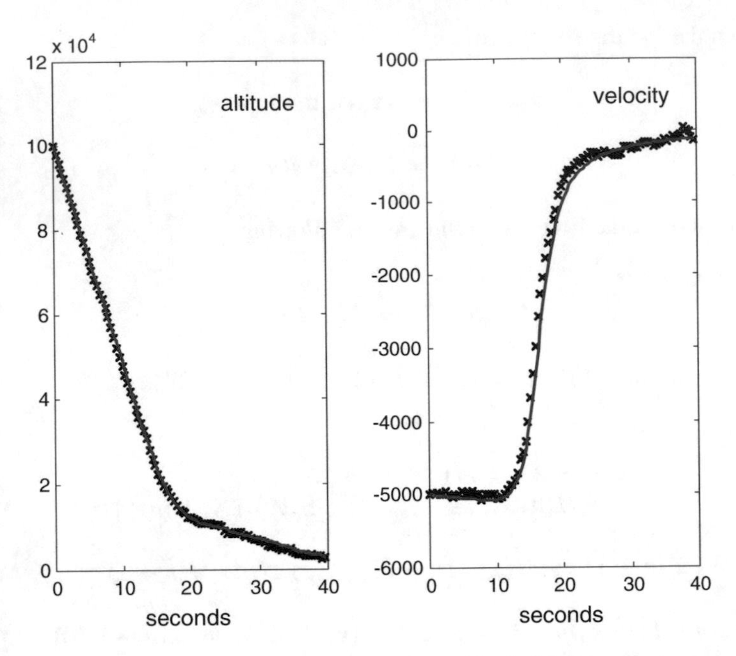

Fig. 1.29 System states (cross marks), and states estimated by the EKF (continuous)

As previously announced, the example of the falling body in noisy conditions is now used to illustrate how EKF works [61]. The Jacobians have been explicited and pictured in subsection (8.4.4.).

The Program 1.11 applies EKF to estimate the states of the system along the fall. Figure 1.29 shows the results.

Program 1.11 is useful to see the details of the EKF algorithm.

Program 1.11 Extended Kalman filter example

```
%Extended Kalman filter example
%Radar monitoring of falling body
%-----------------------------------------
%Prepare for the simulation of the falling body
T=0.4; %sampling period
g=-9.81;
rho0=1.225; %air density, sea level
k=6705.6; %density vs. altitude constant
L=100; %horizontal distance radar<->object
L2=L^2;
Nf=100; %maximum number of samples
rx=zeros(3,Nf); %space for state record
tim=0:T:(Nf-1)*T; %time
%process noise
Sw=[10^5 0 0; 0 10^3 0; 0 0 10^2]; %cov
```

```
bn=randn(3,Nf); sn=zeros(3,Nf);
sn(1,:)=sqrt(Sw(1,1))*bn(1,:); %state noise along simulation
sn(2,:)=sqrt(Sw(2,2))*bn(2,:); %" " "
sn(3,:)=sqrt(Sw(3,3))*bn(3,:); %" " "
%observation noise
Sv=10^6; %cov.
on=sqrt(Sv)*randn(1,Nf); %observation noise along simulation
%----------------------------------------
%Prepare for filtering
%space for matrices
K=zeros(3,Nf); M=zeros(3,3,Nf); P=zeros(3,3,Nf);
%space for recording er(n), xe(n)
rer=zeros(3,Nf); rxe=zeros(3,Nf);
W=eye(3,3); %process noise jacobian
V=1; %observation noise jacobian
%----------------------------------------
%Behaviour of the system and the filter after initial state
x=[10^5; -5000; 400]; %initial state
xe=x; % initial values of filter state
xa=xe; %initial intermediate state
nn=1;
while nn<Nf+1,
%estimation recording
rxe(:,nn)=xe; %state
rer(:,nn)=x-xe; %error
%system
rx(:,nn)=x; %state recording
rho=rho0*exp(-x(1)/k); %air density
d=(rho*(x(2)^2))/(2*x(3)); %drag
%next system state
x(1)=x(1)+(x(2)*T)+sn(1,nn);
x(2)=x(2)+((g+d)*T)+sn(2,nn);
x(3)=x(3)+sn(3,nn);
%system output
y=on(nn)+sqrt(L2+(x(1)^2));
ym=y; %measurement
%Prediction
%a priori state
rho=rho0*exp(-xe(1)/k); %air density
d=(rho*(xe(2)^2))/(2*xe(3)); %drag
xa(1)=xe(1)+(xe(2)*T);
xa(2)=xe(2)+((g+d)*T);
xa(3)=xe(3);
%a priori cov.
f21=-d/k; f22=(rho*xe(2)/xe(3)); f23=-(d/xe(3));
F=[0 1 0; f21 f22 f23; 0 0 0]; %state jacobian
M(:,:,nn+1)=(F*P(:,:,nn)*F')+ (W*Sw*W');
%
%Update
```

```
ya=sqrt(L2+xa(1)^2);
h1=xa(1)/ya;
H=[h1 0 0]; %measurement jacobian
K(:,nn+1)=(M(:,:,nn+1)*H')*inv((H*M(:,:,nn+1)*H')+(V*Sv*V'));
P(:,:,nn+1)=M(:,:,nn+1)-(K(:,nn+1)*H*M(:,:,nn+1));
xe=xa+(K(:,nn+1)*(ym-ya)); %estimated (a posteriori) state
nn=nn+1;
end;
%-------------------------------------------
%display
figure(1)
subplot(1,2,1)
plot(tim,rx(1,1:Nf),'kx'); hold on;
plot(tim,rxe(1,1:Nf),'r');
title('altitude'); xlabel('seconds')
axis([0 Nf*T 0 12*10^4]);
subplot(1,2,2)
plot(tim,rx(2,1:Nf),'kx'); hold on;
plot(tim,rxe(2,1:Nf),'r');
title('velocity'); xlabel('seconds');
axis([0 Nf*T -6000 1000]);
```

An extended version of the Program 1.11 has been included in Appendix B. Next three figures have been obtained with that program.

Figure 1.30 shows the evolution of the state estimation error along an experiment.

Figure 1.31 shows the evolution of the covariance matrix P. Notice the uncertainty increases near the middle of the plots (this is coincident with the drag peak).

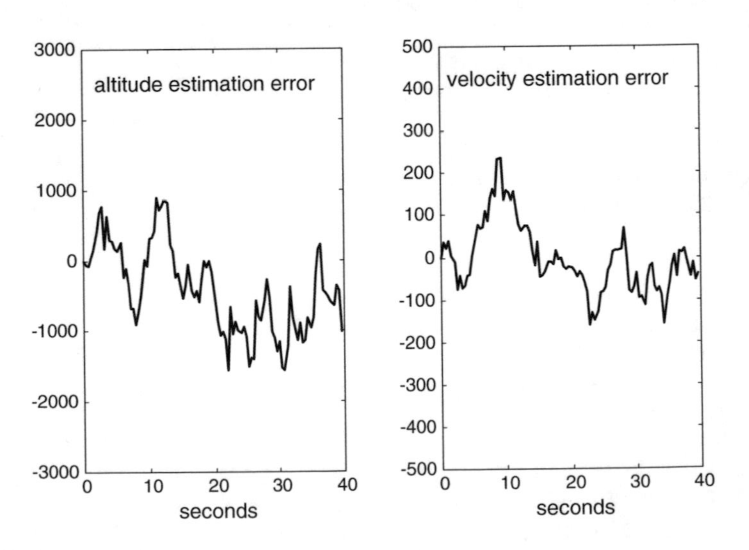

Fig. 1.30 Error evolution

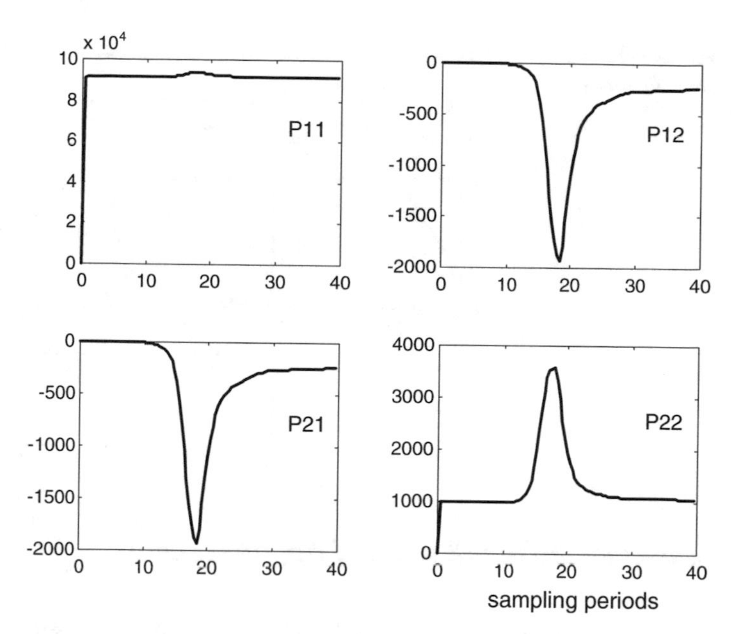

Fig. 1.31 Evolution of matrix P

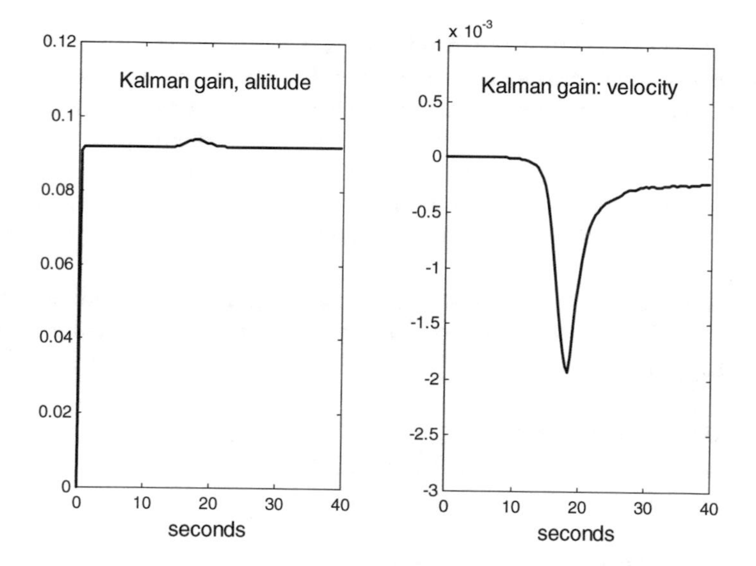

Fig. 1.32 Evolution of the altitude and velocity Kalman gains

Figure 1.32 shows the evolution of the Kalman gains along an experiment. The Kalman gain decreases in the moments of increasing uncertainty.

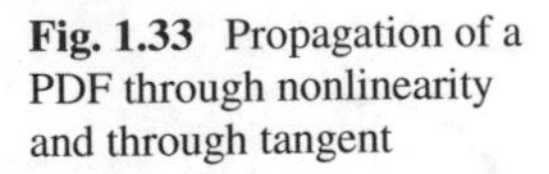

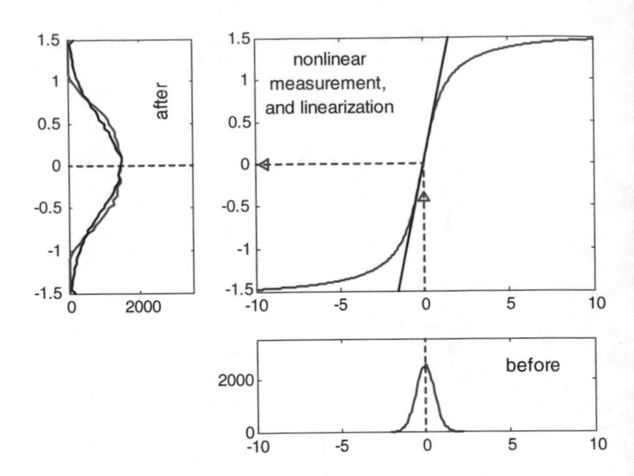

Fig. 1.33 Propagation of a PDF through nonlinearity and through tangent

1.5.2 *Assessment of the Linearized Approximation*

It seems convenient to show some limitations and possible problems associated to the linearization approach. To this purpose the example of the *arctan()* function will be used again.

An important aspect of Kalman filtering is the propagation of Gaussian PDFs through transition or measurement equations. Figure 1.33 shows how a Gaussian PDF propagates through *arctan()*. The linearization uses a tangent to the nonlinear function at (0, 0); this tangent has been plotted inside the limits imposed by *arctan()*.

The propagated PDF has been marked with a N. The PDF obtained by linearized propagation (using the tangent) has been also plotted on the propagated (after) side, and marked with a L. A visual comparison detects some differences between the N and the L curves.

Figure 1.33 has been generated with the Program 1.12. The standard deviation of the original PDF has been set to 0.55. The reader is invited to change this value and see what happens.

Program 1.12 Linearized propagation

```
%Propagation through nonlinearity
% and linearized propagation
%
%nonlinear function
vx=-10:0.1:10;
vy=atan(vx);
%gaussian random data
Nd=40000;
bx=-1.5:0.05:1.5; %location of histogram bins
sig=0.55;
adat=sig*randn(1,Nd);
```

```
%histogram of a priori data
Nbins=50;
[ha,hax]=hist(adat,Nbins);
%propagated random data
pdat=atan(adat);
%histogram of posterior data
hpt=hist(pdat,bx);
%data through tangent
ttg=1; tb=0; %the tangent
tdat=ttg*(adat)+tb;
%histogram of data through tangent
htg=hist(tdat,bx);
figure(1)
pl1=0.1; pb1=0.35; pw1=0.2; ph1=0.55;
pl2=0.4; pb2=0.05; pw2=0.55; ph2=0.2;
subplot('position',[pl1 pb1 pw1 ph1]) %left plot
plot(hpt,bx); hold on;
plot(htg,bx,'k'); %data through tangent
plot([0 3500],[0 0],'r--');
axis([0 3500 -1.5 1.5]);
ylabel('after');
subplot('position',[pl2 pb1 pw2 ph1]) %central plot
plot(vx,vy); hold on;
plot([-1.5 1.5],[-1.5 1.5],'k'); %the tangent
plot([-10 0],[0 0],'r--',-9.5,0,'r<');
plot([0 0],[-1.5 0],'r--',0,-0.4,'r^');
title('nonlinear measurement, and linearization');
subplot('position',[pl2 pb2 pw2 ph2]) %bottom plot
plot(hax,ha); hold on;
plot([0 0],[0 3500],'r--');
axis([-10 10 0 3500]);
title('before');
```

In order to see in more detail what happens when the standard deviation of the original PDF changes, another program, Program 1.13, has been developed. Figure 1.34 shows the linearized propagation, and the nonlinear propagation, of a Gaussian PDF, for three values of the standard deviation: *0.2, 0.4, 0.6*.

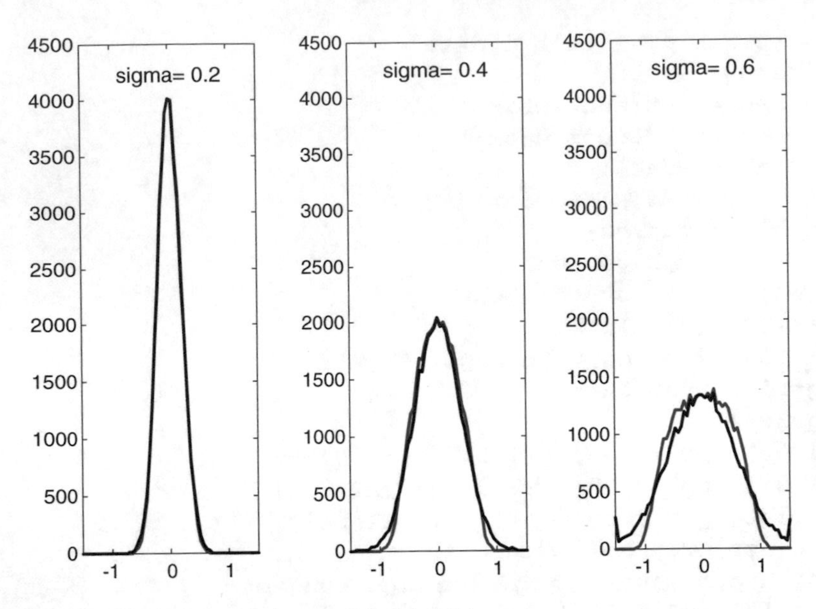

Fig. 1.34 Propagated PDFs for sigma = 0.2, 0.4, 0.6

As the PDF becomes wider the nonlinear propagation (N) diverges more from the linearized propagation (L). If you try to increase the standard deviation above the valuc 0.6, the tails of the original PDF get out from the range (-1.5–1.5) imposed by *arctan()*. This could become a problem in a practical situation.

Program 1.13 Propagation: variances

```
%Propagation through nonlinearity
%Several values of variance
%gaussian random data
Nd=40000;
rdat=randn(1,Nd);
bx=-1.5:0.05:1.5; %location of histogram bins
figure(1)
for nn=1:3,
switch nn,
case 1, sig=0.2;
case 2, sig=0.4;
case 3, sig=0.6;
end;
adat=sig*rdat;
pdat=atan(adat);
hpt=hist(pdat,bx);
%data through tangent
ttg=1; tb=0; %the tangent
```

```
tdat=ttg*(adat)+tb;
%histogram of data through tangent
htg=hist(tdat,bx);
subplot(1,3,nn)
plot(bx,hpt,'b'); hold on;
plot(bx,htg,'k');
tit=['sigma= ',num2str(sig)];
title(tit);
axis([-1.5 1.5 0 4500]);
end;
```

Another problem with the saturation limits imposed by *arctan()* becomes apparent if the original PDF is shifted. For instance, consider the case of the original Gaussian PDF being shifted to the right by 0.8. Figure 1.35, which has been generated with the Program 1.14, compares the nonlinear propagation (N) and the linearized propagation (L). The standard deviation has been set to 0.5, in order to avoid saturation (although a little can be observed at the right tail of L). The mean of the data propagated through *arctan()* has been depicted with a cross on the horizontal axis. The peaks of the N and L curves do not coincide. Neither of the peaks coincides with the propagated data mean. The PDF shift causes asymmetry.

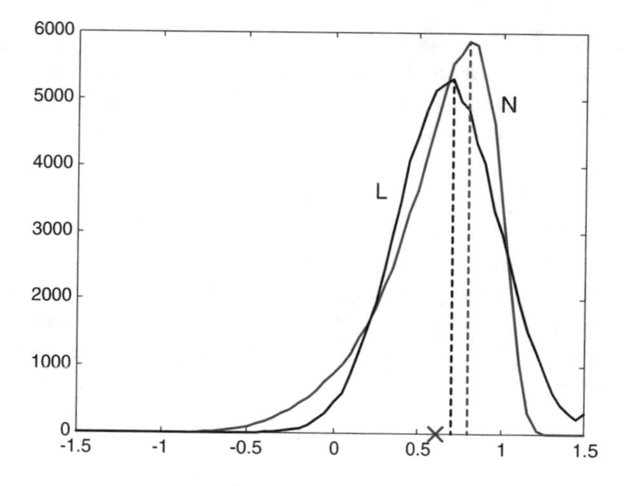

Fig. 1.35 Propagation of a shifted PDF through nonlinearity, and through tangent

Program 1.14 Propagation: asymmetrical result

```
%Propagation through nonlinearity
% and linearized propagation
% Asymmetrical result
%gaussian random data
Nd=80000;
rdat=randn(1,Nd);
bx=-1.5:0.05:1.5; %location of histogram bins
sig=0.5;
hsf=0.8;
adat=hsf+sig*randn(1,Nd); %with horizontal shift
%propagated random data
pdat=atan(adat);
%histogram of posterior data
hpt=hist(pdat,bx);
mup=mean(pdat); %mean of posterior data
[mx,ix]=max(hpt); %posterior histogram peak
ppx=bx(ix); ppy=mx; %" " coordinates
%data through tangent
ttg=1/(1+hsf^2); tb=atan(hsf)-(ttg*hsf); %the tangent
tdat=ttg*(adat)+tb;
%histogram of data through tangent
htg=hist(tdat,bx);
[mx,ix]=max(htg); %tdat histogram peak
ptx=bx(ix); pty=mx; %" " coordinates
%-----------------------------------------
figure(1)
plot(bx,hpt,'b'); hold on;
plot(bx,htg,'k');
plot(mup,0,'rx','Markersize',10); %posterior data mean
plot([ppx ppx],[0 ppy],'b--'); %line to post. hist. peak
plot([ptx ptx],[0 pty],'k--'); %line to tdat. hist. peak
title('nonlinear and linearized measurement');
```

1.6 Unscented Kalman Filter (UKF)

Depending on the particular case, it could be easier to approximate the propagated PDFs than the nonlinear functions.

The issue of PDF approximation is important in the context of Kalman filters, because the Kalman gains are computed in function of covariances of propagated states.

Since computers are more and more powerful, one is tempted to use brute force methods, and this could be reasonable, especially when the alternatives, if any, do not offer good performances. For the approximation of PDFs, it is possible to propagate a lot of samples through the nonlinearities, and then use statistics of the propagated samples. An example of this approach is particle filters, which will be introduced in the next section.

However, it is possible to approximate the propagated PDFs based on the propagation of a few, conveniently selected samples, like for instance the so-called *'sigma-points'* [108].

One of the most cited methods for sigma-point Kalman filtering is the *'Unscented Kalman Filter'* (UKF). With UKF it is not necessary to compute Jabobians nor Hessians, so UKF belongs to the class of *'derivative-free'* Kalman filtering methods. Since UKF uses a few sigma points, it requires moderate computational effort. The propagated PDFs are approximated using the mean and variance obtained from propagated sigma points [46, 53, 96, 115].

This section is devoted to introduce UKF and it is divided into two subsections. The first deals with the *'Unscented Transform'*, which is the basis of UKF. The satellite tracking example will be used to illustrate the main steps of UT. The second subsection describes the UKF algorithm, and includes its application to the falling body example.

1.6.1 The Unscented Transform

Let us start with a nonlinear function

$$\mathbf{y} = \mathbf{f}(\mathbf{x}) \tag{1.168}$$

Suppose that the function is applied to a set of random data with a certain original PDF. Another set of data is obtained, with a propagated PDF. The idea of sigma points is to be able to obtain the mean and covariance matrix of the original data, using a set of M selected data χ_i, as follows:

$$\mu_x = \sum_{i=0}^{M} w_i \chi_i \tag{1.169}$$

$$P_{xx} = \sum_{i=0}^{M} w_i (\chi_i - \mu_x)(\chi_i - \mu_x)^T \tag{1.170}$$

where w_i are suitable weights. They are normalized, so the sum of weights is one.

In addition, the propagated sigma points,

$$\mathbf{Y}_i = \mathbf{f}(\chi_i) \tag{1.171}$$

are used to obtain the mean and covariance matrix of the propagated data:

$$\boldsymbol{\mu}_y = \sum_{i=0}^{M} w_i \mathbf{Y}_i \tag{1.172}$$

$$P_{yy} = \sum_{i=0}^{M} w_i (\mathbf{Y}_i - \boldsymbol{\mu}_y)(\mathbf{Y}_i - \boldsymbol{\mu}_y)^T \tag{1.173}$$

and also:

$$P_{xy} = \sum_{i=0}^{M} w_i (\chi_i - \boldsymbol{\mu}_x)(\mathbf{Y}_i - \boldsymbol{\mu}_y)^T \tag{1.174}$$

A typical selection of sigma points is the following:

$$\begin{aligned} \chi_0 &= \boldsymbol{\mu}_x \\ \chi_i &= \chi_0 + \gamma \boldsymbol{\sigma}_i \ , \ \ i = 1, \ldots, N \\ \chi_i &= \chi_0 - \gamma \boldsymbol{\sigma}_i \ , \ \ i = N+1, \ldots, 2N \end{aligned} \tag{1.175}$$

where γ is a scaling factor, N is the dimension of the vector space, and $\boldsymbol{\sigma}_i$ is the i-th column of the matrix square root of P_{xx} (the covariance matrix):

$$P_{xx} = \boldsymbol{\sigma} \cdot \boldsymbol{\sigma}^T \tag{1.176}$$

The matrix square root can be obtained using the Cholesky decomposition.

In order to determine the weights w_i a set of constraint equations can be established. For example, in the scalar variable case:

$$\left[\sum_{i=0}^{M} w_i\right] - 1 = 0 \tag{1.177}$$

$$\left[\sum_{i=0}^{M} w_i \chi_i\right] - \mu_x = 0 \tag{1.178}$$

$$\left[\sum_{i=0}^{M} w_i (\chi_i - \mu_x)(\chi_i - \mu_x)^T\right] - P_{xx} = 0 \tag{1.179}$$

Suppose that the original PDF is Gaussian $N(0\,,\ 1)$, and that $\gamma = 1$. Then, the constraint equations are:

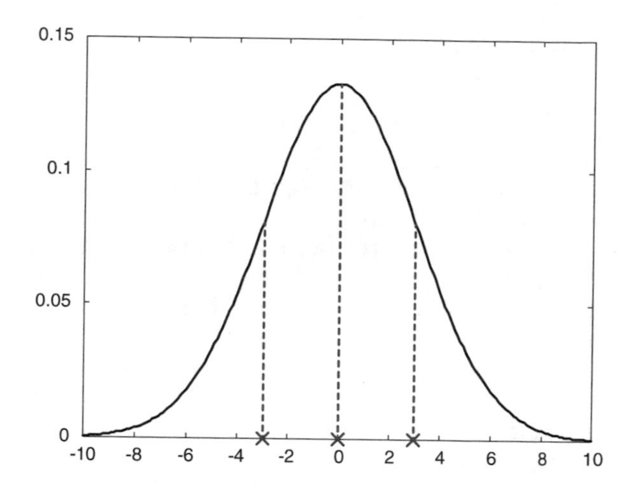

Fig. 1.36 Example of sigma points

$$\begin{array}{l} w_0 + w_1 + w_2 - 1 = 0 \\ w_1 \sigma_1 - w_2 \sigma_2 - 0 = 0 \\ w_1 \sigma_1^2 + w_2 \sigma_2^2 - 1 = 0 \end{array} \tag{1.180}$$

It is natural to choose symmetrical sigma points for a symmetric PDF. Therefore:

$$\sigma_1 = \sigma_2 \tag{1.181}$$

and, from the set of equations:

$$\begin{array}{l} w_1 = w_2 = 1/(2\,\sigma_1^2) \\ w_0 = 1 - 2\,w_1 \end{array} \tag{1.182}$$

A first example of sigma points is illustrated in Fig. 1.36, using a Gaussian PDF. There are three sigma points, represented with cross marks on the horizontal axis. The lateral points are placed at a distance σ (the standard deviation) from the central sigma point.

Program 1.15 Example of sigma points

```
%Example of sigma points
%
x1=-10:0.1:10;
%1D Gaussian---------------------------------
sig=3;
mu=0;
aux=(x1-mu).^2;
y1=(exp(-aux/(2*(sig^2)))/(sig*sqrt(2*pi)));
%sigma points
x1s0=mu;
```

```
x1s1=mu+sig;
x1s2=mu-sig;
%the points on PDF
y1s0=max(y1); %the PDF peak
aux=(x1s1-mu).^2;
y1s1=(exp(-aux/(2*(sig^2)))/(sig*sqrt(2*pi)));
aux=(x1s2-mu).^2;
y1s2=(exp(-aux/(2*(sig^2)))/(sig*sqrt(2*pi)));
figure(1)
plot(x1,y1,'k'); hold on; %the PDF
plot(x1s0,0,'rx','MarkerSize',12); %central sigma point
plot(x1s1,0,'rx','MarkerSize',12); %right sigma point
plot(x1s2,0,'rx','MarkerSize',12); %left sigma point
plot([x1s0 x1s0],[0 y1s0],'b--'); %central sigma point line
plot([x1s1 x1s1],[0 y1s1],'b--'); %right sigma point line
plot([x1s2 x1s2],[0 y1s2],'b--'); %left sigma point line
title('sigma points');
axis([-10 10 0 0.15]);
```

Another example of sigma points is shown in Fig. 1.37 with a bivariate Gaussian PDF and five sigma points.

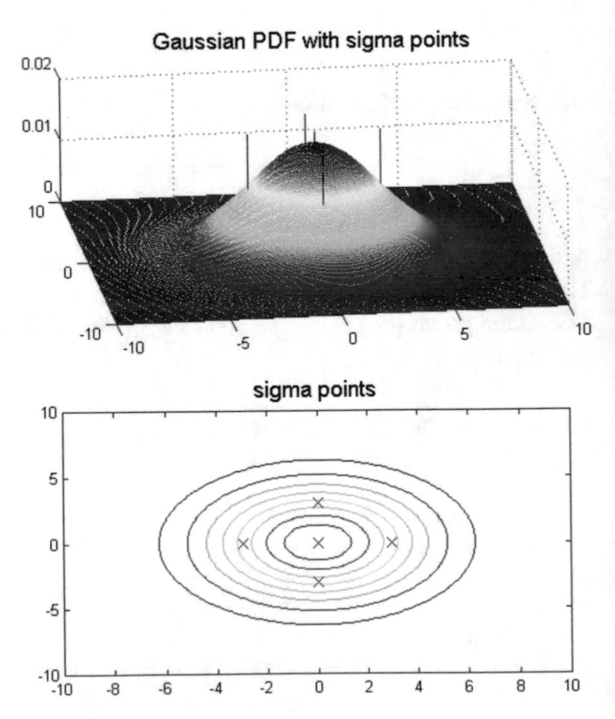

Fig. 1.37 Example of sigma points

Program 1.16 Example of sigma points

```
%Example of sigma points
%
x1=-10:0.1:10;
x2=-10:0.1:10;
pN1=length(x1);
pN2=length(x2);
%2D Gaussian----------------------------------
C=[9 0; 0 9];
mu1=0; mu2=0;
D=det(C);
K=1/(2*pi*sqrt(D)); Q=(C(1,1)*C(2,2))/(2*D);
y=zeros(pN1,pN2); %space for the PDF
for ni=1:pN1,
for nj=1:pN2,
aux1=(((x1(ni)-mu1)^2)/C(1,1))+(((x2(nj)-mu2).^2)/C(2,2));
y(ni,nj)= K*exp(-Q*aux1);
end;
end;
ymax=max(max(y)); %the PDF peak
%sigma points
xs=zeros(2,6); %reserve space
xs0=[mu1; mu2];
xs(:,1)=xs0+[3;0]; xs(:,2)=xs0+[0;3];
xs(:,3)=xs0-[3;0]; xs(:,4)=xs0-[0;3];
figure(1)
subplot(2,1,1)
ypk=ymax+0.002;
mesh(x1,x2,y); hold on; %the PDF
view(-7,45);
%sigma verticals
plot3([mu1 mu1],[mu2 mu2],[0 ypk],'k');
for nn=1:4,
plot3([xs(1,nn) xs(1,nn)],[xs(2,nn) xs(2,nn)],[0 ypk],'k');
end;
title('Gaussian PDF with sigma points');
subplot(2,1,2)
contour(x1,x2,y'); hold on; %the support
%sigma points:
plot(mu1,mu2,'rx','MarkerSize',10);
for nn=1:4,
plot(xs(1,nn),xs(2,nn),'rx','MarkerSize',10);
end;
title('sigma points');
```

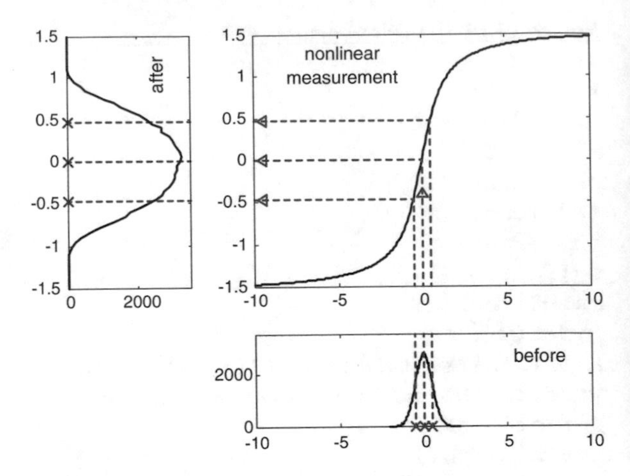

Fig. 1.38 Propagation of sigma points

The sigma points are propagated through the nonlinear function. An example of this is displayed on Fig. 1.38. With the propagated sigma points it is possible to compute the mean and variance of the propagated data. This information could be applied to approximate the propagated PDF with a Gaussian PDF. The figure has been generated with a program that has been included in Appendix B.

It has been noticed that the distance of the lateral sigma points to the central point increases as N increases. This is not convenient, and a scaling scheme has been devised to circumvent this problem:

$$\begin{aligned} \chi_0 &= \boldsymbol{\mu}_x \\ \chi_i &= \chi_0 + (\sqrt{(N+\lambda)\, P_{xx}})_i \;, \quad i = 1, \ldots, N \\ \chi_i &= \chi_0 - (\sqrt{(N+\lambda)\, P_{xx}})_i \;, \quad i = N+1, \ldots, 2N \end{aligned} \tag{1.183}$$

where $(matrix)_i$ means the i-th column of the matrix, and λ is a scaling parameter, such that:

$$\lambda = \alpha^2\,(N+\kappa) - N \tag{1.184}$$

The parameter α determines the spread of the sigma points around the centre, and usually takes a small positive value equal or less than one. The parenthesis term $(N + \kappa)$ is usually equal to 3.

The weighting factors are chosen as follows:

$$\begin{aligned} w_{0m} &= \frac{\lambda}{N+\lambda} \;; \quad w_{0c} = w_{0m} + (1 - \alpha^2 + \beta) \\ w_i &= \frac{1}{2(N+\lambda)} \;, \quad i = 1, \ldots, 2N \end{aligned} \tag{1.185}$$

The parameter β is an added degree of freedom to include a priori knowledge on the original PDF. In the case of a Gaussian PDF, $\beta = 2$ is optimal.

In order to illustrate the UT steps, the example of satellite tracking that was introduced in subsection (8.4.1.) is considered again. A set of radius and angle measurements has been obtained. It is supposed that the measurements obey to Gaussian PDFs. Figure 1.39 shows the bivariate PDF of the data on the angle vs. radius plane.

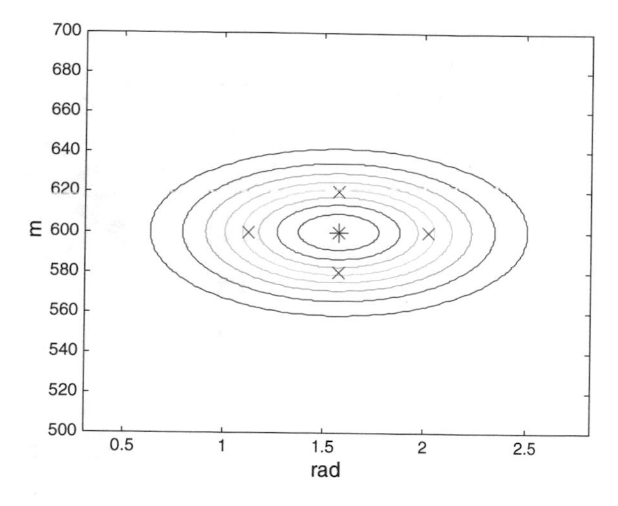

Fig. 1.39 Uncertainty on the angle-radius plane (satellite example), and sigma points

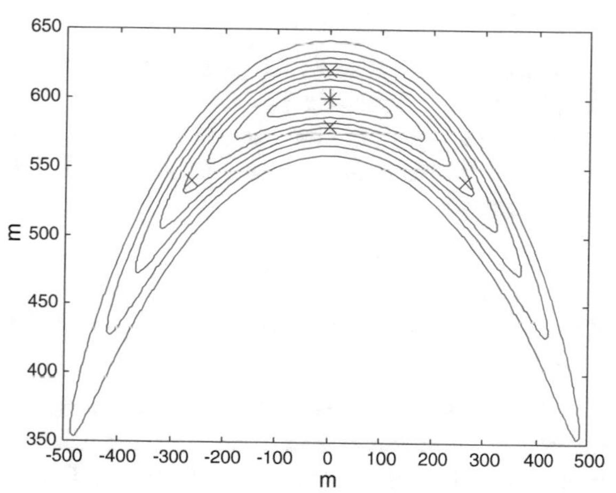

Fig. 1.40 The uncertainty on the Cartesian plane, and sigma points

A set of five sigma points have been selected. They are shown with cross marks.

The data are transformed to Cartesian coordinates (a polar to Cartesian coordinate transformation). Figure 1.40 shows the PDF of the propagated data. In addition, the propagated sigma points have been plotted with cross marks. Clearly, part of the symmetry of the original sigma points is lost on the Cartesian plane.

Now, the weights are applied to obtain the mean and covariance of the propagated data. With this information, a bivariate Gaussian PDF has been obtained (having the computed mean and covariance) as approximation to the propagated PDF. Figure 1.41 shows the cloud of propagated data and a contour plot of the bivariate Gaussian PDF. The mean of propagated data has been plotted with a cross mark, while the centre of the PDF has been plotted with a plus sign. Notice that both points are almost coincident.

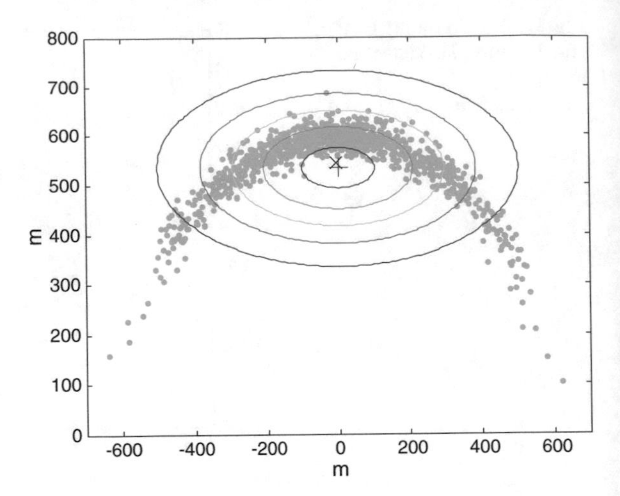

Fig. 1.41 The UT approximation and propagated data points

Figures 1.39, 1.40 and 1.41 have been generated with a long program that has been included in Appendix B.

There are some variations of the Kalman filter that use sigma points. They are denoted as '*sigma point Kalman filters*'. The UKF is an important example of this type of filters.

1.6.2 The Unscented Kalman Filter (UKF)

The UKF takes in a simple way the unscented transformation to obtain the covariance matrices used to compute the Kalman gains. Like the standard Kalman filter, UKF proceeds with a repeated two step algorithm. The details of the UKF algorithm are as follows.

Given the (nonlinear) system:

$$\mathbf{x}(n+1) = \mathbf{f}(\mathbf{x}(n), \mathbf{u}(n), \mathbf{w}(n)) \tag{1.186}$$

$$\mathbf{y}(n) = \mathbf{h}(\mathbf{x}(n), \mathbf{v}(n)) \tag{1.187}$$

Define a set of sigma points $\mathbf{x}_s^{(i)}$ according with the scaled unscented transform previously described.

Now, repeat the following steps.

(a) Prediction

- Use the transition equation to propagate the sigma points:

$$\mathbf{x}_{as}^{(i)}(n+1) = \mathbf{f}(\mathbf{x}_s^{(i)}(n)\,,\ \mathbf{u}(n))\,,\quad i = 0, \ldots, 2N \tag{1.188}$$

- Obtain the mean of the propagated sigma points:

$$\mathbf{x}_a(n+1) = w_{0m}\, x_{as}^{(0)}(n+1) + \sum_{i=1}^{2N} w_i x_{as}^{(i)}(n+1) \tag{1.189}$$

- Compute the a priori covariance matrix:

$$\begin{aligned} M(n+1) &= \Sigma_w + \\ &+ w_{0c}\,(\mathbf{x}_{as}^{(0)}(n+1) - \mathbf{x}_a(n+1))\,(\mathbf{x}_{as}^{(0)}(n+1) - \mathbf{x}_a(n+1))^T + \\ &+ \sum_{i=1}^{2N} w_i(\mathbf{x}_{as}^{(i)}(n+1) - \mathbf{x}_a(n+1))\,(\mathbf{x}_{as}^{(i)}(n+1) - \mathbf{x}_a(n+1))^T \end{aligned} \tag{1.190}$$

(b) Update

- Use the measurement equation to obtain measurements of the propagated sigma points:

$$\mathbf{y}_{as}^{(i)}(n+1) = \mathbf{h}(\mathbf{x}_{as}^{(i)}(n),\ \mathbf{u}(n)),\quad i = 0, ..., 2N \tag{1.191}$$

- Obtain the mean of the measurements:

$$\mathbf{y}_a(n+1) = w_{0m}\, y_{as}^{(0)}(n+1) + \sum_{i=1}^{2N} w_i y_{as}^{(i)}(n+1) \tag{1.192}$$

- Compute the measurement covariance matrix:

$$\begin{aligned} S_{yy}(n+1) &= \Sigma_v + \\ &+ w_{0c}\,(\mathbf{y}_{as}^{(0)}(n+1) - \mathbf{y}_a(n+1))\,(\mathbf{y}_{as}^{(0)}(n+1) - \mathbf{y}_a(n+1))^T + \\ &+ \sum_{i=1}^{2N} w_i(\mathbf{y}_{as}^{(i)}(n+1) - \mathbf{y}_a(n+1))\,(\mathbf{y}_{as}^{(i)}(n+1) - \mathbf{y}_a(n+1))^T \end{aligned} \tag{1.193}$$

- Compute the cross-covariance matrix:

$$\begin{aligned} S_{xy}(n+1) &= \\ &= w_{0c}\,(\mathbf{x}_{as}^{(0)}(n+1) - \mathbf{x}_a(n+1)) \cdot (\mathbf{y}_{as}^{(0)}(n+1) - \mathbf{y}_a(n+1))^T + \\ &+ \sum_{i=1}^{2N} w_i(\mathbf{x}_{as}^{(i)}(n+1) - \mathbf{x}_a(n+1))\,(\mathbf{y}_{as}^{(i)}(n+1) - \mathbf{y}a(n+1))^T \end{aligned} \tag{1.194}$$

- Obtain the Kalman gain as follows:

$$K(n+1) = S_{xy}(n+1) \cdot S_{yy}(n+1)^{-1} \tag{1.195}$$

- And obtain the remaining terms:

$$P(n+1) = M(n+1) - K(n+1)\, S_{yy}(n+1)\, K(n+1)^T \tag{1.196}$$

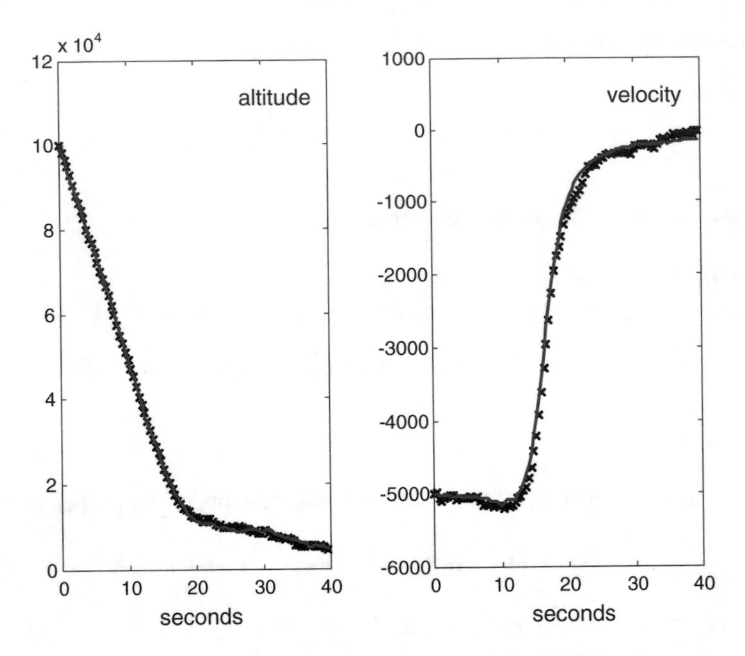

Fig. 1.42 System states (cross marks), and states estimated by the UKF (continuous)

$$\mathbf{x}_e(n+1) = \mathbf{x}_a(n+1) + K(n+1)\,[\mathbf{y}(n+1) - \mathbf{y}_a(n+1)] \tag{1.197}$$

As it was done in the EKF section, the example of the falling body is used to illustrate the UKF algorithm. Figure 1.42 shows the main result, which is the estimation of altitude and velocity of the body along time. The figure has been generated with the Program 1.17.

Notice that the MATLAB function *chol()* has been chosen to compute the matrix square root for the sigma points. This function uses the Cholesky factorization. Some care is needed for the specification of the sigma point parameters, to keep along experiments a positive definite covariance matrix (otherwise MATLAB refuses to apply the factorization). Note also that in the program, the central sigma point has been indexed with number 7, instead of 0.

Program 1.17 Unscented Kalman filter example

```
%Unscented Kalman filter example
%Radar monitoring of falling body
%-----------------------------------------
%Prepare for the simulation of the falling body
T=0.4; %sampling period
g=-9.81;
rho0=1.225; %air density, sea level
k=6705.6; %density vs. altitude constant
L=100; %horizontal distance radar<->object
L2=L^2;
Nf=100; %maximum number of samples
```

```
rx=zeros(3,Nf); %space for state record
rd=zeros(1,Nf); %space for drag record
ry=zeros(1,Nf); %space for measurement record
tim=0:T:(Nf-1)*T; %time
%process noise
Sw=[10^5 0 0; 0 10^3 0; 0 0 10^2]; %cov
bn=randn(3,Nf); sn=zeros(3,Nf);
sn(1,:)=sqrt(Sw(1,1))*bn(1,:); %state noise along simulation
sn(2,:)=sqrt(Sw(2,2))*bn(2,:); %" " "
sn(3,:)=sqrt(Sw(3,3))*bn(3,:); %" " "
%observation noise
Sv=10^6; %cov.
on=sqrt(Sv)*randn(1,Nf); %observation noise along simulation
%----------------------------------------
%Prepare for filtering
%space for matrices
K=zeros(3,Nf); M=zeros(3,3,Nf); P=zeros(3,3,Nf);
%space for recording er(n), xe(n)
rer=zeros(3,Nf); rxe=zeros(3,Nf);
%UKF parameters (to be edited here)
N=3; %space dimension
alpha=0.7; kappa=0; beta=2;
%pre-computation of constants
lambda= ((alpha^2)*(N+kappa))-N;
aab=(1-(alpha^2)+beta);
lN=lambda+N; LaN=lambda/lN; aaN=aab+LaN;
%----------------------------------------
%Behaviour of the system and the filter after initial state
x=[10^5; -5000; 400]; %initial state
xe=x; % initial value of filter state
xa=xe; %initial intermediate state
xs=zeros(3,7); %space for sigma points
xas=zeros(3,7); %space for propagated sigma points
yas=zeros(1,7); %" " "
P(:,:,1)=0.001*eye(3,3); %cov. non-zero init.
nn=1;
while nn<Nf+1,
%estimation recording
rxe(:,nn)=xe; %state
rer(:,nn)=x-xe; %error
%system
rx(:,nn)=x; %state recording
rho=rho0*exp(-x(1)/k); %air density
d=(rho*(x(2)^2))/(2*x(3)); %drag
rd(nn)=d; %drag recording
%next system state
x(1)=x(1)+(x(2)*T)+sn(1,nn);
x(2)=x(2)+((g+d)*T)+sn(2,nn);
x(3)=x(3)+sn(3,nn);
```

```
%system output
y=on(nn)+sqrt(L2+(x(1)^2));
ym=y; %measurement
ry(nn)=ym; %measurement recording
%Prediction
%sigma points
sqP=chol(lN*P(:,:,nn)); %matrix square root
xs(:,7)=xe;
xs(:,1)=xe+sqP(1,:)'; xs(:,2)=xe+sqP(2,:)';
xs(:,3)=xe+sqP(3,:)';
xs(:,4)=xe-sqP(1,:)'; xs(:,5)=xe-sqP(2,:)';
xs(:,6)=xe-sqP(3,:)';
%a priori state
%propagation of sigma points (state transition)
for m=1:7,
rho=rho0*exp(-xs(1,m)/k); %air density
d=(rho*(xs(2,m)^2))/(2*xs(3,m)); %drag
xas(1,m)=xs(1,m)+(xs(2,m)*T);
xas(2,m)=xs(2,m)+((g+d)*T);
xas(3,m)=xs(3,m);
end;
%a priori state mean (a weighted sum)
xa=0;
for m=1:6,
xa=xa+(xas(:,m));
end;
xa=xa/(2*lN);
xa=xa+(LaN*xas(:,7));
%a priori cov.
aux=zeros(3,3); aux1=zeros(3,3);
for m=1:6,
aux=aux+((xas(:,m)-xa(:))*(xas(:,m)-xa(:))');
end;
aux=aux/(2*lN);
aux1=((xas(:,7)-xa(:))*(xas(:,7)-xa(:))');
aux=aux+(aaN*aux1);
M(:,:,nn+1)=aux+Sw;
%Update
%propagation of sigma points (measurement)
for m=1:7,
yas(m)=sqrt(L2+(xas(1,m)^2));
end;
%measurement mean
ya=0;
for m=1:6,
ya=ya+yas(m);
end;
ya=ya/(2*lN);
ya=ya+(LaN*yas(7));
```

```
%measurement cov.
aux2=0;
for m=1:6,
aux2=aux2+((yas(m)-ya)^2);
end;
aux2=aux2/(2*lN);
aux2=aux2+(aaN*((yas(7)-ya)^2));
Syy=aux2+Sv;
%cross cov
aux2=0;
for m=1:6,
aux2=aux2+((xas(:,m)-xa(:))*(yas(m)-ya));
end;
aux2=aux2/(2*lN);
aux2=aux2+(aaN*((xas(:,7)-xa(:))*(yas(7)-ya)));
Sxy=aux2;
%Kalman gain, etc.
K(:,nn+1)=Sxy*inv(Syy);
P(:,:,nn+1)=M(:,:,nn+1)-(K(:,nn+1)*Syy*K(:,nn+1)');
xe=xa+(K(:,nn+1)*(ym-ya)); %estimated (a posteriori) state
nn=nn+1;
end;
%----------------------------------------
%display
figure(1)
subplot(1,2,1)
plot(tim,rx(1,1:Nf),'kx'); hold on;
plot(tim,rxe(1,1:Nf),'r');
title('altitude'); xlabel('seconds')
axis([0 Nf*T 0 12*10^4]);
subplot(1,2,2)
plot(tim,rx(2,1:Nf),'kx'); hold on;
plot(tim,rxe(2,1:Nf),'r');
title('velocity'); xlabel('seconds');
axis([0 Nf*T -6000 1000]);
```

An extended version of the Program 1.17 has been included in Appendix B. Next three figures have been obtained with that program.

Figure 1.43 shows the evolution of the state estimation error along an experiment.

Figure 1.44 shows the evolution of the covariance matrix P. Notice the uncertainty increase near the middle of the plots (this is coincident with the drag peak).

Figure 1.45 shows the evolution of the Kalman gains along an experiment. The evolution of the Kalman gains reflect the changes in the matrix P.

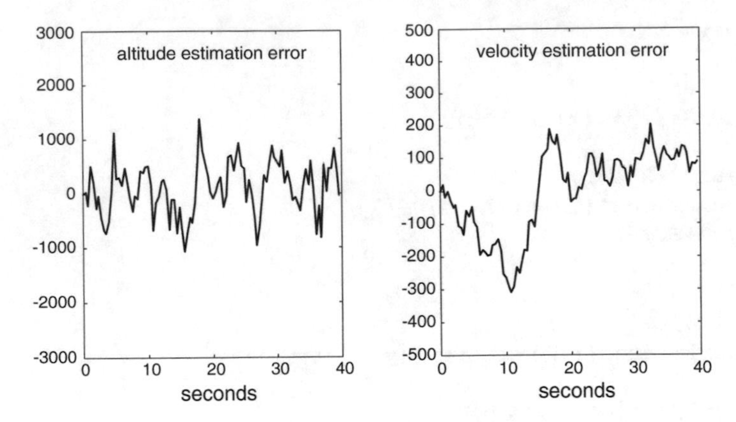

Fig. 1.43 Error evolution

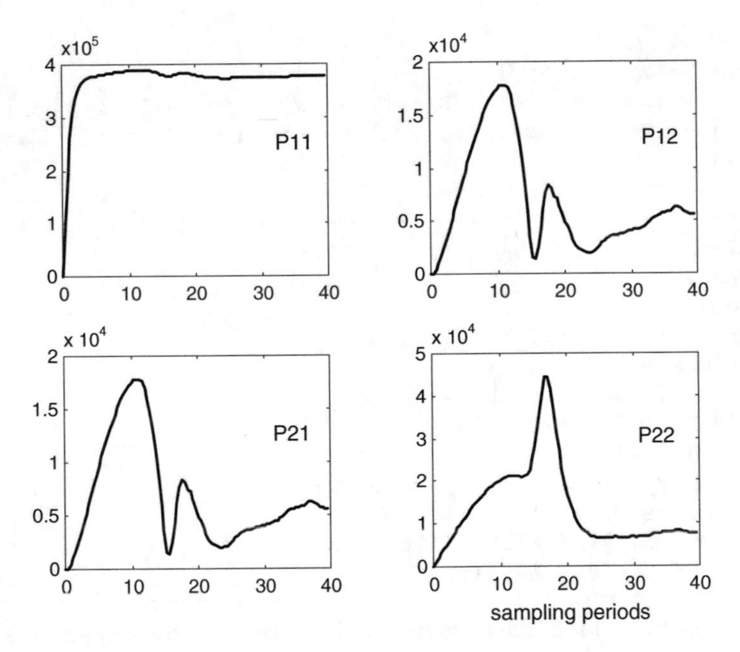

Fig. 1.44 Evolution of matrix P

1.7 Particle Filter

'Particle filters' are applications of Monte Carlo methods to Bayesian estimation. Sets of random states are used to obtain expected values, variances, etc. corresponding to dynamic processes. These random states are called 'particles', then sounding as a statistical mechanics metaphor.

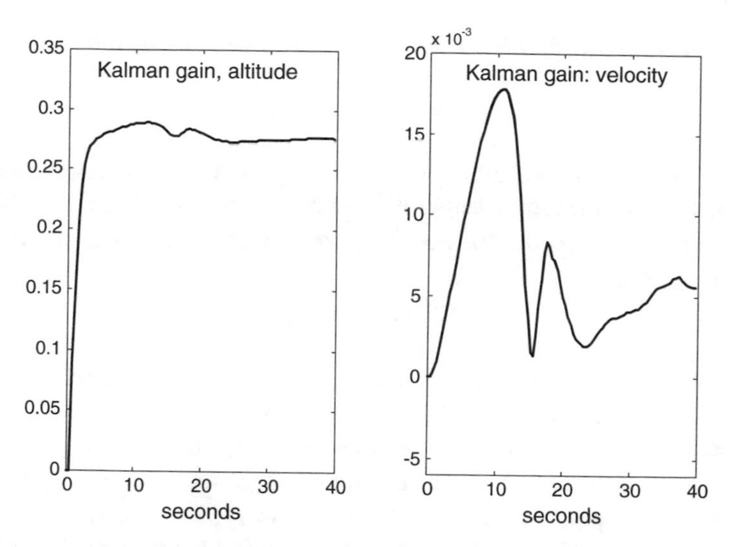

Fig. 1.45 Evolution of the altitude and velocity Kalman gains

The field of particle filters is rapidly expanding, exploring a wide repertory of applications and method formulations. The seminal paper on particle filters was [41]. The background of this section can be found in [7, 45, 96].

It was preferred to start the section with a complete implementation example, using a well-known type of particle filters. This example contains most of the key steps of usual particle filters, including resampling. After the implementation example, the section continues with a description of common resampling schemes. Then, the attention comes to the basic theory of particle filters, which is the root of several algorithm alternatives. The last subsection examines a selection of these alternatives.

The next subsection presents an implementation example, including a MATLAB program and considering again the example of the falling body.

1.7.1 An Implementation of the Particle Filter

Like the other filters already described, the particle filter algorithm repeats prediction and update steps. The algorithm neatly belongs to the Bayesian filtering approach, as it was briefly described in the chapter introduction.

The type of particle filter selected for this subsection is a *sequential importance resampling* (SIR) filter. It has been introduced by several authors with different names, such as *bootstrap* filter, *condensation algorithm*, etc.

Suppose there are N random samples from the posterior PDF at time step $k-1$, $p(\mathbf{x}_{k-1}|\,\mathbf{Y}_{k-1})$. These samples are called 'particles', and will be denoted as $\mathbf{x}_{k-1}^{j}$ with $j = 1, 2 \ldots, N$.

The **prediction** operation is just the propagation of the particles through the system transition equation:

$$\tilde{\mathbf{x}}_k^j = \mathbf{f}(\mathbf{x}_{k-1}^j, \ \mathbf{w}_{k-1}^j) \tag{1.198}$$

In this way, a set of particles from the prior PDF $p(\mathbf{x}_k | \mathbf{Y}_{k-1})$has been obtained.

For the **update** operation a weight W_k^j is computed for each particle, based on the measurements at time step k. The weight W_k^j is the measurement likelihood evaluated at the value of $\tilde{\mathbf{x}}_k^j$:

$$W_k^j = p(\mathbf{y}_k | \tilde{\mathbf{x}}_k^j) \tag{1.199}$$

The weights are then normalized to sum unity. Once the weights are obtained, there is a *resampling* step. The objective is to produce a set of samples from the posterior PDF $p(\mathbf{x}_k | \mathbf{Y}_k)$. These samples are extracted from $\tilde{\mathbf{x}}_k^j$. The resampling procedure is as follows: a particle from $\tilde{\mathbf{x}}_k^j$ is chosen with a probability equal to its weight; the procedure is repeated N times to get a new set $\mathbf{x}_k^j$; the same particle could be chosen several times. The idea is to get more samples from the more plausible states according with their measurement likelihoods.

Let us write in detail these steps, using an easy to program notation.

The program begins with the generation of a set of particles $\bar{px}(0)$. For simpler notation, the index for each particle has been dropped.

(a) Prediction

- Use the transition equation to propagate the particles:

$$a\bar{px}(n+1) = \mathbf{f}(\bar{px}(n), \ \mathbf{w}(n)) \tag{1.200}$$

(b) Update

- Use the measurement equation to obtain measurements of the propagated particles and their standard deviations:

$$\mathbf{y}(n+1) = \mathbf{h}(a\bar{px}(n+1), \ \mathbf{v}(n+1)) \tag{1.201}$$

$$\boldsymbol{\sigma}_y(n+1) = \mathbf{y}_m - \mathbf{y}(n+1) \tag{1.202}$$

  (in the case of our program, $\mathbf{y}_m$ is obtained via simulation of the system, in real-time applications this value could be just the average of measurements).
- Evaluate the measurement PDF at the propagated particles. Then, normalize to obtain the weights. For instance:

$$\boldsymbol{\sigma}_M(n+1) = \max(\boldsymbol{\sigma}_y(n+1)) \, /4 \tag{1.203}$$

$$\mathbf{W}(n+1) = \exp\left(-\left(\boldsymbol{\sigma}_y(n+1)/\boldsymbol{\sigma}_M(n+1)\right)^2\right) \tag{1.204}$$

$$\mathbf{W}(n+1) = \mathbf{W}(n+1)/\sum \mathbf{W}(n+1) \tag{1.205}$$

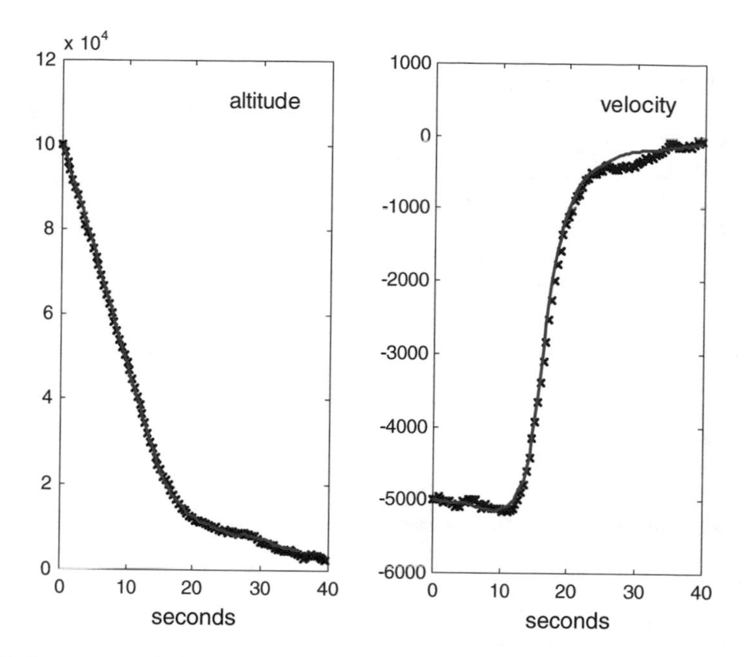

Fig. 1.46 System states (cross marks), and states estimated by the particle filter (continuous)

(the weights have been denoted as *pq(n)* in the program)

- Resampling. Obtain new $\bar{px}(n+1)$ from $a\bar{px}(n+1)$ using the normalized weights.
- Evaluate the mean of the set $\bar{px}(n+1)$. This is the estimated state.

The Program 1.18 provides an implementation of the particle filter for the falling body example. Figure 1.46 shows the results, which are quite good.

Program 1.18 Particle filter example

```
%Particle filter example
%Radar monitoring of falling body
disp('please wait a bit');
%------------------------------------------
%Prepare for the simulation of the falling body
T=0.4; %sampling period
g=-9.81;
rho0=1.225; %air density, sea level
k=6705.6; %density vs. altitude constant
L=100; %horizontal distance radar<->object
L2=L^2;
Nf=100; %maximum number of steps
rx=zeros(3,Nf); %space for state record
tim=0:T:(Nf-1)*T; %time
```

```
%process noise
Sw=[10^5 0 0; 0 10^3 0; 0 0 10^2]; %cov
w11=sqrt(Sw(1,1)); w22=sqrt(Sw(2,2)); w33=sqrt(Sw(3,3));
w=[w11; w22; w33];
%observation noise
Sv=10^6; %cov.
v11=sqrt(Sv);
%------------------------------------------
%Prepare for filtering
%space for recording er(n), xe(n)
rer=zeros(3,Nf); rxe=zeros(3,Nf);
%------------------------------------------
%Behaviour of the system and the filter after initial state
x=[10^5; -5000; 400]; %initial state
xe=x; %initial estimation
%prepare particles
Np=1000; %number of particles
%reserve space
px=zeros(3,Np); %particles
apx=zeros(3,Np); %a priori particles
ny=zeros(1,Np); %particle measurements
vy=zeros(1,Np); %meas. dif.
pq=zeros(1,Np); %particle likelihoods
%particle generation
wnp=randn(3,Np); %noise (initial particles)
for ip=1:Np,
px(:,ip)=x+(w.*wnp(:,ip)); %initial particles
end;
%system noises
wx=randn(3,Nf); %process
wy=randn(1,Nf); %output
nn=1;
while nn<Nf+1,
%estimation recording
rxe(:,nn)=xe; %state
rer(:,nn)=x-xe; %error
%Simulation of the system
%system
rx(:,nn)=x; %state recording
rho=rho0*exp(-x(1)/k); %air density
d=(rho*(x(2)^2))/(2*x(3)); %drag
rd(nn)=d; %drag recording
%next system state
x(1)=x(1)+(x(2)*T);
x(2)=x(2)+((g+d)*T);
x(3)=x(3);
x=x+(w.*wx(:,nn)); %additive noise
%system output
y=sqrt(L2+(x(1)^2))+(v11*wy(nn)); %additive noise
```

```
ym=y; %measurement
%Particle propagation
wp=randn(3,Np); %noise (process)
vm=randn(1,Np); %noise (measurement)
for ip=1:Np,
rho=rho0*exp(-px(1,ip)/k); %air density
d=(rho*(px(2,ip)^2))/(2*px(3,ip)); %drag
%next state
apx(1,ip)=px(1,ip)+(px(2,ip)*T);
apx(2,ip)=px(2,ip)+((g+d)*T);
apx(3,ip)=px(3,ip);
apx(:,ip)=apx(:,ip)+(w.*wp(:,ip)); %additive noise
%measurement (for next state)
ny(ip)=sqrt(L2+(apx(1,ip)^2))+(v11*vm(ip)); %additive noise
vy(ip)=ym-ny(ip);
end;
%Likelihood
%(vectorized part)
%scaling
vs=max(abs(vy))/4;
ip=1:Np;
pq(ip)=exp(-((vy(ip)/vs).^2));
spq=sum(pq);
%normalization
pq(ip)=pq(ip)/spq;
%Prepare for roughening
A=(max(apx')-min(apx'))';
sig=0.2*A*Np^(-1/3);
rn=randn(3,Np); %random numbers
%==================================
%Resampling (systematic)
acq=cumsum(pq);
cmb=linspace(0,1-(1/Np),Np)+(rand(1)/Np); %the "comb"
cmb(Np+1)=1;
ip=1; mm=1;
while(ip<=Np),
if (cmb(ip)<acq(mm)),
aux=apx(:,mm);
px(:,ip)=aux+(sig.*rn(:,ip)); %roughening
ip=ip+1;
else
mm=mm+1;
end;
end;
%==================================
%Results
%estimated state (the particle mean)
xe=sum(px,2)/Np;
nn=nn+1;
```

```
end;
%----------------------------------------
%display
figure(1)
subplot(1,2,1)
plot(tim,rx(1,1:Nf),'kx'); hold on;
plot(tim,rxe(1,1:Nf),'r');
title('altitude'); xlabel('seconds')
axis([0 Nf*T 0 12*10^4]);
subplot(1,2,2)
plot(tim,rx(2,1:Nf),'kx'); hold on;
plot(tim,rxe(2,1:Nf),'r');
title('velocity'); xlabel('seconds');
axis([0 Nf*T -6000 1000]);
```

Figure 1.47 shows the estimation errors recorded during an experiment. The figure has been generated with an extended version of the Program 1.18 that has been included in Appendix B.

1.7.2 Resampling Schemes

According to the literature on particle filters, the target of resampling is to eliminate the particles with low importance weights and multiply particles with high importance weights [50, 80, 96, 104].

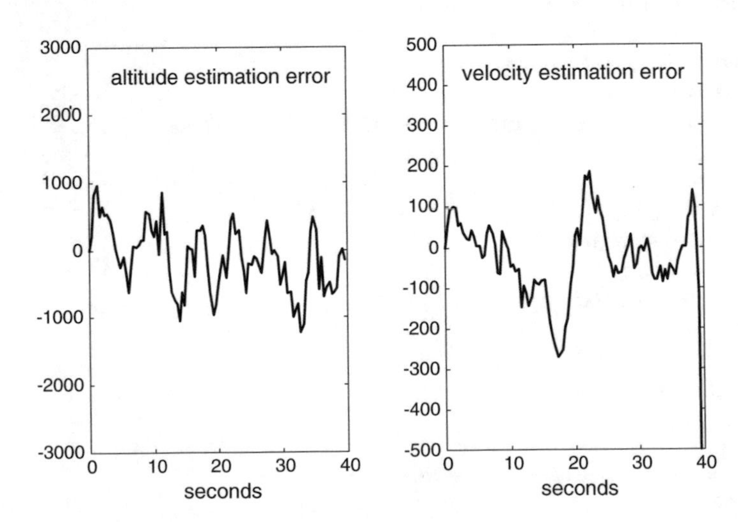

Fig. 1.47 Error evolution

To provide a tool for the reader, which may be used to check the resampling schemes presented below, a program has been developed. This program is included in Appendix B. All the figures in this subsection have been generated with this program.

The four resampling schemes to be described now, have been implemented and included in the same program. The reader could easily examine the results of any of these schemes. The idea of the program is to simulate some first steps (up to three) of the falling body example, and then apply all four resampling schemes to obtain separately the next generation of particles. The results of the different schemes can be compared.

In all simulation experiments, it can be observed that before resampling, many particles have low weights. Figure 1.48 shows, for example, a typical histogram of weights, obtained in a experiment. The left column clearly express the abundance of particles with low weights.

The effect of getting a dispersion of particles (low importance weights) has been denoted as *weight degeneracy.* Resampling mechanisms are intended to counteract this effect.

To further increase the diversity of particles after resampling, some noise can be added: this is denoted as '*roughening*'. Next program fragments give examples of it.

In general, the nucleus of resampling schemes is inversion sampling, where the $F(.)$ is built using the importance weights. The MATLAB *cumsum()* function is employed for this purpose. Figure 1.49 shows the result of this function in a typical experiment.

Figure 1.50 zooms on part of the *cumsum()* plot. When resampling, the number of copies of the particle *apx(mm)* (so a priori particles are denoted in the MATLAB program) should be proportional to the corresponding normalized weight.

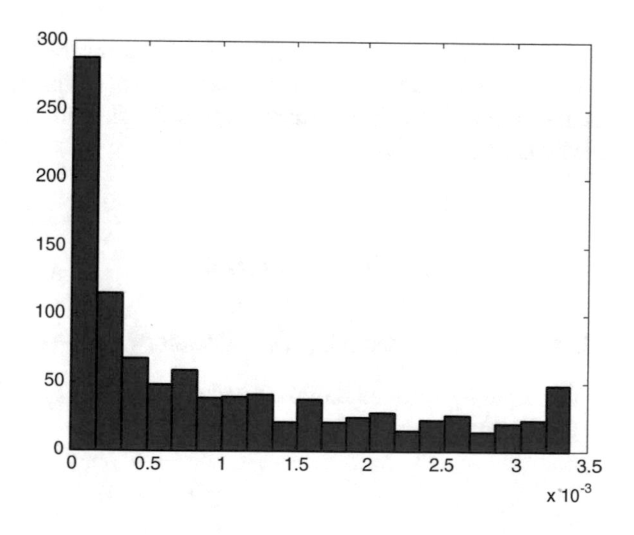

Fig. 1.48 Histogram of weights, resampling example

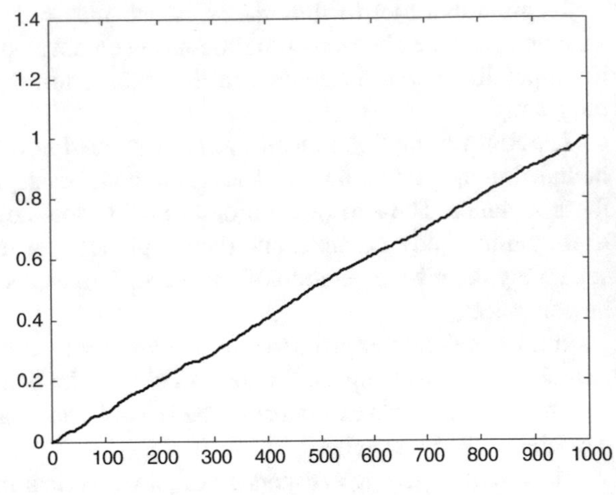

Fig. 1.49 Cumsum() of weights, resampling example

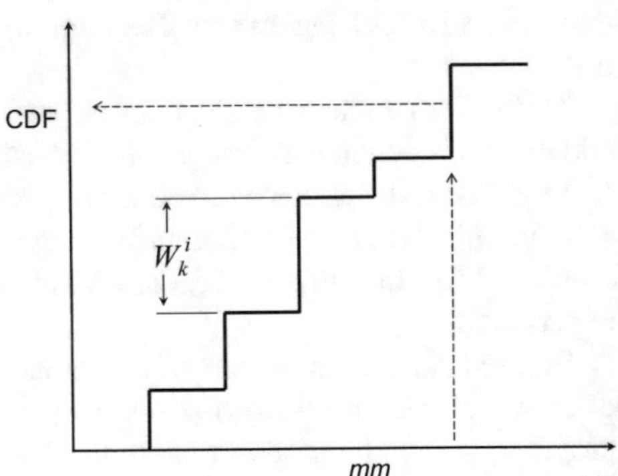

Fig. 1.50 Zoom on the cumsum() plot

Let us proceed with the description of four resampling methods. Pieces of MATLAB code, extracted from the program in Appendix B, are included. These pieces correspond to each of the resampling methods, and serve to give the details of the methods implementation.

1.7.3 Multinomial Resampling

The multinomial resampling algorithm steps are the following:

- Generate *Np* ordered uniform random numbers u_j.
- Use the numbers to select prior particles:

$$x(F^{-1}(u_j)) \tag{1.206}$$

Below is a MATLAB program segment that does multinomial resampling. The first lines compute the cumulative sum *acq* of the normalized weights *pq*, and generate an ordered set of uniform random numbers using the MATLAB *sort()* function. Then, there is a loop that obtains the roughened copies, *Mpx(ip)*, of selected prior particles.

The program use two pointers. The pointer *mm* selects prior particles (the source of copies). The pointer *ip* selects posterior particles (the roughened copies). The internal *while..end* determines values of *mm* trying to select good candidates for replication (prior particles with large weights).

Fragment 1.19 Resampling (multinomial)

```
%Resampling (multinomial)
acq=cumsum(pq);
mm=1;
nr=sort(rand(1,Np)); %ordered random numbers (0, 1]
for ip=1:Np,
while(acq(mm)<nr(ip)),
mm=mm+1;
end;
aux=apx(:,mm);
Mpx(:,ip)=aux+(sig.*rn(:,ip)); %roughening
end;
```

Figure 1.51 shows histograms of the prior particles and the posterior particles obtained by multinomial resampling. The figure corresponds to the third step of the falling body simulation.

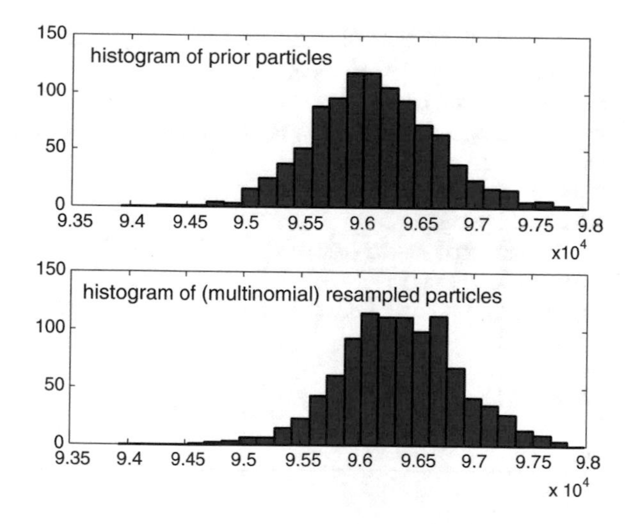

Fig. 1.51 Histograms of prior and resampled particles

1.7.4 Systematic Resampling

Most practical implementations of particle filters use *'systematic resampling'*. It is simple and offers good performances.

The algorithm is like the multinomial already described. Only that the ordered set of random numbers is generated in a different way: first a "comb" of *Np* regularly spaced points is formed; then, the comb is shifted by a uniform random number. This can be observed in the second line of the program fragment below. A simple mathematical description of this generation is:

$$u_j = \frac{(j-1) + \tilde{u}}{N_p}, \quad \tilde{u} \in U[0,1) \tag{1.207}$$

Let us describe in more detail the MATLAB fragment. Like before, two pointers, *mm* and *ip*, are used. There is a main loop that obtains one by one the posterior particles *Spx(:,ip)*. Suppose that, for instance, *ip* = *1*; the first member of the comb *cmb* is compared with the first member of *acq*; the index *mm* is increased until *acq(mm)* becomes larger than *cmb(1)*. Suppose that this happens when *mm* = *3*, the posterior particle *Spx(:,1)* is formed using the particle *apx(:,3)* and some added roughening noise. Now, $ip = 2$. If $cmb(2) < acq(3)$, a second particle *Spx(:,2)* is obtained using *apx(:,3)*. And so on, until there is *n* such that *acq(3)* is not larger than *cmb(n)*. In this case, no more roughened copies of *apx(:,3)*; *mm* is increased until $acq(mm) > cmb(n)$. Then, the same procedure is followed to make roughened copies of *apx(:,mm)*. And so on. The process is repeated until *Np* posterior particles are obtained.

Fragment 1.20 Resampling (systematic)

```
%Resampling (systematic)
acq=cumsum(pq);
cmb=linspace(0,1-(1/Np),Np)+(rand(1)/Np); %the "comb"
cmb(Np+1)=1;
ip=1; mm=1;
while(ip<=Np),
if (cmb(ip)<acq(mm)),
aux=apx(:,mm);
Spx(:,ip)=aux+(sig.*rn(:,ip)); %roughening
ip=ip+1;
else
mm=mm+1;
end;
end;
```

1.7.5 Stratified Resampling

The algorithm is similar to the systematic resampling. Only that the ordered random numbers are obtained as follows:

$$u_j = \frac{(j-1) + \tilde{u}_j}{N_p}, \quad \tilde{u}_j \in U[0,1) \tag{1.208}$$

Below is a MATLAB fragment with the stratified resampling. The posterior particles have been denoted as *Fpx(ip)*.

Fragment 1.21 Resampling (stratified)

```
%Resampling (stratified)
acq=cumsum(pq);
stf=zeros(1,Np);
nr=rand(1,Np)/Np;
j=1:Np;
stf(j)=nr(j)+((j-1)/Np); %(vectorized code)
stf(Np+1)=1;
ip=1; mm=1;
while(ip<=Np),
if (stf(ip)<acq(mm)),
aux=apx(:,mm);
Fpx(:,ip)=aux+(sig.*rn(:,ip)); %roughening
ip=ip+1;
else
mm=mm+1;
end;
end;
```

1.7.6 Residual Resampling

The residual resampling could be described as a two pass process. In the first, deterministic pass, *na(j)* copies of the particle *apx(j)* are obtained, where *na(j)* is computed as *floor(Np*pq(j))* (recall that *pq()* are the normalized weights). Suppose that *NR* copies have been obtained. The second pass gets the remaining number of particles, *Np-NR*, considering that the probability for selecting *apx(i)* is proportional to *rpq(i)* (a modified weight: see program).The second pass could use any of the previous resampling schemes.

Fragment 1.22 Resampling (residual)

```
%Resampling (residual)
acq=cumsum(pq);
mm=1;
```

```
%preparation
na=floor(Np*pq); %repetition counts
NR=sum(na); %total count
Npr=Np-NR; %number of non-repeated particles
rpq=((Np*pq)-na)/Npr; %modified weights
acq=cumsum(rpq); %for the monomial part
%deterministic part
mm=1;
for j=1:Np,
for nn=1:na(j),
Rpx(:,mm)=apx(:,j);
mm=mm+1;
end;
end;
%multinomial part:
nr=sort(rand(1,Npr)); %ordered random numbers (0, 1]
for j=1:Npr,
while(acq(mm)<nr(j)),
mm=mm+1;
end;
aux=apx(:,mm);
Rpx(:,NR+j)=aux+(sig.*rn(:,j)); %roughening
end;
```

1.7.7 Comparison

Some comparison studies conclude that stratified and systematic resampling are better than multinomial resampling in terms of better filter estimates and computational complexity. In the case of residual resampling, approximately half of the copies would be deterministic, and the other half stochastic: this means less computations, but it could be compensated by recalculation of weights and other aspects of the algorithm.

Figure 1.52 shows histograms of resampled particles corresponding to third step of the falling body simulation. Each histogram has been obtained with one of the four resampling methods. Notice that residual resampling results in less variance.

1.7.8 Roughening

In order to prevent sample impoverishment, roughening is applied to the posterior particles [21, 96]. A zero mean noise is added, with variance proportional to:

$$k\,\mathbf{M}\,N_p^{-1/n} \tag{1.209}$$

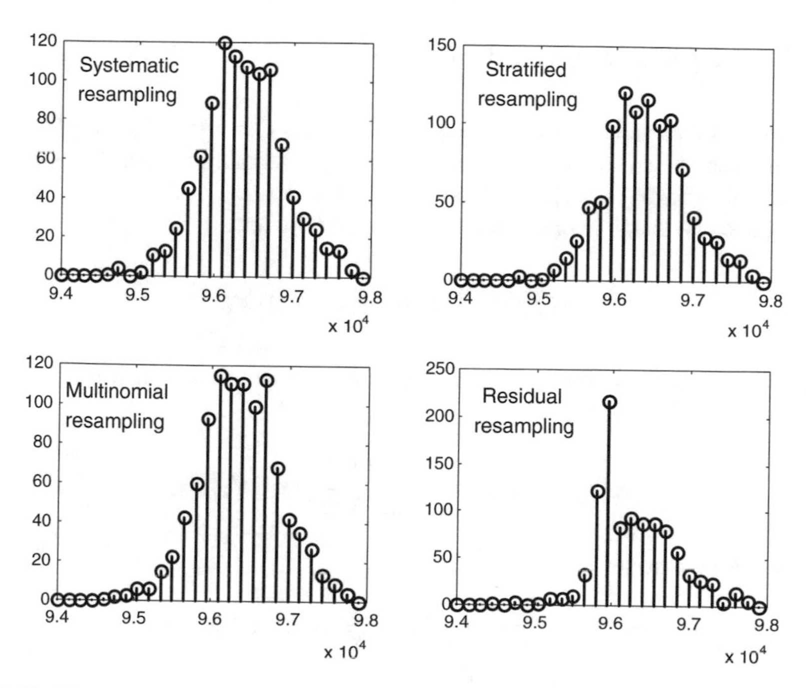

Fig. 1.52 Histograms of systematic, stratified, multinomial, residual resampling example

where n is the dimension of the space, k is a tuning parameter, and the *j-th* element of the vector **M** is:

$$\mathbf{M}(j) = \max_{m,n}(\mathbf{x}_k^m(j) - \mathbf{x}_k^n(j)) \tag{1.210}$$

where $\mathbf{x}_k^m(j)$ is the *j-th* component of the posterior particle $\mathbf{x}_k^m$ (which is a vector).

1.7.9 Basic Theory of the Particle Filter

Particle filters are applied to different kinds of applications, in economy, experimental sciences, sociology, etc. In a number of cases, analytical solutions for the Bayesian estimation equations are difficult to obtain. In such cases, it is reasonable to put Monte Carlo methods in action.

1.7.10 Sequential Monte Carlo (SMC)

A majority of particle filters use Sequential Monte Carlo (SMC) sampling; only in some cases batch mode is chosen. The type of sampling is usually *importance sampling*, employing a proposal importance function [7].

Suppose one wants to obtain an expectation of the form:

$$E(f) = \int f(\mathbf{x}_k)\, p(\mathbf{x}_k \mid \mathbf{Y}_k)\, d\mathbf{x}_k \tag{1.211}$$

According with the importance sampling approach, select a proposal importance function $q(\mathbf{x}_k \mid \mathbf{Y}_k)$, and write:

$$E(f) = \int f(\mathbf{x}_k)\, \frac{p(\mathbf{x}_k \mid \mathbf{Y}_k)}{q(\mathbf{x}_k \mid \mathbf{Y}_k)} q(\mathbf{x}_k \mid \mathbf{Y}_k)\, d\mathbf{x}_k \tag{1.212}$$

Invoking Bayes' rule:

$$\begin{aligned} E(f) = & \int f(\mathbf{x}_k)\, \tfrac{p(\mathbf{Y}_k|\mathbf{x}_k)\,p(\mathbf{x}_k)}{p(\mathbf{Y}_k) q(\mathbf{x}_k \mid \mathbf{Y}_k)} q(\mathbf{x}_k \mid \mathbf{Y}_k)\, d\mathbf{x}_k = \\ = & \int f(\mathbf{x}_k)\, \tfrac{w_k(\mathbf{x}_k)}{p(\mathbf{Y}_k)} q(\mathbf{x}_k \mid \mathbf{Y}_k)\, d\mathbf{x}_k \end{aligned} \tag{1.213}$$

where $w_k(\mathbf{x}_k)$ are the unnormalized importance weights:

$$w_k(\mathbf{x}_k) = \tfrac{p(\mathbf{Y}_k|\mathbf{x}_k)\,p(\mathbf{x}_k)}{q(\mathbf{x}_k \mid \mathbf{Y}_k)} \tag{1.214}$$

We obtain the following convenient expression:

$$E(f) = \frac{\int f(\mathbf{x}_k) w_k(\mathbf{x}_k) q(\mathbf{x}_k \mid \mathbf{Y}_k)\, d\mathbf{x}_k}{\int p(\mathbf{Y}_k|\mathbf{x}_k) p(\mathbf{x}_k) \frac{q(\mathbf{x}_k \mid \mathbf{Y}_k)}{q(\mathbf{x}_k \mid \mathbf{Y}_k)} d\mathbf{x}_k} = \frac{E(w_k(\mathbf{x}_k)\, f(\mathbf{x}_k))}{E(w_k(\mathbf{x}_k))} \tag{1.215}$$

Now, the expectations could be approximated by drawing N samples from $q(\mathbf{x}_k \mid \mathbf{Y}_k)$ and averaging:

$$E(f) \approx \frac{(1/N) \sum_{i=1}^{N} w_k(x_k^i)\, f(x_k^i)}{(1/N) \sum_{i=1}^{N} w_k(x_k^i)} = \sum_{i=1}^{N} W_k(x_k^i)\, f(x_k^i) \tag{1.216}$$

where

$$W_k(x_k^i) = \frac{w_k(x_k^i)}{\sum_{i=1}^{N} w_k(x_k^i)} \tag{1.217}$$

are the normalized weights.

In order to get an iterative version, it is convenient to choose a proposal distribution that could be factorized as follows:

$$q(\mathbf{x}_k \mid \mathbf{Y}_k) = q(\mathbf{x}_k|\mathbf{x}_{k-1}, \mathbf{Y}_k)q(\mathbf{x}_{k-1} \mid \mathbf{Y}_{k-1}) \quad (1.218)$$

Coming back to the unnormalized weights, one has:

$$w_k(\mathbf{x}_k) = \frac{p(\mathbf{Y}_k|\mathbf{x}_k)\,p(\mathbf{x}_k)}{q(\mathbf{x}_k|\mathbf{Y}_k)} = \frac{p(\mathbf{x}_k|\mathbf{Y}_k)p(\mathbf{Y}_k)}{q(\mathbf{x}_k|\mathbf{Y}_k)} \quad (1.219)$$

Now, factorize the posterior:

$$p(\mathbf{x}_k \mid \mathbf{Y}_k) = \frac{p(\mathbf{x}_{k-1} \mid \mathbf{Y}_{k-1})p(\mathbf{y}_k|\mathbf{x}_k)p(\mathbf{x}_k|\mathbf{x}_{k-1})}{p(\mathbf{y}_k|\mathbf{Y}_{k-1})} \quad (1.220)$$

Consider the factorization of the posterior and the proposal distribution:

$$w_k(\mathbf{x}_k) = \frac{p(\mathbf{x}_k|\mathbf{Y}_k)p(\mathbf{Y}_k)}{q(\mathbf{x}_k|\mathbf{Y}_k)} = \\ = \frac{p(\mathbf{x}_{k-1}|\mathbf{Y}_{k-1})}{q(\mathbf{x}_{k-1}|\mathbf{Y}_{k-1})} \cdot \frac{p(\mathbf{y}_k|\mathbf{x}_k)p(\mathbf{x}_k|\mathbf{x}_{k-1})}{q(\mathbf{x}_k|\mathbf{x}_{k-1}, \mathbf{Y}_k)} \cdot \frac{p(\mathbf{Y}_k)}{p(\mathbf{y}_k|\mathbf{Y}_{k-1})} \quad (1.221)$$

Therefore:

$$w_k(\mathbf{x}_k) = w_{k-1}(\mathbf{x}_{k-1}) \cdot \frac{p(\mathbf{y}_k|\mathbf{x}_k)p(\mathbf{x}_k|\mathbf{x}_{k-1})}{q(\mathbf{x}_k|\mathbf{x}_{k-1}, \mathbf{Y}_k)} \quad (1.222)$$

With this last expression, all the elements for the sequential importance sampling (SIS) algorithm have been obtained. Let us summarize the algorithm:

SIS Algorithm:

(the next 3 steps to be repeated for k = 0, 1, 2...)

1. Draw N samples from a proposal importance function $q(\mathbf{x}_k \mid \mathbf{Y}_k)$
2. Compute the importance weights:

$$w_k(\mathbf{x}_k^i) = w_{k-1}(\mathbf{x}_{k-1}^i) \cdot \frac{p(\mathbf{y}_k|\mathbf{x}_k^i)p(\mathbf{x}_k^i|\mathbf{x}_{k-1}^i)}{q(\mathbf{x}_k^i|\mathbf{x}_{k-1}^i, \mathbf{Y}_k)} \quad (1.223)$$

3. Obtain the normalized weights

$$W_k(x_k^i) = \frac{w_k(\mathbf{x}_k^i)}{\sum_{i=1}^{N} w_k(\mathbf{x}_k^i)} \quad (1.224)$$

On the basis of the normalized weights, one could approximate the posterior distribution as follows:

$$p(\mathbf{x}_k|\mathbf{Y}_k) \approx \sum_{i=1}^{N} W_k(x_k^i)\,\delta(\mathbf{x}_k - \mathbf{x}_k^i) \tag{1.225}$$

One could also approximate the mean of the posterior distribution:

$$\mu_k \approx \sum_{i=1}^{N} W_k(x_k^i) \cdot \mathbf{x}_k^i = \frac{1}{N}\sum_{i=1}^{N} \mathbf{x}_k^{i\,*} \tag{1.226}$$

and the covariance:

$$cov(\mathbf{x}_k) \approx \frac{1}{N}\sum_{i=1}^{N} (\mathbf{x}_k^{i\,*} - \mu_k)(\mathbf{x}_k^{i\,*} - \mu_k)^T \tag{1.227}$$

Notice that when the posterior distribution is multimodal and/or skewed the mean and the covariance may not be sufficient for statistical characterization.

A serious problem with the SIS algorithm is the degeneracy phenomenon: after a few iterations all but one particle have insignificant weight. A theorem establishes that the unconditional variance of the '*importance ratio*' increases over time (as the algorithm is repeated). The importance ratio is:

$$\frac{p(\mathbf{x}_k^i|\mathbf{Y}_k)}{q(\mathbf{x}_k^i|\mathbf{Y}_k)} \tag{1.228}$$

(the weights are proportional to this ratio)

A measure of the degeneracy is the *'effective sample size'*,N_{eff}. This parameter can be estimated with:

$$\hat{N}_{eff} = \frac{1}{\sum_{i=1}^{N} (W_k^i)^2} \tag{1.229}$$

Small values of N_{eff} indicates significant degeneracy.

Degeneration can be reduced with a suitable choice of the proposal distribution, and using resampling.

However, since resampling only keeps and replicates some particles, and eliminates the rest, the number of distinct samples tends to decrease. This is called *'sample impoverishment'*. This represents a problem.

Some algorithms use resampling in every cycle. Other algorithms monitor the degeneracy, using N_{eff}, and apply from time to time resampling (when $N_{eff} < threshold$).

Sequential importance sampling algorithms that include resampling are denoted as *'sequential importance resampling'* (SIR) algorithms. Therefore:

SIR Algorithm:

(the next 4 steps to be repeated for k = 0, 1, 2...)

1. Draw N samples from a proposal importance function $q(\mathbf{x}_k \mid \mathbf{Y}_k)$
2. Compute the importance weights:

$$w_k(\mathbf{x}_k^i) = w_{k-1}(\mathbf{x}_{k-1}^i) \cdot \frac{p(\mathbf{y}_k|\mathbf{x}_k^i)\,p(\mathbf{x}_k^i|\mathbf{x}_{k-1}^i)}{q(\mathbf{x}_k^i|\mathbf{x}_{k-1}^i,\, \mathbf{Y}_k)} \tag{1.230}$$

3. Compute the normalized weights

$$W_k(x_k^i) = \frac{w_k(\mathbf{x}_k^i)}{\sum\limits_{i=1}^{N} w_k(\mathbf{x}_k^i)} \tag{1.231}$$

4. Apply resampling to obtain a set of equally weighted particles $\mathbf{x}_k^{i\,*}$, which can be used for estimation of the posterior PDF, mean, etc.

1.7.11 *Proposal Importance Functions*

As said before, it is essential to select a good proposal importance function. It has been observed in practical applications that situations as represented in Fig. 1.53 may appear. The problem that could rise is that, depending on the chosen proposal function and the resampling scheme, most of the resampled particles would fall in the small overlapping zone as represented in the figure. This sample impoverishment problem may be caused by good sensors (peaky likelihoods). Also, any outlier pushing the likelihood to the right would cause the collapse of the algorithm.

The research is active in suggesting good proposal functions. Let us briefly review some important alternatives [11, 15, 29, 45, 125].

1.7.11.1 Optimal Proposal

It has been shown that the optimal importance function that minimizes the variance of the weights, conditioned upon $\mathbf{x}_{k-1}^i$ and $\mathbf{y}_k$, is the following:

$$q(\mathbf{x}_k \mid \mathbf{x}_{k-1}^i,\, \mathbf{y}_k) = p(\mathbf{x}_k \mid \mathbf{x}_{k-1}^i,\, \mathbf{y}_k) = \frac{p(\mathbf{y}_k|\,\mathbf{x}_k)\; p(\mathbf{x}_k \mid \mathbf{x}_{k-1}^i)}{p(\mathbf{y}_k|\,\mathbf{x}_{k-1}^i)} \tag{1.232}$$

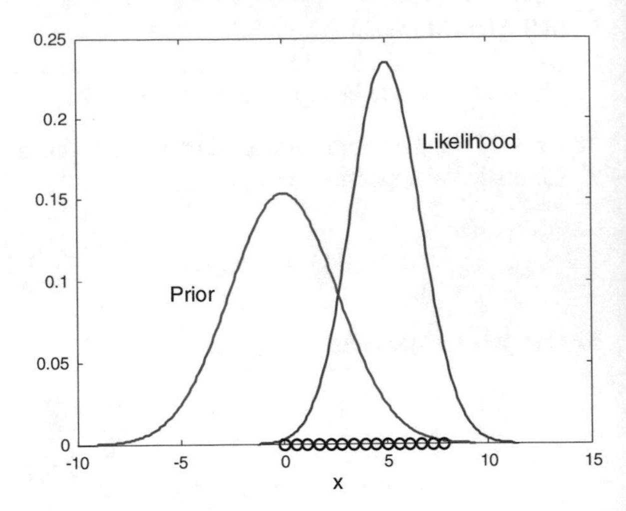

Fig. 1.53 A possible situation of prior and likelihood PDFs

This variance minimization implies that N_{eff} is maximized. Notice that the importance function combines the prior and the likelihood.

With this choice, the weight update is:

$$w_k(\mathbf{x}_k^i) = w_{k-1}(\mathbf{x}_{k-1}^i) \cdot p(\mathbf{y}_k|\mathbf{x}_{k-1}^i) \tag{1.233}$$

1.7.11.2 Standard Proposal:

Most practical applications use the prior as proposal importance function, and so it is named the standard proposal:

$$q(\mathbf{x}_k \,|\, \mathbf{x}_{k-1}^i, \mathbf{y}_k) = p(\mathbf{x}_k \,|\, \mathbf{x}_{k-1}^i) \tag{1.234}$$

The weight update is simple:

$$w_k(\mathbf{x}_k^i) = w_{k-1}(\mathbf{x}_{k-1}^i) \cdot p(\mathbf{y}_k|\mathbf{x}_k^i) \tag{1.235}$$

This proposal is the core of the bootstrap particle filter. The program 1.18 provides an implementation example of it.

1.7.11.3 Likelihood Proposal:

In certain cases, the predominant factor in the optimal proposal could be the likelihood, so the (scaled) likelihood could be chosen as proposal:

$$q(\mathbf{x}_k \,|\, \mathbf{x}_{k-1}^i, \mathbf{y}_k) \propto p(\mathbf{y}_k|\, \mathbf{x}_k) \tag{1.236}$$

In this case, the weight update is:

$$w_k(\mathbf{x}_k^i) = w_{k-1}(\mathbf{x}_{k-1}^i) \cdot p(\mathbf{x}_k^i | \mathbf{x}_{k-1}^i) \tag{1.237}$$

It is opportune to remark that in many cases it is not easy to find a good proposal. An avenue to get a solution is provided by factorization of the proposal, so it can be sequentially constructed as the filter algorithm runs. This is a main feature of sequential importance sampling.

Bootstrap techniques are commonly used in statistics, and are based on resampling from observed data. The basic idea is to get estimations using cumulative distributions of samples.

Before introducing other proposal function alternatives, it is convenient to consider the case of Gaussian $p(\mathbf{x}_k \,|\, \mathbf{x}_{k-1}^i,\, \mathbf{y}_k)$. This may happen, for instance, when the transition equation is nonlinear and the measurement equation is linear:

$$\mathbf{x}_k = \mathbf{f}(\mathbf{x}_{k-1}) + \mathbf{w}_k \tag{1.238}$$

$$\mathbf{y}_k = C\,\mathbf{x}_k + \mathbf{v}_k \tag{1.239}$$

where the noises are zero-mean Gaussian.

Define:

$$\Sigma^{-1} = \Sigma_w^{-1} + C^T \Sigma_v^{-1} C \tag{1.240}$$

$$\mathbf{m}_k = \Sigma\,(\,\Sigma_w^{-1}\,\mathbf{f}(\mathbf{x}_{k-1}) + C^T \Sigma_v^{-1}\,\mathbf{y}_k) \tag{1.241}$$

Then,

$$p(\mathbf{x}_k \,|\, \mathbf{x}_{k-1},\, \mathbf{y}_k) = N(\mathbf{m}_k,\, \Sigma) \tag{1.242}$$

$$p(\mathbf{y}_k \,|\, \mathbf{x}_{k-1}) = N(C\,\mathbf{f}(\mathbf{x}_{k-1}),\; \Sigma_w + C\,\Sigma_v\,C^T) \tag{1.243}$$

This lucky Gaussian circumstance is not found in many applications. In general there is no easy analytical way to get the posterior PDF, and you should try something related with proposal importance functions.

Part of the research has proposed to use EKF, or UKF, to obtain before each resampling step a proposal importance function. This idea has produced several combinations of particle and Kalman filter versions. Some authors use the term 'adaptation' for that mechanism. However, the term adaptation is also used for certain particle filter algorithms that change the number of particles along time, depending on the range to explore in each moment. A reason for the adaptation of the number of particles is to get fast particle filters, by using small filter populations (as small as possible in each step).

In general, you should decide which type of proposal function would be adequate for the process at hand, more or less fixed or adaptive.

Particles could also be used to estimate a good proposal before any resampling step. It may be compared to a scouting party. The prediction effort could be addressed to estimate the likelihood. In this context, it is illustrative to consider the *'Auxiliary Particle Filter'* introduced in [84], which has been used in several reported applications with good results [113, 120].

1.7.11.4 Auxiliary Particle Filter (APF)

Recall that the filtering target is to obtain the posterior in each step. Using Bayes' rule:

$$p(\mathbf{x}_{k+1}|\mathbf{Y}_{k+1}) \propto p(\mathbf{y}_{k+1}|\mathbf{x}_{k+1})p(\mathbf{x}_{k+1}|\mathbf{Y}_k) \tag{1.244}$$

And the prior $p(\mathbf{x}_{k+1} \mid \mathbf{Y}_k)$ is:

$$p(\mathbf{x}_{k+1}|\mathbf{Y}_k) = \int p(\mathbf{x}_{k+1}|\mathbf{x}_k)p(\mathbf{x}_k|\mathbf{Y}_k)\,d\mathbf{x}_k \approx \sum_{i=1}^{N} W_k(\mathbf{x}_k^i)\; p(\mathbf{x}_{k+1}|\mathbf{x}_k^i) \tag{1.245}$$

Observe that, according with the right-hand side of the last equation, the prior could be regarded as a mixture density. Samples could be drawn as follows: select the *i-th* component $p(\mathbf{x}_{k+1} \mid \mathbf{x}_k^i)$ with probability $W_k(\mathbf{x}_k^i)$, and then take a sample from it.

Combining the previous two equations, one has:

$$p(\mathbf{x}_{k+1}|\mathbf{Y}_{k+1}) \propto p(\mathbf{y}_{k+1}|\mathbf{x}_{k+1})\sum_{i=1}^{N} W_k(\mathbf{x}_k^i)\; p(\mathbf{x}_{k+1}|\mathbf{x}_k^i) \tag{1.246}$$

But this could be written as follows:

$$p(\mathbf{x}_{k+1}|\mathbf{Y}_{k+1}) \propto \sum_{i=1}^{N} W_k(\mathbf{x}_k^i)\; p(\mathbf{y}_{k+1}|\mathbf{x}_{k+1})\; p(\mathbf{x}_{k+1}|\mathbf{x}_k^i) \tag{1.247}$$

Again, this can be regarded as a mixture. Therefore, to sample from the posterior:

- select the *i-th* component $p(\mathbf{x}_{k+1} \mid \mathbf{x}_k^i)$ with probability:

  $$W_k(\mathbf{x}_k^i)\; p(\mathbf{y}_{k+1}|\mathbf{x}_{k+1})$$

- and take a sample from it

Another way to express this idea is by using an auxiliary variable, the index i, and write:

$$p(\mathbf{x}_{k+1},\; i|\mathbf{Y}_{k+1}) \propto W_k(\mathbf{x}_k^i)\; p(\mathbf{y}_{k+1}|\mathbf{x}_{k+1})\; p(\mathbf{x}_{k+1}|\mathbf{x}_k^i)\,, \quad i = 1, \ldots, N \tag{1.248}$$

One could draw, from each $p(\mathbf{x}_{k+1}, i \mid \mathbf{Y}_{k+1})$ several samples, up to R. This could be done using SIR, also obtaining R weights.

It has been suggested in [84] to use the following approximation for $p(\mathbf{x}_{k+1}, i \mid \mathbf{Y}_{k+1})$:

$$q(\mathbf{x}_{k+1},\ i|\mathbf{Y}_{k+1}) \ \propto \ W_k(\mathbf{x}_k^i)\, p(\mathbf{y}_{k+1}|\boldsymbol{\mu}_{k+1}^i)\, p(\mathbf{x}_{k+1}|\mathbf{x}_k^i) \tag{1.249}$$

where $\boldsymbol{\mu}_{k+1}^i$ is any likely value, the mean, the mode, etc., associated with the transition PDF. Suitable approximations are of the form:

$$q(i|\mathbf{Y}_{k+1}) \ \propto \ W_k(\mathbf{x}_k^i)\, p(\mathbf{y}_{k+1}|\boldsymbol{\mu}_{k+1}^i) \tag{1.250}$$

Then, samples would be drawn from $p(\mathbf{x}_{k+1} \mid \mathbf{x}_k^i)$ with probability $\lambda_i = q(i|\mathbf{Y}_{k+1})$. These λ_i are called first-stage weights, and correspond to likelihoods. The purpose of this sampling is to select particles with large predictive likelihoods.

Having obtained R samples $\mathbf{x}_k^j$, a reweighting is done, using second-stage weights:

$$w_j = \frac{p(\mathbf{y}_{k+1}|\mathbf{x}_{k+1}^j)}{p(\mathbf{y}_{k+1}|[\boldsymbol{\mu}_{k+1}^i]^j)}, \quad j = 1, \ldots, R \tag{1.251}$$

APF Algorithm:

Begin by generating N samples $\mathbf{x}_0^i$. from $p(\mathbf{x}_0)$, set $\boldsymbol{\mu}_0^i = \mathbf{x}_0^i$.

(the next 4 steps to be repeated for k = 0, 1, 2...)

1. For *i=1,...,N*, compute $\boldsymbol{\mu}_{k+1}^i = E(p(\mathbf{x}_{k+1} \mid \mathbf{x}_k^i)$
2. For *i=1,...,N*, calculate the first-stage weights $\lambda_i = W_k(\mathbf{x}_k^i)\, p(\mathbf{y}_{k+1}|\boldsymbol{\mu}_{k+1}^i)$, and normalize them.
3. Resample using the first-stage weights, and obtain $\mathbf{x}_k^j$particles
4. Compute the second-stage weights.

When the process noise is small, APF is usually better than SIR. For large noise, $\boldsymbol{\mu}_{k+1}^i$ would not give enough information about $p(\mathbf{x}_{k+1} \mid \mathbf{x}_k^i)$ and APF could degrade.

1.7.12 *Particle Filter Variants*

Many particle filter variants have been and are being proposed. For evident reasons of space here it is only possible to introduce, in a few words, some representative examples.

1.7.13 Marginalized Particle Filter (Rao-Blackwellized Particle Filter)

In certain cases it may be possible to divide the process at hand into linear-Gaussian and non-linear parts, so the state vector may be partitioned:

$$\mathbf{x}_k = \begin{pmatrix} \mathbf{x}_k^L \\ \mathbf{x}_k^N \end{pmatrix} \tag{1.252}$$

So the posterior, the target of the filter, may be factorized as follows:

$$p(\mathbf{x}_k|\mathbf{Y}_k) = p(\mathbf{x}_k^L|\mathbf{x}_k^N, \mathbf{Y}_k)\, p(\mathbf{x}_k^N|\, \mathbf{Y}_k) \tag{1.253}$$

where $p(\mathbf{x}_k^L|\mathbf{x}_k^N, \mathbf{Y}_k)$ is Gaussian and $p(\mathbf{x}_k^N|\, \mathbf{Y}_k)$ is non-Gaussian.

The linear part $\mathbf{x}_k^L$ of the state vector has been 'marginalized out'. The Gaussian term could be obtained using a Kalman filter, and the non-Gaussian term using a particle filter [44, 67]. This procedure is called *Rao-Blackwellization*.

The distinctive advantage of the method is that the dimension of the nonlinear part is smaller, so fewer particles would be required. However, the method requires a Kalman filter update for each of the $\mathbf{x}_k^N$ particles. Actually, the method is also called *'mixture Kalman filter'*.

It can be shown that with respect to the standard particle filter, the quality of the estimation, in terms of less variance, is better.

A state space model for the case considered could be:

$$\mathbf{x}_{k+1}^N = f_k^N(\mathbf{x}_k^N) + A_k^N(\mathbf{x}_k^N)\,\mathbf{x}_k^L + G_k^N(\mathbf{x}_k^N)\,\mathbf{w}_k^N \tag{1.254}$$

$$\mathbf{x}_{k+1}^L = f_k^L(\mathbf{x}_k^N) + A_k^L(\mathbf{x}_k^N)\,\mathbf{x}_k^L + G_k^L(\mathbf{x}_k^N)\,\mathbf{w}_k^L \tag{1.255}$$

$$\mathbf{y}_k = h_k(\mathbf{x}_k^N) + C_k(\mathbf{x}_k^N)\,\mathbf{x}_k^L + \mathbf{v}_k \tag{1.256}$$

After initialization, the algorithm repeats the update and the prediction step:
(a) Update

- Particle filter: obtain and normalize the importance weights corresponding to the $\mathbf{x}_k^{(i)}$ particles, using the likelihood:

$$p(\mathbf{y}_k|\mathbf{x}_k^{N(i)}) = N(\hat{y}_k^{(i)}, \Sigma_{y,k}^{(i)}) \tag{1.257}$$

where:

$$\hat{y}_k^{(i)} = h_k(\mathbf{x}_k^{N(i)}) + C_k(\mathbf{x}_k^{N(i)})\,\hat{x}_k^{(i)} \tag{1.258}$$

$$\Sigma_{y,k}^{(i)} = C_k(\mathbf{x}_k^{N(i)})\,P_k^{(i)}\,C_k(\mathbf{x}_k^{N(i)})^T + \Sigma_{v,k}^{(i)} \tag{1.259}$$

- Kalman filter update for each particle $\mathbf{x}_k^{(i)}$:

$$K_k^{(i)} = M_k^{(i)} C_{k-1}(\mathbf{x}_{k-1}^{N(i)}) \cdot (L_k^{(i)})^{-1} \tag{1.260}$$

$$P_k^{(i)} = M_k^{(i)} - K_k^{(i)} L_k^{(i)} (K_k^{(i)})^T \tag{1.261}$$

$$\hat{x}_k^{L(i)} = \mathbf{x}_{a,k}^{L(i)} + K_k^{(i)}(\mathbf{y}_k - h_k(\mathbf{x}_k^{N(i)}) - C_k(\mathbf{x}_k^{N(i)})\, \mathbf{x}_{a,k}^{L(i)}) \tag{1.262}$$

- Resampling if required

(b) Prediction

- Particle filter prediction using the first equation of the state space model (the term with $\mathbf{x}_k^L$ is interpreted as process noise)
- Kalman filter prediction for each particle using the second equation of the state space model:

$$\mathbf{x}_{a,k+1}^{L(i)} = f_k^L(\mathbf{x}_k^{N(i)}) + A_k^L(\mathbf{x}_k^{N(i)})\, \hat{x}_k^{L(i)} \tag{1.263}$$

$$M_{k+1}^{(i)} = A_k^L(\mathbf{x}_k^{N(i)})\, P_k^{(i)}\, A_k^L(\mathbf{x}_k^{N(i)})^T + \Sigma_w \tag{1.264}$$

$$L_{k+1}^{(i)} = C_k(\mathbf{x}_k^{N(i)})\, M_{k+1}^{(i)}\, C_k(\mathbf{x}_k^{N(i)})^T + \Sigma_v \tag{1.265}$$

A typical application of Rao-Blackwellized filters is object tracking with a camera. Given an object, the size of its captured image depends on the distance object-camera. The distance could be estimated by a particle filter, and its corresponding size, and the rate of size changing, could be estimated with a Kalman filter.

The article [45] presents examples of Rao-Blackwellized filters related with positioning of different vehicles: aircrafts, cars, ships, and submersibles.

1.7.14 Regularized Particle Filters

Given a set of particles, PDFs could be approximated via the Dirac delta, like for instance the posterior approximation mentioned in the SIS algorithm, which is reproduced below:

$$p(\mathbf{x}_k|\mathbf{Y}_k) \approx \sum_{i=1}^{N} W_k(x_k^i)\, \delta(\mathbf{x}_k - \mathbf{x}_k^i) \tag{1.266}$$

The idea of regularization is to employ a continuous function, a kernel, instead of the Dirac delta [7, 17, 80]. This is analogous to the Parzen estimation, [126]. Hence, continuing with the example of the posterior approximation, one could write:

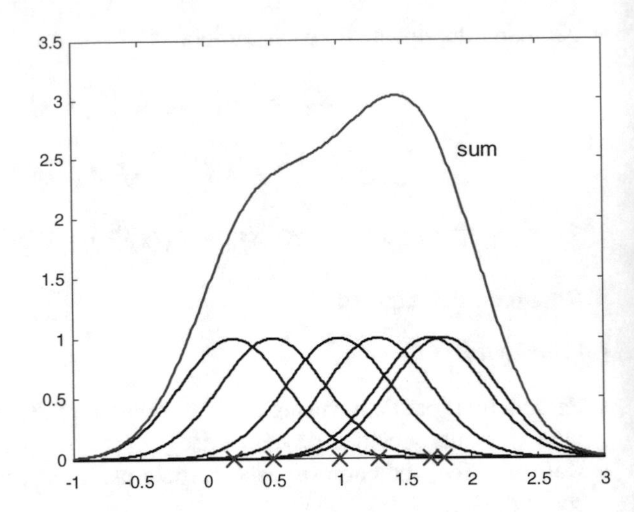

Fig. 1.54 Using Gaussian kernel for filter regularization

$$p(\mathbf{x}_k|\mathbf{Y}_k) \approx \sum_{i=1}^{N} W_k(x_k^i)\; K(\mathbf{x}_k, \mathbf{x}_k^i) \tag{1.267}$$

where the kernel $K(\mathbf{x}_k, \mathbf{x}_k^i)$ is symmetric, unimodal and smooth, with a limited bandwidth. Good candidates for the kernel are Gaussian or Epanechnikov kernels. In essence, the kernel is used for interpolation between particles. Figure 1.54 illustrates the approach. Observe that the sum of kernels, which are placed around the particles, is a continuous approximation of the PDF.

The width of the kernel is chosen to minimize the difference between the posterior PDF and the kernel estimate. Once the continuous approximation is obtained, a new generation of particles, with the desired variance, could be extracted from it.

The regularization could be applied at the prediction step, or at the updating step.

Some authors propose to use the posterior estimation to estimate its gradient and move the particles, following the gradient direction, toward the modes of the posterior [18].

Heuristic optimization is being applied to obtain best kernels or to drive particles to their neighboring maximum of the posterior [118].

1.8 The Perspective of Numerical Integration

Many of the new Bayesian filtering approaches are based on different ways of handling the integrals that appear in the estimation methodology.

Then, it is convenient to look at the matter from the perspective of numerical integration. It is good to see the place occupied by filters already described in this chapter, and to open the floor for other filters that have not been mentioned yet.

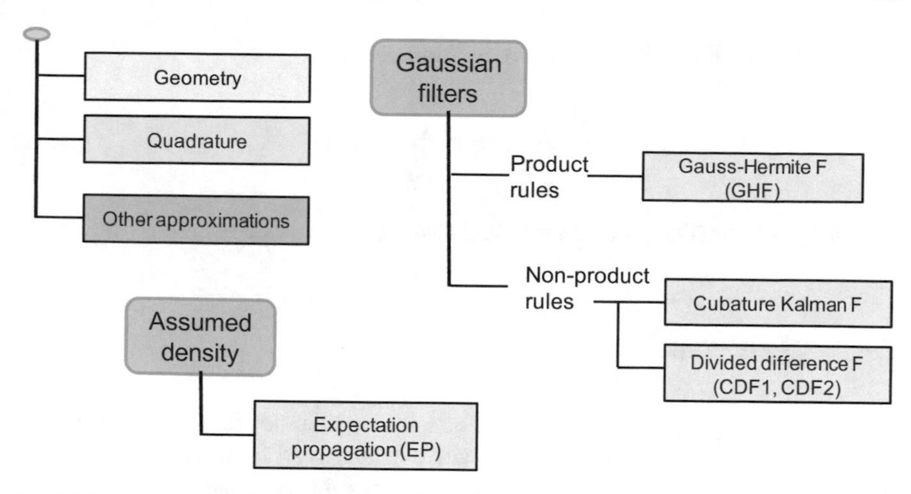

Fig. 1.55 Organigram of the section

This section begins with a rapid overview of numerical integration methods. Then, it comes to important examples of their use for Bayesian filters (Fig. 1.55).

Let us come back to the roots. Many aspects of filtering involve integrals. For instance, a typical target is to obtain the mean and the variance of the posterior:

$$E(\mathbf{x}) = \int \mathbf{x}\, p(\mathbf{x} \,|\, \mathbf{Y})\, d\mathbf{x} \tag{1.268}$$

$$E(\mathbf{x}^2) = \int \mathbf{x}^2\, p(\mathbf{x} \,|\, \mathbf{Y})\, d\mathbf{x} \tag{1.269}$$

$$var(\mathbf{x}) = E(\mathbf{x}^2) - E(\mathbf{x})^2 \tag{1.270}$$

There are deterministic or non-deterministic numerical integration algorithms. The archetype of non-deterministic methods is Monte-Carlo estimation of integrals, which has been already described. Now, let us focus on deterministic integration algorithms.

1.8.1 Geometry

Intuitive geometrical considerations could be applied for problems in one dimension. The integral is an area. The numerical integration could be done by adding rectangles or other suitable polygons. A similar point of view could be applied for two dimensions, computing volumes by adding convenient polyedra.

Thus, in one dimension, the approximation could be:

$$\int_a^b f(x)\,dx \approx \sum_{i=1}^{N} w_i\, f(x_i) \tag{1.271}$$

The integrand function $f(x)$ is evaluated at N equally spaced locations (knots).

1.8.2 Quadrature

There is a more general framework: the *quadrature* methods for numerical integration. The approximation just described, with evenly spaced knots, is a particular case of quadrature. But in general, quadrature methods do not require the knots to be equally spaced, and this can lead to more exact, and even computationally cheaper, integration.

Actually, the mean value theorem says that:

$$\int_a^b f(x)\,dx = (b-a)\, f(c), \quad c \in [a,\, b] \tag{1.272}$$

So it is possible to evaluate the integral by determining the opportune knot c. This is an extreme case of quadrature.

Newton-Cotes methods choose the knots to be evenly spaced. *Gaussian quadrature* methods do not require evenly spaced knots. In general the results obtained with Gaussian quadrature are better than the results of Newton-Cotes methods.

Examples of Newton-Cotes methods are the trapezoid rule, with two knots, the Simpson's rule, with three knots, and the Milne's rule with four knots.

When used for multidimensional integrals, the quadrature methods can be called *cubature* methods.

A quadrature formula:

$$\int_a^b f(x)\,dx \approx \sum_{i=1}^{n} w_i\, f(x_i) \tag{1.273}$$

is said to be *exact to degree n*if it yields exact equality when $f(x)$ is any polynomial of degree n or less.

Just for illustration purposes, let us include a simple example. One tries the following approximation:

$$\int_{-1}^{1} f(x)\,dx \approx\; a\, f(x_1) + b\, f(x_2) \tag{1.274}$$

The approximation should b exact for any polynomial of degree 3. Hence:

$$\int_{-1}^{1} dx = 2 = a + b$$

$$\int_{-1}^{1} x\, dx = 0 = a\, x_1 + b\, x_2$$

$$\int_{-1}^{1} x^2\, dx = \frac{2}{3} = a\, x_1^2 + b\, x_2^2$$

$$\int_{-1}^{1} x^3\, dx = 0 = a\, x_1^3 + b\, x_2^3$$

From this system of equations, one obtains that:

$$a = b = 1; \;\; x_1 = -x_2 = -\sqrt{\frac{1}{3}}$$

Therefore:

$$\int_{-1}^{1} f(x)\, dx \approx \; f(-\sqrt{1/3}) + f(\sqrt{1/3}) \qquad (1.275)$$

In Gaussian quadrature scenarios with n knots, it is usual to employ orthogonal polynomials in order to determine the knots and the weights w_i. Let us advance that the suitable knots are precisely the zeros of these polynomials.

A set $p_1(x),\; p_1(x),\; \ldots,\; p_n(x)$ of polynomials is orthogonal on an interval (a, b) if:

$$\int_{a}^{b} p_j(x)\, p_k(x)\, dx = 0 \qquad (1.276)$$

for all $j \neq k$.

All the zeros of $p_m(x)$ are real, with multiplicity 1, and lie in (a, b). Let us denote these zeros as: $x_1,\; x_2, \ldots,\; x_m$.

Now, consider the following case:

$$\int_{a}^{b} f(x)\, w(x)\, dx \approx \sum_{i=1}^{n} w_i\, f(x_i) \qquad (1.277)$$

where the knots are the zeros of an orthogonal polynomial $p_n(x)$, and the weights are given by:

$$w_i = \int_{a}^{b} l_i(x)\, w(x)\, dx \qquad (1.278)$$

with $l_i(x)$ the i-th Lagrange interpolating polynomial for these knots.

A theorem shows that the approximation (1.275) is exact if $f(x)$ is a polynomial of degree $\leq 2n - 1$. This is the fundament of Gaussian quadrature.

The idea of Gaussian quadrature is to integrate an interpolation that passes through the roots of an orthogonal polynomial. This polynomial, in turn, can be expressed in the Lagrange basis as follows:

$$p_n(x) = \sum_{i=1}^{n} f(x_i)\, l_i(x) \tag{1.279}$$

Then, the integral of the interpolation is:

$$\begin{aligned} \int_a^b p_n(x)\, w(x)\, dx &= \int_a^b \sum_{i=1}^{n} f(x_i)\, l_i(x)\, w(x)\, dx = \\ &= \sum_{i=1}^{n} f(x_i) \int_a^b l_i(x) w(x)\, dx = \sum_{i=1}^{n} f(x_i)\, w_i \end{aligned} \tag{1.280}$$

There are several sets of orthogonal polynomials, together with the corresponding knots and weights for Gaussian quadrature, published in books and papers. The orthogonal polynomials can be constructed with three term recurrence relations.

For the case of *(a, b) = (-1, 1)* and $w(x) = 1$ the orthogonal polynomials are the Legendre polynomials.

Here is a table concerning most common orthogonal polynomials:

Name	$w(x)$	Interval
Legendre	1	(-1, 1)
Hermite	$\exp(-x^2)$	$(-\infty, \infty)$
Laguerre	$x^{\alpha} \exp(-x^2)$	$(0, \infty)$
Chebyshev	$1/\sqrt{(1 - x^2)}$	$(-1, 1)$
Jacobi	$(1 - x)^{\alpha} (1 + x)^{\beta}$	$(-1, 1)$

Changes of variables could be applied to extend the application of $(-1, 1)$ intervals. Also, the interval of integration could be decomposed into smaller regions, over which quadrature can be applied.

The Gaussian quadrature is very appropriate when the function to integrate can be expressed as the product of a polynomial and a weight $w(x)$ of a tabulated type. In particular, Hermite weights could fit well with Bayesian problems.

When the integrand is not a polynomial, quadrature could still be applied for approximation. The error is bounded according with the following result:

$$\int_a^b f(x)\, w(x)\, dx = \sum_{i=1}^{n} w_i\, f(x_i) + \frac{1}{(2n)!} \frac{d^{2n} f(\xi)}{dx^{2n}} \int_a^b \psi^2(x)\, w(x)\, dx \tag{1.281}$$

where $\psi(x) = \Pi_{i=1}^{n}(x - x_i)$, and ξ is some number in the interval (a, b).

Finding the roots of polynomials may face numerical difficulties. For this reason, conventional simple approaches are discouraged for quadrature purposes. Instead there are algorithms based on finding the eigenvalues of a specific matrix (called the Jabobi matrix). This matrix is built based on the recurrence relation of the chosen orthogonal polynomial. A QR iteration could be used to find the eigenvalues and eigenvectors.

MATLAB provides a series of functions, *quad(), quadl(), quadgk(), quadv(),* etc., for numerical integration using different types of quadrature.

1.8.3 Other Approximations

A simple idea is to try approximations of the integrand, with easy to integrate components. For instance, logarithms could be applied to convert products into sums.

Of course, series expansions around a certain point could be a convenient approach. In certain cases it would be opportune to decompose the integral into several regions, applying Taylor series in each one.

Consider a certain Taylor expansion:

$$f(h) = f(0) + b_1 h + b_2 h^2 + b_3 h^3 + \ldots \tag{1.282}$$

where the coefficients b_i are values of derivatives. The error of the approximation would be a certain $O(h)$. Now, let us halve h:

$$f(h/2) = f(0) + b_1 \frac{1}{2} h + b_2 \frac{1}{4} h^2 + b_3 \frac{1}{8} h^3 + \ldots \tag{1.283}$$

The error would be approximately $O(h)/2$. Then, combining the two expansions:

$$g(h) = 2\, f(h/2) - f(h) = f(0) - b_2 \frac{1}{2} h^2 - b_3 \frac{3}{4} h^3 + \ldots \tag{1.284}$$

the error would be $O(h^2)$.

This is the basis of *Richardson extrapolation*, which repeatedly halves h to increase the accuracy. The extrapolation can be applied to the composite trapezoid rule, obtaining the *Romberg integration* method, which is widely used.

When the integrand is well approximated with a Gaussian curve, it can be convenient to use the *Laplace's method*. Let us introduce this method. Consider for instance that one has:

$$\int f(x)\, dx = K \tag{1.285}$$

where $f(x)$ has a peak at a point x_0. Take the logarithm of the integrand and apply a Taylor expansion around the peak:

$$\ln f(x) \approx \ln f(x_0) - \frac{b}{2}(x - x_0)^2 + \ldots \tag{1.286}$$

with:

$$b = -\frac{\partial^2}{\partial x^2} \ln f(x)|_{x=x_0} \tag{1.287}$$

Then, it is possible to approximate the integrand with a Gaussian:

$$\exp(\ln f(x)) = f(x) \approx f(x_0) \exp(-\frac{b}{2}(x - x_0)^2) \tag{1.288}$$

And the value of the integral:

$$K \approx f(x_0)\sqrt{2\pi/b} \tag{1.289}$$

This approach is also called *saddle-point approximation.*

From a more general point of view, if the idea is to approximate the integrand, it is pertinent to have a measure of the approximation quality. The research frequently uses the *Kullback-Leibler (KLD) divergence* for this purpose (KLD has been already mentioned in the previous book).

For example, suppose that the integrand is a chi-square distribution, and one wants to approximate it with a Gaussian PDF multiplied by a constant. A simple method could be to try several values of the constant and the PDF variance, until a minimum value of thc Kullback-Leibler divergence (or any other valid measure) is reached: this is the basis of the *variational* approach.

Next figure compares the results of the Laplace's method and the variational method for the approximation of the Student's T PDF (Fig. 1.56).

Let us add some comments about the variational method for numerical integration. A basic approach is to obtain a computable lower bound of the integral. The bound depends on some parameters, and the parameters are tuned for the tightest approximation (reaching equality when possible). Of course it could also be done with an upper bound. Expressing this in mathematical terms, one wants to evaluate an integral:

$$I = \int_a^b f(x)\,dx \tag{1.290}$$

This could be written as follows:

$$I = \int_a^b q(x,\theta)\frac{f(x)}{q(x,\theta)}\,dx \tag{1.291}$$

where $q(x, \theta)$ is an arbitrary function.

When trying to establishing bounds, the *Jensen's inequality* for convex functions is often a good starting point. This inequality can be expressed in several forms. In the case of logarithms of PDFs, the inequality is:

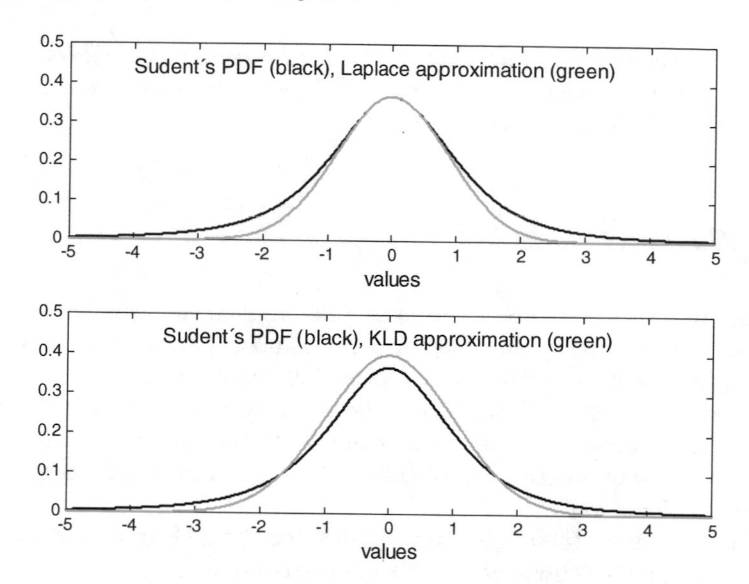

Fig. 1.56 Approximations of Student's T PDF: (*top*) Laplace's method, (*bottom*) KLD minimization

$$\ln \int_a^b q(x) \frac{f(x)}{q(x)} dx \geq \int_a^b q(x) \ln \frac{f(x)}{q(x)} dx \tag{1.292}$$

Therefore, a lower bound B could be:

$$I \geq B = \exp\{ \int_a^b q(x, \theta) \ln \frac{f(x)}{q(x, \theta)} dx \} \tag{1.293}$$

This lower bound is obviously related with the KLD from $q(x, \theta)$ to $f(x)$, which is:

$$D_{KL}(q \,||f) = \int_a^b q(x, \theta) \ln \frac{q(x, \theta)}{f(x)} dx = -B \tag{1.294}$$

The approximation target would be to minimize the KLD, by suitable changes in the parameters θ.

Another procedure for the variational approach is the following. An easy- to-integrate lower bound $g(x, \theta)$ is found such that:

$$f(x) \geq g(x, \theta), \ \forall \theta \tag{1.295}$$

Hence,

$$I \geq G(\theta) = \int_a^b g(x, \theta)\, dx \tag{1.296}$$

And the parameters θ are optimized to get $G(\theta)$ as close to I as possible.

By the way, there is an interesting paper, [34], which proposes the adaptation of the sample size in particle filters using KLD.

1.8.4 Gaussian Filters

Gaussian expressions are present in many filtering methods. It is not only when the noises and perturbations are Gaussian. In addition, Gaussian functions may be used to approximate unimodal distributions, and even multimodal distributions using Gausian sums. Therefore, Gaussian filters cover a large assortment of methods. The aim of this subsection is to offer a certain organized view of Gaussian filters, from the point of view of numerical integration, and to introduce some filters that have not been mentioned yet.

One of the reasons for the preference of theorists about Gaussian distributions is that they are uniquely characterized by mean and covariance.

When dealing with Bayesian filtering in Gaussian conditions, it is typical to deal with multiple integrals like the following:

$$I = \frac{1}{\sqrt{((2\pi)^d \det \Sigma)}} \int \ldots \int F(\mathbf{x}) \exp\left(\frac{1}{2} (\mathbf{x} - \boldsymbol{\mu})^T \Sigma^{-1} (\mathbf{x} - \boldsymbol{\mu})\right) dx \quad (1.297)$$

where d is thc space dimension.

Decomposing $\Sigma = \zeta^T \zeta$ and using a change of variables $\mathbf{x} = \boldsymbol{\mu} + \zeta^T \mathbf{z}$, the above integral can be expressed as:

$$I = \frac{1}{((2\pi)^{d/2}} \int \ldots \int G(\mathbf{z}) \exp\left(\frac{1}{2} \mathbf{z}^T \mathbf{z}\right) d\mathbf{z} \quad (1.298)$$

1.8.4.1 Product Rules

A filter based on the application of Gauss-Hermite quadrature for this integral has been proposed. The acronym of the filter is *GHF (Gaussian-Hermite filter)*. The main idea is to decompose the multiple integral into a product of univariate integrals, and apply Gauss-Hermite quadrature for each one. It is, then, a *product rule*. Weights and knots are taken from standard tables.

1.8.4.2 Non-product Rules

Product rules suffer from the curse of dimensionality problems: as space dimension increases linearly the number of knots increases exponentially (it is a multidimensional grid).

There are alternatives to product rules that obtain similar precision results with less knots. For instance, *lattice rules* that transform the grid of knots according with the integration domain; or *sparse grids* that concentrate most knots in suitable areas. There are also rules that exploit the symmetries of the problem; let us focus on this alternative.

In a *fully symmetric* set of cubature knots, equally weighted knots are symmetrically distributed around origin. The set $[\mathbf{p}]$ can be obtained from a *generator* knot $\mathbf{p}$ by sign changes and coordinate permutations. For example, given $\mathbf{p} = (1, 0)$, the set $[\mathbf{p}] = (1, 0), (-1, 0), (0, 1), (0, -1)$ can be generated.

Based on the invariant theory of Sobolev, it is possible to approximate the integral with a third degree cubature rule:

$$I = w_0\, G(0) + w_1\, G[\mathbf{p}] \tag{1.299}$$

where the set $[\mathbf{p}]$ contains $2d$ knots.

This cubature rule is exact for $G(\mathbf{z})$ being a monomial up to degree three. The monomials in this case have the form $z_1^{d1} \cdot z_2^{d2} \cdots z_n^{dn}$, with the sum of exponents equal to the monomial degree n. Thus, considering that the approximation should be exact for $n = 0$ and $n = 2$, the weights can be obtained from two equations, one for $G(\mathbf{z}) = 1$ (the integral I is 1), and one for $G(\mathbf{z}) = z_1^2$ (again the integral I is 1). The result is:

$$w_0 = 1 - \left(\frac{d}{u^2}\right), \quad w_1 = \frac{1}{2\,u^2} \tag{1.300}$$

where u is a free parameter.

Therefore, the following approximation could be used:

$$I = w_0 F(0) + w_1 \sum_{i=1}^{d} F(\mu + \zeta^T u\, \mathbf{e}_i) + w_1 \sum_{i=1}^{d} F(\mu - \zeta^T u\, \mathbf{e}_i) \tag{1.301}$$

where the weights are as in (1.298), and $\mathbf{e}_i$ has d components with the *i-th* component equal to 1 and the rest 0.

Notice that this approximation is the basis of the UKF filter.

More precision could be attained using:

$$I = w_0\, G(0) + w_1\, G\,[\mathbf{p}] + w_2\, G\,[\mathbf{p}, \mathbf{p}] \tag{1.302}$$

with:

$$u = \sqrt{3}\,, w_0 = 1 + \left(\frac{d^2 - 7d}{18}\right), \quad w_1 = \frac{4 - d}{18}, \; w_2 = \frac{1}{36} \tag{1.303}$$

By considering radial and spherical symmetries, it is possible to deduce other cubature rules that can be combined for our integration problem. In this way, the ***cubature***

Kalman filter *(CKF)* has been proposed. The CKF third-degree approximation is:

$$I = w_1 \sum_{i=1}^{d} F(\mu + \zeta^T \mathbf{e}_i) + w_1 \sum_{i=1}^{d} F(\mu - \zeta^T \mathbf{e}_i) \tag{1.304}$$

where the weight is $w_1 = 1/2d$.

The CKF is coincident with UKF for certain values of the UKF parameters [99, 123].

Here is a **CKF algorithm**, as described in [99].

Repeat the following steps:

(a) Prediction

- Draw points $\boldsymbol{\xi}^{(i)}$ from the intersection of the *n-dimensional* unit sphere and the Cartesian axes; scale them by $\sqrt{n}$:

$$\boldsymbol{\xi}^{(i)} = \begin{cases} \sqrt{n}\, e_i, & i = 1, \ldots, n \\ -\sqrt{n}\, e_{i-n}, & i = n+1, \ldots, 2N \end{cases} \tag{1.305}$$

- Obtain cubature points:

$$\mathbf{x}_c^{(i)}(n) = \sqrt{\Sigma(n)}\xi^{(i)} + \boldsymbol{\mu}_x, \quad i = 1, ..., 2N \tag{1.306}$$

- Use the transition equation to propagate the cubature points:

$$\mathbf{x}_{ac}^{(i)}(n+1) = \mathbf{f}(\mathbf{x}_c^{(i)}(n)), \quad i = 1, ..., 2N \tag{1.307}$$

- Obtain the mean of the propagated cubature points:

$$\mathbf{x}_a(n+1) = \frac{1}{2N} \sum_{i=1}^{2N} x_{ac}^{(i)}(n+1) \tag{1.308}$$

- Compute the a priori covariance matrix:

$$M(n+1) = \Sigma_w - \mathbf{x}_a(n+1) \cdot \mathbf{x}_a(n+1)^T + \sum_{i=1}^{2N} \mathbf{x}_{ac}^{(i)}(n+1) \cdot x_{ac}^{(i)}(n+1)^T \tag{1.309}$$

(b) Update

- Use the measurement equation to obtain measurements of the propagated sigma points:

$$\mathbf{y}_{ac}^{(i)}(n+1) = \mathbf{h}(\mathbf{x}_{ac}^{(i)}(n)), \quad i = 1, ..., 2N \tag{1.310}$$

- Obtain the mean of the measurements:

$$\mathbf{y}_a(n+1) = \frac{1}{2N}\sum_{i=1}^{2N} y_{ac}^{(i)}(n+1) \tag{1.311}$$

- Compute the measurement covariance matrix:

$$S_{yy}(n+1) = \Sigma_v - \mathbf{y}_a(n+1)\cdot\mathbf{y}_a(n+1)^T + \frac{1}{2N}\sum_{i=1}^{2N}\mathbf{y}_{ac}^{(i)}(n+1)\cdot\mathbf{y}_{ac}^{(i)}(n+1)^T \tag{1.312}$$

- Compute the cross-covariance matrix:

$$S_{xy}(n+1) = -\mathbf{x}_a(n+1)\cdot\mathbf{y}_a(n+1)^T + \frac{1}{2N}\sum_{i=1}^{2N}\mathbf{x}_{ac}^{(i)}(n+1)\cdot\mathbf{y}_{ac}^{(i)}(n+1)^T \tag{1.313}$$

- Obtain the Kalman gain as follows:

$$K(n+1) = S_{xy}(n+1)\cdot S_{yy}(n+1)^{-1} \tag{1.314}$$

- And obtain the remaining terms:

$$P(n+1) = M(n+1) - K(n+1)\,S_{yy}(n+1)\,K(n+1)^T \tag{1.315}$$

$$\mathbf{x}_e(n+1) = \mathbf{x}_a(n+1) + K(n+1)\,[\mathbf{y}(n+1) - \mathbf{y}_a(n+1)\,] \tag{1.316}$$

Go to (a)

Other non-product rules are now briefly introduced.

On the basis of a Taylor expansion of $F(\mathbf{x})$, numerical integration rules corresponding to EKF can be obtained [123].

The derivatives of the Taylor expansion can be replaced by central divided differences. This is the approach of the *divided difference filter (DDF)*, with its two versions: DDF1 based on first-order Taylor expansion, and DDF2 based on second-order Taylor expansion [82].

The *central difference filter (CDF)* has also been proposed, with two versions CDF1 and CDF2 [51]. It uses quadratic interpolation polynomials in the numerical integration. It has been shown that CDF1 is like DDF1.

1.8.4.3 Comparison

Many research contributions are devoted to compare different filters [26], but it is not easy to get a clear view. The next comments are just a general prospect based on [123].

In terms of precision one could take the integration of monomials as reference. For instance, UKF3 tends to be better than the degree-2 GHF (let us denote it as GHF(2)) because UKF3 is exact for monomials up to degree 4 (with $u = \sqrt{3}$). According with this precision criterion, the filters could be ordered as follows:

$EKF < DDF1 < GHF(2) < UKF3 < DDF2\, or\, CDF1 < CDF2$
$< GHF(3), UKF5$

There is some controversy about computational costs, in special when comparing EKF –which has to compute the Jacobian- and the UKF3. The cost of DDF2 or CDF1 is similar to UKF3. UKF5 has higher cost. GHF has the highest cost, while DDF1 has the lowest.

In terms of numerical stability, it is desired that all weights were positive and all knots were inside the integration limits. UKF3 would have a negative weight when $d > 3$. GHF is the only one that is completely stable. Because of unstability, it may happen that covariance becomes nonpositive in a filtering step, which means the collapse of the algorithm. Some measures could be taken to prevent this problem.

1.8.5 Assumed Density. Expectation Propagation

Suppose you want to approximate an integrand. Let us introduce an iterative procedure [74]. First you write the integrand as a product of terms:

$$f(x) = \prod_{i=1}^{n} t_i(x) \tag{1.317}$$

The idea is to approximate with a $q(x)$ the integrand term by term. The function $q(x)$ has an *assumed PDF* with some parameters to adjust. You start with the first term and adjust $q(x)$to fit this term using a divergence measurement D (for instance the KLD). Denote $q_1(x)$ this first approximation. Now incorporate the second term and get $a(x) = q_1(x)\, t_2(x)$. Adjust $q(x)$ to fit $a(x)$ using D, obtain a new $q_2(x)$. Next, get $a(x) = q_2(x)\, t_3(x)$, adjust $q(x)$ to fit $a(x)$ using D, obtain $q_3(x)$...And so on, till all terms have been incorporated. The final $q_n(x)$ is the approximation to the integrand.

In general the fitting steps are solved by moment matching.

There is a problem with this procedure: the result depends on the order by which terms are incorporated.

A solution for this problem is *expectation propagation (EP)*. The idea is to approximate the factorized integrand by a factorized $q(x)$:

$$q(x) = \prod_{i=1}^{n} \hat{t}_i(x) \tag{1.318}$$

where all the factors belong to the same exponential family.

Each term is chosen for:

$$q(x) = \hat{t}_j(x) \prod_{i \neq j} \hat{t}_i(x) \tag{1.319}$$

to be close to

$$t_j(x) \prod_{i \neq j} \hat{t}_i(x) \tag{1.320}$$

minimizing the divergence.

1.9 Other Bayesian Filters

The number of Bayesian filter variants and new proposals is large and continuously increasing. This section intends to show the trends in this area, and to give the distinctive ideas of algorithms that have found application success (Fig. 1.57).

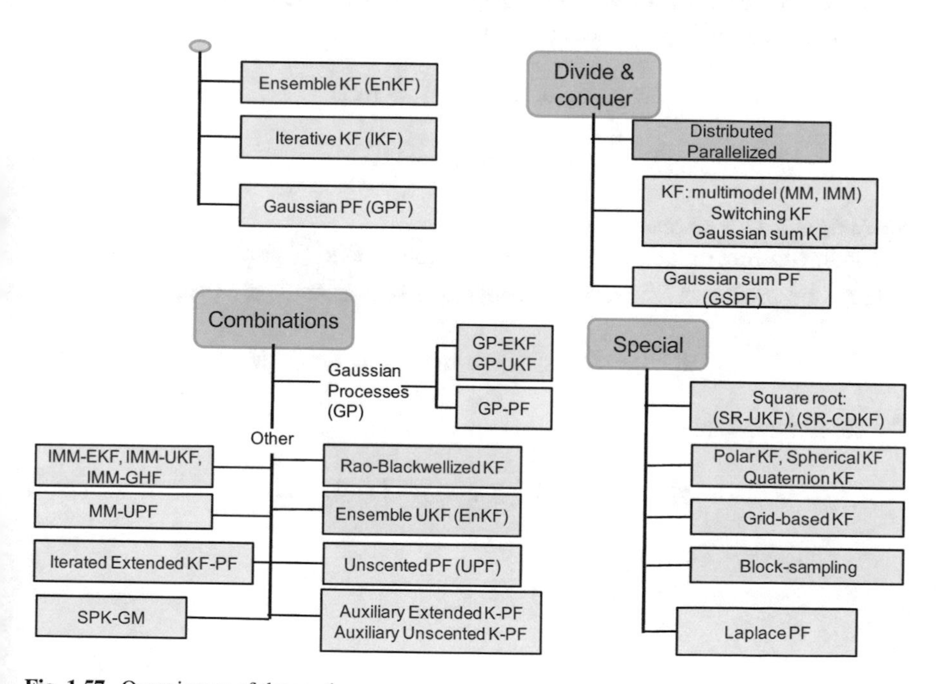

Fig. 1.57 Organigram of the section

Some key references are given for each identified activity nucleus, so the reader is invited to explore the specific topics from these starting points.

1.9.1 Ensemble Kalman Filter (EnKF)

Like particle filters the ensemble Kalman filter (EnKF) uses a random set of states: an ensemble. The algorithm follows a cycle of prediction and update steps, on the basis of a state space model, which usually has nonlinearities. The distinctive aspects are that the error covariance matrix is computed from the ensemble (empirical estimation), and that a common Kalman gain is used to update each ensemble member. See [62] and references therein for more details.

This type of filter is widely used in numerical weather prediction (NWP), where the dimension of the state space may be 10^7. The use of EKF would involve $10^7 \times 10^7$ matrices, this is clearly excessive. Instead, the EnKF would require just some hundreds of particles.

Other typical applications of EnKF are related to monitoring of ocean states and currents, ecology, geophysics, etc.

The mentioned application contexts have a different perspective, and use a different terminology. The filter is regarded as a data assimilation tool. Data assimilation obtains agreements combining model predictions and observations. The prediction step is called the forecast step; the update step is called the analysis step. The state space model is usually written as follows:

$$\mathbf{x}_k = M_k(\mathbf{x}_{k-1}) + \mathbf{u}_k \tag{1.321}$$

$$\mathbf{y}_k = H_k(\mathbf{x}_k) + \mathbf{v}_k \tag{1.322}$$

where $\mathbf{u}_k$ and $\mathbf{v}_k$ are noises.

The forecasting (transition equation) equation takes $\mathbf{x}_{k-1}^{a\,(i)}$ particles from the analysis step, and predicts $\mathbf{x}_k^{b\,(i)}$ particles (denoted as background particles).

$$\mathbf{x}_k^{b(i)} = M_k(\mathbf{x}_{k-1}^{a\,(i)}) + \mathbf{u}_k; \quad i = 1, \ldots, N \tag{1.323}$$

The observations are:

$$\mathbf{y}_k^{b\,(i)} = H_k(\mathbf{x}_{k-1}^{b\,(i)}) + \mathbf{v}_k; \quad i = 1, \ldots, N \tag{1.324}$$

Now, it is possible to compute the mean and the pertinent covariance matrices:

$$\hat{y}_k^b = \frac{1}{N} \sum_{i=1}^{N} \mathbf{y}_k^{b\,(i)} \tag{1.325}$$

$$P_{xy}^k = \frac{1}{N-1} \sum_{i=1}^{N} (\mathbf{x}_k^{b\,(i)} - \hat{x}_k^b)\,(\mathbf{y}_k^{b\,(i)} - \hat{y}_k^b)^T \tag{1.326}$$

$$P_{yy}^k = \frac{1}{N-1} \sum_{i=1}^{N} (\mathbf{y}_k^{b\,(i)} - \hat{y}_k^b)\,(\mathbf{y}_k^{b\,(i)} - \hat{y}_k^b)^T \tag{1.327}$$

where $\hat{x}_k^b$ is the mean of the background particles.

The Kalman gain is approximated as follows:

$$K_k = P_{xy}^k\,(P_{yy}^k + \Sigma_v)^{-1} \tag{1.328}$$

Using incoming observations, the background ensemble is updated as follows:

$$\mathbf{x}_k^{a(i)} = \mathbf{x}_k^{b(i)} + K_k\,(\mathbf{y}_k^{s(i)} - H_k(\mathbf{x}_k^{b\,(i)}));\quad i = 1, \ldots, N \tag{1.329}$$

where $\mathbf{y}_k^{s\,(i)}$ are surrogate observations, obtained by sampling from a normal distribution with mean $\hat{y}_k^b$ and covariance Σ_v.

1.9.2 *Iterative Kalman Filter*

Recall the update step of the extended Kalman filter:

$$K(n+1) = M(n+1)\;H(n)^T\;[H(n)\,M(n+1)\,H(n)^T + V(n)\,\Sigma_v\,V(n)^T]^{-1}$$

$$P(n+1) = M(n+1) - K(n+1)\;H(n)\;M(n+1)$$

$$\mathbf{x}_e(n+1) = \mathbf{x}_a(n+1) + K(n+1)\,[\mathbf{y}(n+1) - h(\mathbf{x}_a(n+1,\,0)]$$

where the Jacobian H(n) is evaluated at $\mathbf{x}_a(n+1)$.

The *Iterative Kalman Filter (IKF)* iteratively improves the update step. It does so between samples using some intermediate states $\mathbf{x}_i$.

For any n, the iterations are as follows:

$$i = 0; \quad \mathbf{x}_0 = \mathbf{x}_a(n+1) \tag{1.330}$$

1. increase i
2. compute:

$$K_i = M(n+1)\, H_i(n)^T\, [H_i(n)\, M(n+1)\, H_i(n)^T + V(n)\, \Sigma_v\, V(n)^T]^{-1} \tag{1.331}$$

$$\mathbf{x}_{i+1} = \mathbf{x}_i + K_i\, [\mathbf{y}(n+1) - h(x_i) - H_i(n)(\mathbf{x}_0 - \mathbf{x}_i)] \tag{1.332}$$

go to (1) unless a convergence criterion is fulfilled.

The convergence criterion could be that the difference $(\mathbf{x}_i - \mathbf{x}_{i-1})$ is sufficiently small (a certain threshold is used).

Notice that $H_i(n)$ is evaluated at $\mathbf{x}_i$.

Once the iterations have converged for a certain $i = m$, then the estimation is updated: $\mathbf{x}_e(n+1) = \mathbf{x}_m$.

There are nonlinear cases where EKF fails, but an adequately designed IKF can be used. See for example [106].

It has been shown in [10] that the iterated update of IKF is an instance of the Gauss-Newton optimization method.

1.9.3 Gaussian Particle Filter

There are particle filter applications where the approximation of the posterior by a single Gaussian PDF is adequate.

Two versions of *'Gaussian particle filter' (GPF)* were introduced in [59]. Resampling is not required. The noise can be non-Gaussian and non-additive. The filtering algorithm is parallelizable. Two standard examples were shown with GPF results being compared to EKF, UKF, and SIS (particle filter). GPF outperforms EKF and UKF. On the other hand SIS is slightly better than GPF, but it requires much more computation.

1.9.4 Divide and Conquer

Specific filtering circumstances that require large computational effort and/or short time response have been tackled by using several processors. *Distributed archi-*

tectures have been proposed. *Parallelization* techniques have been introduced. The filtering algorithms have been modified in order to share the processing work.

Also, the characteristics of some applications suggest the use of algorithms with some kind of partition inside. This could be the use of several models pertaining the dynamics and/or the perturbations. How to manage several models give birth to a number of alternatives.

In this context, the word 'mixture' has been used with several meanings. One has been already mentioned, when describing the Rao-Blackwellized particle filter. Another meaning is connected with sums of Gaussians to approximate PDFs. In addition the use of several filters could be named as mixtures.

As a way to organize this subsection, it has divided into two parts. The first is devoted to Kalman filters and has relatively old roots. The second looks at particle filters, where a lot of exploration activity exists.

1.9.4.1 Divide and Conquer with Kalman Filters

It is typical in flight control systems to use a gain scheduling approach: the control constants are switched according with the flight state. This is because the aircraft dynamics is not the same when taking off, cruising, or landing. It is also natural to use several models, and several Kalman filters. *Multi-model Kalman filters* have been proposed from time ago. In the case of examples like the flight control, it is also natural to switch from one filter to another (*'switching Kalman filter'*) [77], and one of the problems is the switching mechanism, which must be bump-free.

In the switching Kalman filters a set of models is sequentially used, one after one. There are other methods that use several models (or filters) in parallel. In this case a mechanism is included in order to extract the output information, perhaps by selecting the most successful filter in each moment, or by some kind of information fusion. In the case of multiple target tracking it is natural to use several filters in parallel, and the output information concerns all targets. For systems with a set of interacting components, it has been proposed to use multiple interacting models [12].

In particular the so-called Gaussian sum filters [3] can be regarded as banks of Kalman filters in parallel. The Gaussian sum is used to approximate the PDF of interest: the transition PDF or the posterior PDF. Each Gaussian component of the sum has an associated Kalman filter.

More in general, given a bank of Kalman filters, which could be structured as a tree, a management unit could be added to adapt, by pruning or merging, the size and structure of the bank to current signal processing needs.

1.9.4.2 Divide and Conquer with Particle Filters

Gaussian sum are 'universal approximators' for PDFs. The idea of using Gaussian sums in particle filters has been already introduced [60] as banks of particle filters. Actually, three types of '*Gaussian sum particle filter' (GSPF)* are described in [60], depending on the type of noise considered: general, additive Gaussian, or additive non-Gaussian.

1.9.5 Combinations

1.9.5.1 Filters Based on Gaussian Processes

The prediction capability of Gaussian processes (GP) could be used to predict the next state and the observations in the context of a Kalman filter. Some authors [58] express it as that GP can be applied directly to learn the transition and observation models, in such a way that GP can be integrated into Kalman filters or particle filters.

Actually, the literature shows examples of Gaussian process extended Kalman filters, Gaussian process UKF, Gaussian process particle filters, etc.

1.9.5.2 Other Combinations

The scientific literature offers a rich variety of combinations of the described filtering methods. Some of them are briefly cited below.

A Rao-Blackwellised unscented Kalman filter is introduced in [14], variance reduction and lower computational cost is achieved.

A modification of the ensemble Kalman filter based on the unscented transform is introduced in [68] and denoted as EnUKF.

The unscented particle filter was described in [109] (it was previously introduced in a Cambridge University technical report). The idea was to use a bank of unscented filters to obtain the importance proposal distribution.

In [98] the auxiliary extended and the auxiliary unscented Kalman particle filters were introduced. This publication includes extensive comparisons with other filters.

A set of combinations of interacting multiple model (IMM) and other methods is studied in [25]. In particular this article compares IMM-EKF with IMM-UKF, IMM-GHF, and multiple model unscented particle filter (MM-UPF).

In [65] an iterated extended Kalman particle filter was introduced.

In the context of particle filters, it was proposed in [111] to use a bank of sigma-point Kalman filters for the proposal distribution, and a Gaussian mixture for the posterior, so there is no resampling.

1.9.6 Algorithms with Special Characteristics

Instead of calculating the matrix square-root of the covariance at each filtering iteration, square-root forms of UKF and Central Difference KF directly propagate and update the square root of the covariance using the Cholesky factors. The acronyms are SR-UKF and SR-CDKF. They are better concerning numerical stability and computational cost [110].

Polar coordinates are more adequate than other coordinates for certain applications (for instance, target tracking). There have been several efforts to formulate Kalman filtering in polar [2, 116], or spherical coordinates [71].

When measuring 3D angular magnitudes, using inertial units, accelerometers, magnetometers, etc. It is convenient to use quaternions in order to avoid singularities. There are formulations of the Kalman filter using quaternions, like in [90] and references therein.

In mobile robotics and other applications, spatial grids are used. Some adaptations of the filtering algorithms have been proposed to include grids, like in [43] and references therein.

In [78] the use of Laplace method for the approximation of integrals in particle filters has been proposed.

It has been proposed [28] to use block sampling strategies instead of one-at-a-time sampling. This results in efficiency improvement.

1.10 Smoothing

In some situations it is possible to improve state estimation, obtaining smoothed results. The idea is to use the Kalman filter methodology and combine it with data from future. In some way it is similar to the *filtfilt()* MATLAB function, which uses past and future data for filtering. Of course, "future" data can be easily handled in off-line cases, with all data already recorded. For on-line cases, a future window could be used, by allowing a delayed output, like in the case of some TV programs that insert a short delay between reality and transmission.

Smoothing is gaining importance in practical life. For instance, it is being applied in vehicular GPS.

This section starts with a short look at prediction, to pick some convenient concepts. Then, it develops in some detail one-stage smoothing, attaining an important expression. Then, in the third subsection, it expands the study to the most relevant types of smoothers. A last subsection is devoted to a more general, Bayesian view of the topic. The section is based on [8, 40, 72, 96].

1.10.1 Optimal Prediction

Based on the dynamic model of the system, it is possible to predict future states. This prediction will be as good as permitted by the process noise. The basic mathematical aspects about prediction can be expressed using the transition matrix, which is introduced next.

The transition matrix $\Phi(k)$ of the (noiseless) system $\mathbf{x}(n+1) = A\,\mathbf{x}(n)$ is such that:

$$\Phi(j+1) = A\,\Phi(j),\ \ j = 0,\ 1,\ 2, \ldots \tag{1.333}$$

with $\Phi(0) = 1$.

Now let us affirm that, based on measurements and model, the best prediction of the state, for $i < n$ is the following:

$$\mathbf{x}_p(n|i) = \hat{\mathbf{x}}(n|i) = \Phi(n, i)\,\hat{\mathbf{x}}(i) \tag{1.334}$$

where $\Phi(n, i)$ is equivalent to $\Phi(n - i)$.

Evidently: $\Phi(n+1, n) = A$.

The prediction error can be expressed as follows:

$$p\bar{e}r\ (n|i) = \mathbf{x}(n) - \mathbf{x}_p(n|i) \tag{1.335}$$

The error covariance would be:

$$\Sigma_p\,(n|i) = \Phi(n, i)\,\Sigma_p\,(i|i)\,\Phi(n, i)^T + \sum_{j=i+1}^{n} \Phi(n, i)\,\Sigma_w\,\Phi(n, i)^T \tag{1.336}$$

In this last expression one can observe the effect of the process noise.

1.10.2 One-Stage Smoothing

As a first step into smoothing theory, the simplest case is now considered: to obtain a smoother estimation using the next measurement. The following mathematical development is based on the innovations process.

The innovations could be written as follows:

$$\eta(n+1) = \mathbf{y}(n+1) - C\,\hat{\mathbf{x}}(n+1|n) = C\,p\bar{e}r(n+1) + \mathbf{v} \tag{1.337}$$

For brevity purposes, let us denote the series of obtained measurements as:

$$Y(n) = \{\mathbf{y}(1),\ \mathbf{y}(2),\ \ldots,\ \mathbf{y}(n)\} \tag{1.338}$$

The target of one-stage smoothing, is to get the following estimate:

$$\begin{aligned} &\mathbf{x}_S(n|n+1) = E(\mathbf{x}(n)|\mathbf{y}(1),\ \mathbf{y}(2),\ \ldots,\ \mathbf{y}(n+1)) = \\ &= E(\mathbf{x}(n)\,|Y(n),\ \eta(n+1)) = E(\mathbf{x}(n)\,|Y(n)) + E(\mathbf{x}(n)|\ \eta(n+1)) \end{aligned} \tag{1.339}$$

Since $\mathbf{x}(n)$ and the innovations $\eta(n+1)$ are jointly Gaussian with zero mean, the following partition could be considered:

$$\begin{pmatrix} \mathbf{x}(n) \\ \eta(n+1) \end{pmatrix} \tag{1.340}$$

and then:

$$\mathbf{x}_S(n|n+1) = \hat{x}(n) + \frac{cov(\mathbf{x}(n), \eta(n+1)}{\Sigma_{in}(n+1)} \cdot \eta(n+1) \tag{1.341}$$

Let us focus on the numerator of the fraction.

$$cov(\mathbf{x}(n), \eta(n+1)) = cov(\mathbf{x}(n), p\bar{e}r(n+1)) \cdot C^T + E(\mathbf{x}(n)|\mathbf{v}) \tag{1.342}$$

The second term is zero. Also:

$$p\bar{e}r(n+1) = A\, \bar{er}(n) + \mathbf{w} \tag{1.343}$$

Then:

$$\begin{aligned} cov(\mathbf{x}(n), \eta(n+1)) &= cov(\mathbf{x}(n), \bar{er}(n)) \cdot A^T C^T = \\ &= cov(\bar{er}(n), \bar{er}(n)) \cdot A^T C^T = \Sigma(n) \cdot A^T C^T \end{aligned} \tag{1.344}$$

In consequence:

$$\mathbf{x}_S(n|n+1) = \hat{\mathbf{x}}(n) + F(n|n+1)[\mathbf{y}(n+1) - C\, A\, \hat{\mathbf{x}}(n)] \tag{1.345}$$

with:

$$F(n|n+1) = \Sigma(n) \cdot \frac{A^T C^T}{\Sigma_{in}(n+1)} \tag{1.346}$$

Let us show that that the optimal smoothing includes the Kalman gain $K(n+1)$. From AKF filter, one has that:

$$\frac{C^T}{\Sigma_{in}(n+1)} = \frac{K(n+1)}{M(n+1)} \tag{1.347}$$

Therefore:

$$F(n|n+1) = \frac{\Sigma(n) A^T}{M(n+1)} \cdot K(n+1) = G(n) \cdot K(n+1) \tag{1.348}$$

A convenient expression of the one-stage smoother is the following:

$$\mathbf{x}_S(n|n+1) = \hat{\mathbf{x}}(n) + G(n)[\hat{\mathbf{x}}(n+1) - \hat{\mathbf{x}}(n+1|n)] \tag{1.349}$$

where $\hat{\mathbf{x}}(n)$ and $\hat{\mathbf{x}}(n+1)$ should be computed with the Kalman filter.

1.10.3 Three Types of Smoothers

There are three main types of smoothers [72], which could be briefly introduced as follows:

- *Fixed-interval*
  It is desired to obtain:

$$\mathbf{x}_S(n|N)\,,\; n = 0,\, 1,\, \ldots,\, N-1 \tag{1.350}$$

  The data processing is done after all data are gathered. It is an off-line task. For each n within $0..N-1$, one wishes to obtain the optimal estimate of $\mathbf{x}(n)$ using all the available measurements $\mathbf{y}(n),\; n = 0,\, 1, \ldots, N$.
- *Fixed-point*
  It is desired to obtain:

$$\mathbf{x}_S(n|j)\,,\; j = n+1,\, n+2,\, \ldots \tag{1.351}$$

  with n a fixed integer.

  The state estimate is improved using future measurements. This can be done on-line, but the result of say $\mathbf{x}_s(n|n+d)$ is subject to a delay of d samples.
- *Fixed-lag*
  It is desired to obtain:

$$\mathbf{x}_S(n|n+d)\,,\; n = 0,\, 1,\, \ldots \tag{1.352}$$

  with d a fixed integer.

  The data processing can be done on-line, being subject to a delay of d samples.

The algorithms for these types of smoothers can be deduced from the one-stage smoothing, or by augmentation of the state vector.

1.10.3.1 Algorithms Based on Previous Expressions

Let us continue with the expressions obtained for one-stage smoothing [8]. The three types of smoothers can be implemented as follows:

- *Fixed-interval:*

$$\mathbf{x}_S(n|N) \;=\; \hat{\mathbf{x}}(n) \;+\; G(n)[\hat{\mathbf{x}}(n+1|N) - \hat{\mathbf{x}}(n+1|n)] \tag{1.353}$$

  This must be solved backwards in time, starting from $\mathbf{x}_s(N|N)$.

Some authors call this algorithm as the *'two-pass smoother'*, since there is a forward pass with the Kalman filter and a backward pass using the expression above. The algorithm is also known as the *'RTS (Rauch-Tung-Striebel)'* smoother [87].

Mendel proposed in [72] a version of the two-pass smoother that has better computational characteristics. A *residual state vector* is defined as follows:

$$\mathbf{r}(n|N) = M(n)^{-1}\,[\hat{\mathbf{x}}(n|N) - \hat{\mathbf{x}}(n|n-1)] \tag{1.354}$$

Using this vector, the RTS expression is transformed to:

$$\mathbf{x}_S(n|N) = \hat{\mathbf{x}}(n|n-1) + M(n)\,\mathbf{r}(n|N) \tag{1.355}$$

where $\mathbf{r}(n|N)$ can be obtained with a backward recursive equation:

$$\mathbf{r}(j|N) = A_p(j)^T\,\mathbf{r}(j+1|N) + C^T\,[C\,M(j)\,C^T + \Sigma_v]^{-1}\,\eta(j) \tag{1.356}$$

$$j = N-1,\ N-2,\ \ldots,\ 1 \tag{1.357}$$

with: $A_p(j) = A\,[I - K(j)\,C]$

Another fixed-interval algorithm, less convenient for computation, is the *'forward-backward filter'* proposed by Fraser and Potter [36]. Suppose you want to get a smoothed value at m: $\mathbf{x}_s(m)$. The algorithm runs the Kalman filter up to m, to obtain a forward estimate, and runs a reverse system down to m,

$$\mathbf{x}(n-1) = A^{-1}\,\mathbf{x}(n) - A^{-1}\,\mathbf{w}(n-1) \tag{1.358}$$

to obtain a backward estimate. Both estimates are combined to yield the smoothed value.

- *Fixed-point:*

$$\mathbf{x}_S(n|j) = \hat{\mathbf{x}}(n|j-1) + B(j)[\hat{\mathbf{x}}(j) - \hat{\mathbf{x}}(j|j-1)] \tag{1.359}$$

where: $B(j) = B(j-1)\,G(j-1)$.

- *Fixed-lag:*

$$\begin{aligned} \mathbf{x}_S(n+1|n+1+N) = \; & A\,\hat{\mathbf{x}}(n|n+N) + \\ & + B(n+1+N)\,K(n+1+N)\,\eta(n+1+N|n+N) + \\ & + U(n+1)[\hat{\mathbf{x}}(n|n+N) - \hat{\mathbf{x}}(n)] \end{aligned} \tag{1.360}$$

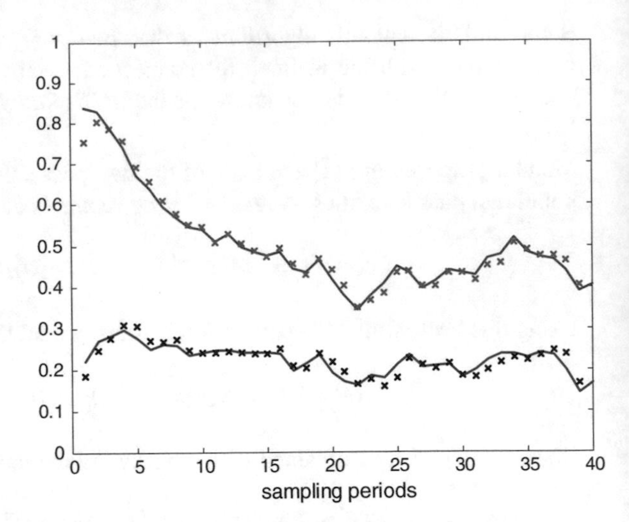

Fig. 1.58 System estimated states (*cross marks*), and fixed-interval smoothed states (*continuous*)

with:

$$B(n+1+N) = \prod_{i=n+1}^{n+N} G(i) \tag{1.361}$$

$$U(n+1) = \Sigma_w\, A^T\, \Sigma^{-1}(n) \tag{1.362}$$

Program 1.23 brings an example of fixed-interval smoothing, implementing the algorithm just described. The same two-tank system treated in section 8.3 has been selected. Figure 1.58 compares the smoothed state estimates, continuous curves, against the state estimation given by the Kalman filter, cross marks. The effect of smoothing is less variance.

Notice in the program that there are two steps: first to run the Kalman filter along all samples, and second to run the smoothing procedure, which only requires a few code lines. The Kalman filter implementation is a more compact version of Program 1.3.

Program 1.23 Fixed-interval smoothing example

```
%Fixed-interval smoothing example
%state space system model (2 tank system):
A1=1; A2=1; R1=0.5; R2=0.4;
cA=[-1/(R1*A1) 1/(R1*A1); 1/(R1*A2) -(1/A2)*((1/R1)+(1/R2))];
cB=[1/A1; 0]; cC=[1 0; 0 1]; cD=0;
Ts=0.1; %sampling period
%discrete-time model:
csys=ss(cA,cB,cC,cD); dsys=c2d(csys,Ts,'zoh');
[A,B,C,D]=ssdata(dsys); %retrieves discrete-time model matrices
%simulation horizon
Nf=40;
```

```
%process noise
Sw=[12e-4 0; 0 6e-4]; %cov
sn=zeros(2,Nf);
sn(1,:)=sqrt(Sw(1,1))*randn(1,Nf);
sn(2,:)=sqrt(Sw(2,2))*randn(1,Nf);
%observation noise
Sv=[6e-4 0; 0 15e-4]; %cov.
on=zeros(2,Nf);
on(1,:)=sqrt(Sv(1,1))*randn(1,Nf);
on(2,:)=sqrt(Sv(2,2))*randn(1,Nf);
% system simulation preparation
x=[1;0]; % state vector with initial tank levels
u=0.4; %constant input
% Kalman filter simulation preparation
%space for matrices
K=zeros(2,2); M=zeros(2,2); P=zeros(2,2);
xe=[0.5; 0.2]; % filter state vector with initial values
%space for recording xa(n), xe(n)
rxa=zeros(2,Nf-1);rxe=zeros(2,Nf-1);
%behaviour of the system and the Kalman filter
% after initial state
% with constant input u
for nn=1:Nf-1,
%system simulation
xn=(A*x)+(B*u)+sn(nn); %next system state
x=xn; %system state actualization
ym=(C*x)+on(:,nn); %output measurement
%Prediction
xa=(A*xe)+(B*u); %a priori state
M=(A*P*A')+ Sw;
%Update
K=(M*C')*inv((C*M*C')+Sv);
P=M-(K*C*M);
xe=xa+(K*(ym-(C*xa))); %estimated (a posteriori) state
%recording xa(n), xe(n)
rxa(:,nn)=xa;
rxe(:,nn)=xe;
end;
%Smoothing----------------------------
xs=zeros(2,Nf);
xs(:,Nf)=xe; %final estimated state
for nn=(Nf-1):-1:1,
G=(P*A')*inv(M);
xs(:,nn)=rxe(:,nn)+(G*(xs(:,nn+1)-rxa(:,nn)));
end;
%----------------------------------------
% display of state evolution
figure(3)
plot(xs(1,:),'r'); %plots xs1
hold on;
```

```
plot(xs(2,:),'b'); %plots xs2
plot(rxe(1,:),'mx'); %plots xe1
plot(rxe(2,:),'kx'); %plots xe2
axis([0 Nf 0 1]);
xlabel('sampling periods');
title('Kalman filter states(x) and Smoothed states(-)');
```

1.10.3.2 Algorithms Based on State Augmentation

There are some smoothing algorithms that are based on augmentation of the state vector. Here are two notable examples, as described in [40] and detailed in [96].

- *Fixed-lag:*
  Let us augment the state vector by delayed versions of the state:

$$\mathbf{x}_A(n+1) = \begin{bmatrix} \mathbf{x}(n+1) \\ \mathbf{x}^1(n+1) \\ \mathbf{x}^2(n+1) \\ - - - \\ \mathbf{x}^l(n+1) \end{bmatrix} = \begin{bmatrix} A\ 0 \ldots 0 \\ I\ 0 \ldots 0 \\ 0\ I \ldots 0 \\ - - - \\ 0 \ldots I\ 0 \end{bmatrix} \cdot \begin{bmatrix} \mathbf{x}(n) \\ \mathbf{x}^1(n) \\ \mathbf{x}^2(n) \\ - - - \\ \mathbf{x}^l(n) \end{bmatrix} + \begin{bmatrix} I \\ 0 \\ 0 \\ - - - \\ 0 \end{bmatrix} \cdot \mathbf{w} \tag{1.363}$$

$$\mathbf{y}(n) = [C\ 0\ 0 \ldots 0] \cdot \begin{bmatrix} \mathbf{x}(n) \\ \mathbf{x}^1(n) \\ \mathbf{x}^2(n) \\ - - - \\ \mathbf{x}^l(n) \end{bmatrix} + \mathbf{v} \tag{1.364}$$

  where: $\mathbf{x}^j(n) = \mathbf{x}(n-j)$

  The OPKF filter can be used to obtain smoothed estimates:

$$\hat{x}_A(n+1) = \begin{bmatrix} \hat{x}(n+1) \\ \hat{x}^1(n+1) \\ \hat{x}^2(n+1) \\ - - - \\ \hat{x}^l(n+1) \end{bmatrix} =$$

$$= \begin{bmatrix} A\ 0 \ldots 0 \\ I\ 0 \ldots 0 \\ 0\ I \ldots 0 \\ - - - \\ 0 \ldots I\ 0 \end{bmatrix} \cdot \begin{bmatrix} \hat{x}(n) \\ \hat{x}^1(n) \\ \hat{x}^2(n) \\ - - - \\ \hat{x}^l(n) \end{bmatrix} + \begin{bmatrix} K(n) \\ K^1(n) \\ K^2(n) \\ - - - \\ K^l(n) \end{bmatrix} \cdot [\mathbf{y}(n) - C\,\hat{x}(n)] \tag{1.365}$$

The state error covariance can be expressed as:

$$\Sigma_A(n) = \begin{bmatrix} \Sigma(n) & \Sigma^1(n)^T & \ldots \Sigma^l(n)^T \\ \Sigma^1(n) & \Sigma^{11}(n) & \ldots \Sigma^{1l}(n) \\ - & - & - \\ \Sigma^l(n) & \Sigma^{l1}(n) & \ldots \Sigma^{ll}(n) \end{bmatrix} \tag{1.366}$$

The Kalman gains can be computed as follows:

$$K^j(n) = \Sigma^{j-1}(n) \cdot C^T [C\,\Sigma(n)\,C^T + \Sigma_v]^{-1} \tag{1.367}$$

where:

$$\Sigma^j(n+1) = \Sigma^{j-1}(n)\,[A - K(n)C]^T \tag{1.368}$$

$$\Sigma^{jj}(n+1) = \Sigma^{j-1,j-1}(n) - \Sigma^{j-1}(n)\,[K^j(n)C]^T \tag{1.369}$$

and:

$$\Sigma^0(n) = \Sigma(n) \tag{1.370}$$

$$K(n) = A\,\Sigma(n)\,C^T[C\,\Sigma(n)\,C^T + \Sigma_v]^{-1} \tag{1.371}$$

As lag increases the estimation error variance decreases. For $l \to \infty$ the filter approaches the non-causal Wiener filter. Nearly optimal performance is obtained for l two or three times the dominant time constants of the system.

In order to give an implementation example, a program has been developed.This program is included in Appendix B. The first part of the program is very similar to the previous program, since it includes the system simulation and a complete run of the Kalman filter for state estimation. The second part of the program is the code for fixed-lag smoothing. This part of the code has been included below.

Figure 1.59 compares as before the smoothed state estimates, continuous curves, against the state estimation given by the Kalman filter, cross marks. The lag has been set to $L = 5$ samples. The continuous curves have been left shifted to compensate for the delay.

Fragment 1.24 Smoothing

```
% Smoothing-----------------------------
% Smoothing preparation
N=zeros(2,2); P=zeros(2,2);
% augmented state vectors
axa=zeros(2*(L+1),1);
axp=zeros(2*(L+1),1);
% augmented input
bu=zeros(2*(L+1),1); bu(1:2,1)=B*u;
% augmented A matrix
aA=diag(ones(2*L,1),-2); aA(1:2,1:2)=A;
```

```
% augmented K
aK=zeros(2*(L+1),2);
% set of covariances
Pj=zeros(2,2,L);
%space for recording xs(n)
rxs=zeros(2,Nf-1);
jed=(2*L)+1; %pointer for last entries
%jed=1;
%action:
axa(1:2,1)=rxe(:,1); %initial values
for nn=1:Nf,
M=(A*P*A')+Sw;
N=(C*P*C')+Sv;
ivN=inv(N);
K=(A*P*C')*ivN;
aK(1:2,:)=K;
aK(3:4,:)=(P*C')*ivN;
for jj=1:L,
bg=1+(jj*2); ed=bg+1;
aK(bg:ed,:)=(Pj(:,:,jj)*C')*ivN;
end;
aux=[A-K*C]';
Pj(:,:,1)=P*aux;
for jj=1:L-1,
Pj(:,:,jj+1)=Pj(:,:,jj)*aux;
end;
axp=(aA*axa)+bu+aK*(rym(:,nn)-C*axa(1:2,1));
P=M-(K*N*K');
rxs(:,nn)=axp(jed:jed+1);
axa=axp; %actualization (implies shifting)
end;
%-------------------------------------------
% display of state evolution
figure(3)
plot(rxs(1,L:end),'r'); %plots xs1
hold on;
plot(rxs(2,L:end),'b'); %plots xs2
plot(rxe(1,:),'mx'); %plots xe1
plot(rxe(2,:),'kx'); %plots xe2
axis([0 Nf 0 1]);
xlabel('sampling periods');
title('Kalman filter states(x) and Smoothed states(-)');
```

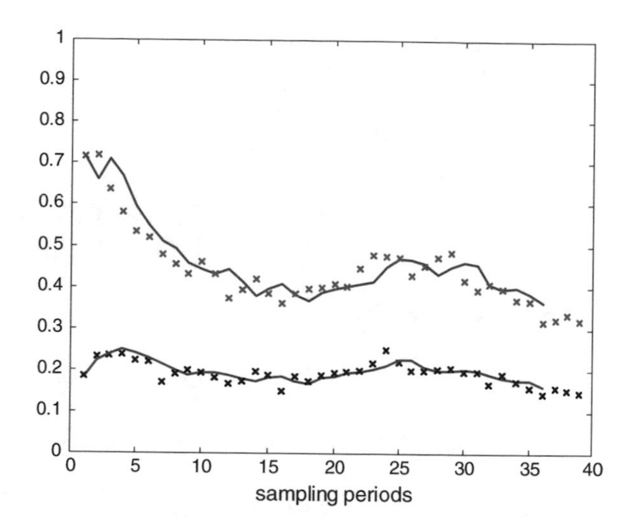

Fig. 1.59 System estimated states (*cross marks*), and fixed-lagl smoothed states (*continuous*)

- *Fixed-point:*

A different augmented state vector is used.
Define:

$$\mathbf{x}'(n) = \begin{cases} \mathbf{x}'(n) \; = \; \mathbf{x}(n), \;\; n \leq n_f \\ \mathbf{x}'(n+1) \; = \; \mathbf{x}(n), \;\; n > n_f \end{cases} \tag{1.372}$$

Notice that:

$$\mathbf{x}'(n) = \mathbf{x}(n_f), \quad n \geq n_f \tag{1.373}$$

As before, the augmented state vector and the OPKF filter can be used to obtain the smoothed estimates:

$$\begin{bmatrix} \hat{\mathbf{x}}(n+1) \\ \hat{\mathbf{x}}'(n+1) \end{bmatrix} = \begin{bmatrix} A & 0 \\ 0 & I \end{bmatrix} \begin{bmatrix} \hat{\mathbf{x}}(n) \\ \hat{\mathbf{x}}'(n) \end{bmatrix} + \begin{bmatrix} K(n) \\ K'(n) \end{bmatrix} [\mathbf{y}(n) \; - \; C\,\hat{\mathbf{x}}(n)] \tag{1.374}$$

where:

$$\begin{bmatrix} K(n) \\ K'(n) \end{bmatrix} = \begin{bmatrix} A & 0 \\ 0 & I \end{bmatrix} \begin{bmatrix} \Sigma_{11}(n) & \Sigma_{12}(n) \\ \Sigma_{21}(n) & \Sigma_{22}(n) \end{bmatrix} \begin{bmatrix} C^T \\ 0 \end{bmatrix} [C\,\Sigma_{11}(n)\,C^T \; + \; \Sigma_v]^{-1} \tag{1.375}$$

Denote:

$$\Sigma_A(n) = \begin{bmatrix} \Sigma_{11}(n) & \Sigma_{12}(n) \\ \Sigma_{21}(n) & \Sigma_{22}(n) \end{bmatrix} \tag{1.376}$$

One has that:

$$\Sigma_A(n+1) = \begin{bmatrix} A & 0 \\ 0 & I \end{bmatrix} \Sigma_A(n) \begin{bmatrix} A^T & 0 \\ 0 & I \end{bmatrix} + \begin{bmatrix} \Sigma_w & 0 \\ 0 & 0 \end{bmatrix} - \\ - \begin{bmatrix} K(n) \\ K'(n) \end{bmatrix} \cdot [C\, \Sigma_{11}(n)\, C^T \,+\, \Sigma_v] \cdot \begin{bmatrix} K(n) \\ K'(n) \end{bmatrix} \tag{1.377}$$

with:

$$\Sigma_A(n_f) = \begin{bmatrix} \Sigma_{11}(n_f) & \Sigma_{11}(n_f) \\ \Sigma_{11}(n_f) & \Sigma_{11}(n_f) \end{bmatrix} \tag{1.378}$$

It can be deduced that:

$$K'(n) = \, \Sigma_{21}(n)\, C^T\, N(n+1)^{-1} \tag{1.379}$$

$$\Sigma_{11}(n+1) = \, M(n+1) \,-\, K(n)\;N(n+1)\, K(n)^T \tag{1.380}$$

$$K(n) = \, A\, \Sigma_{11}(n)\, C^T\, N(n+1)^{-1} \tag{1.381}$$

$$\Sigma_{21}(n+1) = \, \Sigma_{21}(n)\, [A \,-\, K(n)\, C]^T \tag{1.382}$$

$$\Sigma_{22}(n+1) = \, \Sigma_{22}(n) - \Sigma_{21}(n) C^T\, K'(n)^T \tag{1.383}$$

with:

$$\Sigma_{21}(n_f) = \, \Sigma_{11}(n_f) \tag{1.384}$$

A program has been developed to provide an example of implementation. The program is included in Appendix B. It is very similar to the previous program, with the main difference being the smoothing code. This code has been listed below.

The program considers the smoothing of state estimate at a fixed point, the $10th$ sample. Figure 1.60 shows the improvement of estimation as more future samples are taken into account. Figure 1.61 shows the corresponding evolution of covariances; clearly the effect of more than ten future samples is not important.

In principle it is not necessary to run the Kalman filter along all samples. It suffices with running it up to the fixed point.

Fragment 1.25 Smoothing

```
%Smoothing----------------------------
% Smoothing preparation
Nfix=10; %the fixed point
%space for matrices
N=zeros(2,2); P11=zeros(2,2); P21=zeros(2,2);
% augmented state vectors
axa=zeros(4,1);
axp=zeros(4,1);
% augmented input
bu=zeros(4,1); bu(1:2,1)=B*u;
```

```
% augmented A matrix
aA=diag(ones(4,1)); aA(1:2,1:2)=A;
% augmented K
aK=zeros(4,2);
%space for recording xs(Nfix), P11(n)
rxs=zeros(2,Nf);
rP11=zeros(2,2,Nf);
%action:
P11=rP(:,:,Nfix); P21=P11; %initial values
axa(1:2,1)=rxe(:,Nfix); %initial values
axa(3:4,1)=rxe(:,Nfix); %initial values
for nn=Nfix:Nf,
M=(A*P11*A')+Sw;
N=(C*P11*C')+Sv;
ivN=inv(N);
K=(A*P11*C')*ivN;
Ka=(P21*C')*ivN;
aK(1:2,:)=K; aK(3:4,:)=Ka;
axp=(aA*axa)+bu+aK*(rym(:,nn)-C*axa(1:2,1));
axa=axp; %actualization
rP21(:,:,nn)=P21; %recording
rxs(:,nn)=axp(3:4,1);
P11=M-(K*N*K');
P21=P21*(A-(K*C))';
end;
%------------------------------------------
% display of smoothed state at Nfix
figure(3)
plot(rxs(1,Nfix:end),'r'); %plots xs1
hold on;
plot(rxs(2,Nfix:end),'b'); %plots xs2
axis([0 Nf 0 0.6]);
xlabel('sampling periods');
title('State smoothing at Nfix');
% display of Covariance evolution
figure(4)
subplot(2,2,1)
plot(squeeze(rP21(1,1,Nfix:end)),'k');
title('Evolution of covariance');
subplot(2,2,2)
plot(squeeze(rP21(1,2,Nfix:end)),'k');
subplot(2,2,3)
plot(squeeze(rP21(2,1,Nfix:end)),'k');
subplot(2,2,4)
plot(squeeze(rP21(2,2,Nfix:end)),'k');
```

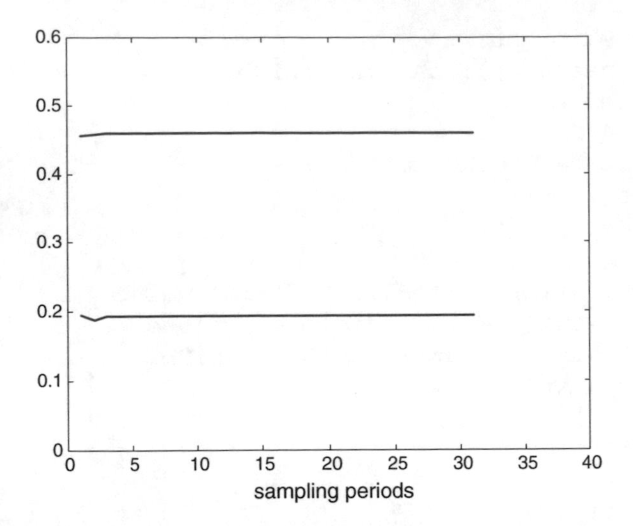

Fig. 1.60 Smoothing of states with fixed-point smoothing

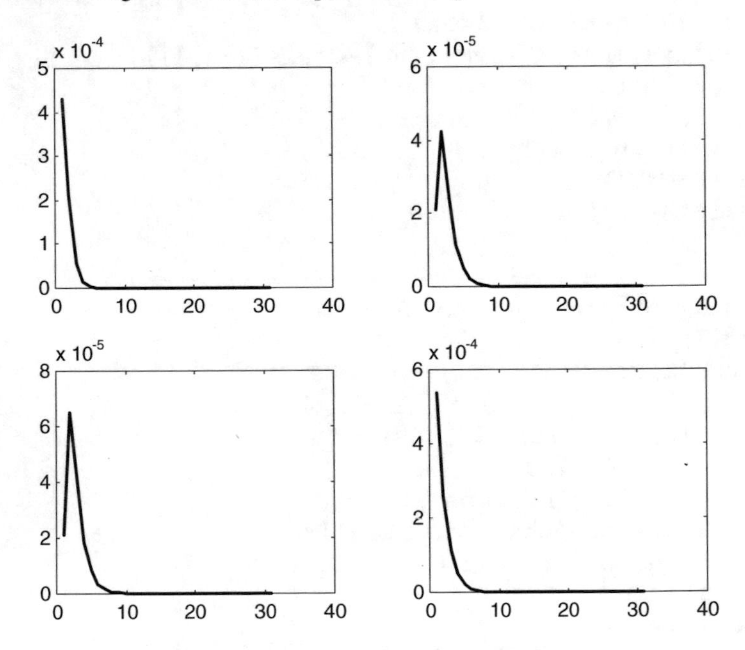

Fig. 1.61 Evolution of covariances in the fixed-point smoothing example

1.10.4 Bayesian Smoothing

Smoothing techniques are rapidly extending to more general contexts, not only the linear system with Gaussian noises.

One of the approaches for the application of Bayes to smoothing is given by the following equation, due to Kitagawa [57]:

$$\begin{aligned} p(\mathbf{x}_n|\mathbf{Y}_N) &= \int p(\mathbf{x}_n,\, \mathbf{x}_{n+1}|\mathbf{Y}_N)\, d\,\mathbf{x}_{n+1} = \\ &= \int p(\mathbf{x}_{n+1}|\mathbf{Y}_N) p(\mathbf{x}_n|\,\mathbf{x}_{n+1},\, \mathbf{Y}_N)\, d\,\mathbf{x}_{n+1} = \\ &= \int p(\mathbf{x}_{n+1}|\mathbf{Y}_N) p(\mathbf{x}_n|\,\mathbf{x}_{n+1},\, \mathbf{Y}_n)\, d\,\mathbf{x}_{n+1} = \\ &= p(\mathbf{x}_n|\mathbf{Y}_n) \int \frac{p(\mathbf{x}_{n+1}|\mathbf{Y}_N)\, p(\mathbf{x}_{n+1}|\mathbf{x}_n)}{p(\mathbf{x}_{n+1}|\mathbf{Y}_n)}\, d\mathbf{x}_{n+1} \end{aligned} \tag{1.385}$$

With a forward recursion you can compute $p(\mathbf{x}_n|\mathbf{Y}_n)$; and with a backward recursion $p(\mathbf{x}_n|\mathbf{Y}_N)$ from $p(\mathbf{x}_{n+1}|\mathbf{Y}_N)$.

The same methods already introduced, like EKF or UKF, are being applied for smoothing in the presence of nonlinearities. Likewise, based on Bayesian expressions, particle filters are being adapted for this scenario [13, 29]; one of the ways to do so is the following approximations:

$$\begin{aligned} &\int \frac{p(\mathbf{x}_{n+1}|\mathbf{Y}_N)\, p(\mathbf{x}_{n+1}|\mathbf{x}_n)}{p(\mathbf{x}_{n+1}|\mathbf{Y}_n)}\, d\mathbf{x}_{n+1} \approx \sum_i W_{n+1|N}(\mathbf{x}^i_{n+1})\, \frac{p(\mathbf{x}^i_{n+1}\,|\mathbf{x}_n)}{p(\mathbf{x}^i_{n+1}\,|\mathbf{Y}_n)} = \\ &= \sum_i W_{n+1|N}(\mathbf{x}^i_{n+1})\, \frac{p(\mathbf{x}^i_{n+1}\,|\mathbf{x}_n)}{\sum_j W_n(\mathbf{x}^j_n) p(\mathbf{x}^i_{n+1}\,|\mathbf{x}^j_n)} \end{aligned} \tag{1.386}$$

and:

$$p(\mathbf{x}_n|\mathbf{Y}_n) \approx \sum_k W_n(\mathbf{x}^k_n)\;\; \delta(\mathbf{x}_n\; - \mathbf{x}^k_n\,) \tag{1.387}$$

These approximations can be combined to obtain the next compact expression:

$$p(\mathbf{x}_n|\mathbf{Y}_N) \approx \sum_i W_{n|N}(\mathbf{x}^i_n)\;\; \delta(\mathbf{x}_n\; - \mathbf{x}^i_n\,) \tag{1.388}$$

with:

$$W_{n|N}(\mathbf{x}^i_n) = W_n(\mathbf{x}^i_n) \sum_k W_{n+1|N}(\mathbf{x}^k_{n+1})\, \frac{p(\mathbf{x}^k_{n+1}\,|\mathbf{x}^i_n)}{\sum_j W_n(\mathbf{x}^j_n) p(\mathbf{x}^k_{n+1}\,|\mathbf{x}^j_n)} \tag{1.389}$$

The smoothing algorithm would have a first forward pass, running the particle filter up to N and for $k = 1 : N$ store the sets $\{\mathbf{x}^i_k,\; W_k(\mathbf{x}^i_k)\}$. The second pass run backwards from N to n, recursively computing the importance weights.

1.11 Applications of Bayesian Filters

This section intends to give an idea of the varied applications of Bayesian filters. Of course it is not a complete review. Anyway, the literature that has been selected for this section offers a lot of references and hints, to help the reader to explore the area. For this reason, most of the cited works are academic dissertations or papers with large bibliographies.

In addition to the mentioned literature, there are books, like [48, 76], that include a set of chapters with different Bayesian filter applications.

1.11.1 Navigation

Interplanetary satellites use a Guidance-Navigation-Control (GNC) system. Knowing where to go, and its position and attitude, the satellite performs controlled actions.

First applications of Kalman filter focused on aerospace navigation, providing information on the state (position and attitude). An extensive and very interesting account of this application from 1960 is given by [42]. The Apollo mission to the Moon thrusted significant research efforts, including the development of EKF and the use of Monte Carlo techniques for testing. The numerical difficulties of the conventional Kalman filter were overcome with square-root filtering. On board implementation with 15 bit arithmetic was achieved.

Also during the 1960's the Kalman filter was introduced for aircraft inertial navigation systems.

See [49] for a comprehensive review of navigation systems.

Nowadays the navigation theme is attracting a lot of attention for cars, mobile robots, and autonomous vehicles (air, land, water, underwater). In each case two main aspects are what sensors are you using, and what is the influence of the environment. Other important aspects are what kind of maneuvering is involved, and how is the on-board processing system (weight, energy needs, etc.).

1.11.2 Tracking

Like navigation, tracking is a typical application of Kalman filtering. One can easily imagine scenarios related to radar and aircrafts, airports, missiles, air traffic, ships and maritime traffic, etc.

Most examples that usually appear in the scientific literature are related to navigation or tracking. The use of angles involves trigonometric functions; and the use of distances involves square roots. Then, there are nonlinearities.

In addition to radar there are other tracking devices. Microphones, or ultrasonic transducers could be used for air or underwater applications. Devices based on infrared, temperature, pressure, etc., could be used for certain types of detection and tracking. Cameras could be employed for monitoring and security, face tracking, pedestrian tracking, etc.

An extensive review of maneuvering target tracking is [63], [64]. Another review, more related to people tracking, is [124]. An interesting article about Bayesian location estimation is [35]. Particle filters are applied for positioning problems in [45].

Tracking is being extended to several targets at the same time. Proposed solutions involve multiple models, banks of filters, partitioning of particles, etc.

1.11.3 Information Fusion

Suppose you are driving a futuristic car with a lot of sensors, including artificial vision. Ultrasonic devices detect collision risk within 0.01 s. The vision device needs 1 s. processing time to detect an object. The car moves at 15 m/s, which is a moderate speed. A dog crosses at 10 m. from the car. Should the on-board computer wait for the vision?

Another scenario. This is a ship in a harbour. The ship requires protection. Sonar detectors show that something is moving underwater towards the ship. Underwater microphones detect the sound of a propeller. It would be important to know if this sound comes from that thing. Also, it would be relevant to estimate if the motion of that thing could correspond to a machine or something else.

Fusion problems deal with information from several sensors, perhaps with different delays. It is relevant to determine if different types of signals come from the same object. Notice that sometimes it is difficult to locate an airplane based on its sound, instinctively you look at the wrong place in the sky.

Lately, a typical fusion problem concerning several types of vehicles, is to combine inertial units and GPS to determine own position [24].

Most of the reported fusion applications use Bayesian estimation. The main idea is to combine sensors and mathematical models.

1.11.4 SLAM

Where in the world am I in this moment? If you are on the street you look around, recognize some familiar buildings, and this is enough to know your position. If you were a tourist some years ago, a typical procedure was to use a map and read street names. Nowadays GPS solves the problem when you don't know the place. However, it is not possible to use GPS under certain circumstances.

Suppose your mission is to study the sea floor in a certain ocean zone. There is no GPS underwater. What you can do is to recognize some possible landmarks on the floor, and then determine your position with respect to the landmarks. Starting from this, it would be possible to obtain a map of other new landmarks, expanding the exploration field and being able to move knowing where you are on this map.

This is the simultaneous localization and mapping (SLAM) problem. It is a main research topic in the field of mobile robots, with many applications in perspective. Bayesian state estimation is extensively used.

Several types of maps could be considered based on landmarks, features, occupancy grids, topological descriptions, etc.

In mathematical terms, when using a landmark map, the problem includes a state vector $\mathbf{x}(n)$ with the location and attitude of the vehicle, a control vector $\mathbf{u}(n)$ for driving the vehicle, a vector $\overleftarrow{m}$ with the locations of the landmarks, and a vector $\mathbf{z}(n)$ with observations of the landmarks from the vehicle. The following joint distribution must be computed for all times [31]:

$$p((\mathbf{x}(n),\ \mathbf{m}\,|\,\mathbf{Z}_n,\ \mathbf{U}_n,\ \mathbf{x}_0) \tag{1.390}$$

where $\mathbf{Z}_n,\ \mathbf{U}_n$ are histories of observations and control inputs.

There are two models:

- The observation model:

$$p((\mathbf{z}(n)\,|\,\mathbf{x}(n),\ \mathbf{m}) \tag{1.391}$$

- The motion model:

$$p((\mathbf{x}(n)\,|\,\mathbf{x}(n-1),\ \mathbf{u}(n)) \tag{1.392}$$

A prediction-correction **recursive algorithm for SLAM** is:

Prediction

$$p(\mathbf{x}(n),\ \mathbf{m}\,|\,\mathbf{Z}_{n-1},\ \mathbf{U}_n,\ \mathbf{x}_0) = \\ = \int p(\mathbf{x}(n)|\,\mathbf{x}(n-1),\ \mathbf{u}(n)) \cdot p(\mathbf{x}(n-1),\ \mathbf{m}|\mathbf{Z}_{n-1},\ \mathbf{U}_{n-1},\ \mathbf{x}_0)\, d\,\mathbf{x}(n-1) \tag{1.393}$$

Update

$$p(\mathbf{x}(n),\ \mathbf{m}\,|\,\mathbf{Z}_n,\ \mathbf{U}_n,\ \mathbf{x}_0) = \frac{p(\mathbf{z}(n)|\,\mathbf{x}(n),\mathbf{m})\,p(\mathbf{x}(n),\ \mathbf{m}|\mathbf{Z}_{n-1},\ \mathbf{U}_n,\ \mathbf{x}_0)}{p(\mathbf{z}(n),\ \mathbf{m}|\mathbf{Z}_{n-1},\ \mathbf{U}_n)}\,. \tag{1.394}$$

Important contributions to solve the SLAM problem use EKF. More recently the FastSLAM algorithm was introduced, based on particle filters. Bayesian information filters have been also proposed [114].

SLAM is a core topic of the so-called probabilistic robotics [105].

1.11.5 Speech, Sounds

One of the most important signals for humans is speech. The research has contributed with algorithms for speech enhancement. Increase of intelligibility, quality improvement, noise rejection, etc., are objectives of such enhancement. The use of iterative and sequential Kalman filter has been proposed in [38, 39]. A special modification of the Kalman filter that considers properties of the human auditory system has been investigated by [69]. A more recent contribution, using Bayesian inference is [70].

Speech enhancement is used in mobile phones, hearing aids, speech recognition, etc.

Another important area is vibration monitoring, which concerns machinery and structures –like for instance, bridges-. The use of Bayesian filters, together with signal classification methods, is an active research field. An example of this research is [117].

1.11.6 Images

The Kalman filtering approach to image restoration has received a lot of attention. Extended Kalman filters have been used for parameter estimation of the PSF. Also, the degradation model has been put as a Gauss-Markov model, so the Kalman filter can be applied to estimate the image to be recovered [23], [79]. In the recent years the use of particle filters for image recovery is being considered [91].

Another application is the use of Kalman filtering for image stabilization using models of the camera motion [32].

1.11.7 Earth Monitoring and Forecasting

Many forecasting applications are based on the use of the ensemble Kalman filter. For example in aspects of ocean dynamics [33], or concerning weather prediction [4].

Other papers about Bayesian filters and weather prediction are [37]. And about ocean dynamics [89].

Seismic signals are important for quake monitoring, analysis of civil structures, and for oil reservoirs monitoring [54]. Bayesian recursive estimation for seismic signals is investigated in [9].

1.11.8 Energy and Economy

Nowadays a lot of different electrical energy sources are being combined for daily power supply. Fossil or nuclear power plants, wind, solar panels, water, waves, tides...Good management systems are needed to coordinate sources, to obtain alternate electrical power with correct voltage and frequency. Bayesian filtering is being proposed to guarantee this behaviour [107, 122].

Other facets of electrical energy are consumption and price. This is treated in [47, 75] with Bayesian methods.

1.11.9 Medical Applications

Bayesian filtering methods can be applied for biosignal monitoring and analysis [92, 102]. In [94] Bayesian filters are used for arrhythmia verification in ICU patients.

Automated health monitoring systems propose to monitor patients during their activity at home. Bayesian methods for this application are investigated in [121].

A switching Kalman filter is proposed in [85] for neonatal intensive care.

1.11.10 Traffic

Vehicle traffic is an important part of modern life. Step by step, information technologies are coming into this scene through several entrances. One of the first contributions was related to traffic prediction, based on models [22]. This was used for semaphore managing. In certain countries this was also used to inform drivers about the *future* status of parking facilities or road crowding.

Currently the target is to devise cars with no human driver. The transition to this could be through intelligent highways and automatic convoys, perhaps with the help of some kind of beacons.

Particle filters adapt quite naturally to traffic applications. See for instance [30, 73].

In addition, Bayesian filters could help for robotized cars [5].

But it is not only a matter of cars. There are also ships and flying vehicles. Bayesian methods are being applied for sea and air traffic [81]. The future would mix manned and unmanned vehicles; it would be important to take into account, with adequate models, the type of responses that could be expected from robots or from humans.

1.11.11 *Other Applications*

It has been proposed to train neural networks with non-linear variants of the Kalman filter. In [83] this is done with good results for price prediction using ARMA models.

1.12 Frequently Cited Examples

The purpose of this section is to collect typical examples used in the scientific literature. Frequently these examples are employed to compare different filtering algorithms. One of these typical examples, the body falling towards Earth, has already been used in previous sections and will not be included here.

1.12.1 *Bearings-Only Tracking*

A typical scenario is a tracking device, perhaps based on circular scanning with a laser or with an ultrasonic transducer, focusing on a mobile robot. The information obtained is just an angle, corresponding to the line from the device to the robot. In the simplest case, the device is fixed. In other cases the device could be mounted on a moving platform.

Some of the scientific papers that study in particular this case are [1, 2, 6, 41, 59, 86]

Figure 1.62 shows a tracking scenario on the X-Y plane, with the tracking device at the origin.

A simple discrete-time model of the robot motion on the X-Y plane, including some process noise and supposing constant velocity, could be the following:

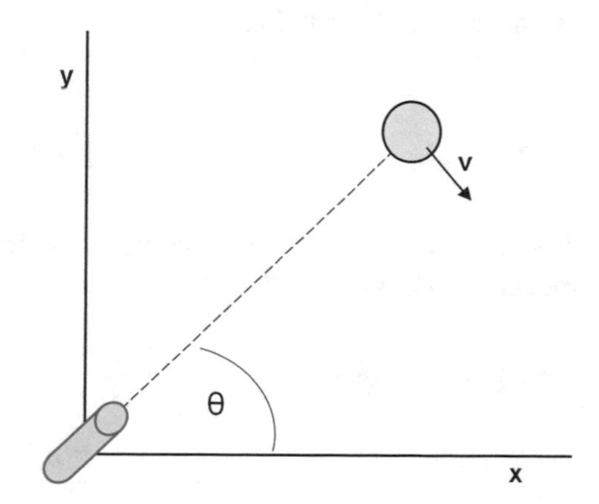

Fig. 1.62 Bearings only tracking scenario

$$\begin{bmatrix} x(n+1) \\ \dot{x}(n+1) \\ y(n+1) \\ \dot{y}(n+1) \end{bmatrix} = \begin{pmatrix} 1 & T & 0 & 0 \\ 0 & 1 & 0 & 0 \\ 0 & 0 & 1 & T \\ 0 & 0 & 0 & 1 \end{pmatrix} \begin{bmatrix} x(n) \\ \dot{x}(n) \\ y(n) \\ \dot{y}(n) \end{bmatrix} + \begin{pmatrix} 0.5 & 0 \\ 1 & 0 \\ 0 & 0.5 \\ 0 & 1 \end{pmatrix} \begin{bmatrix} w_x(n) \\ w_y(n) \end{bmatrix} \tag{1.395}$$

where T is the sampling period, and:

$$w_x(n) \propto N(0,\ 1)\ ,\ \ w_y(n) \propto N(0,\ 1) \tag{1.396}$$

The bearings only measurement induces a nonlinearity in the model:

$$\theta(n) = \arctan(\frac{y(n)}{x(n)}) + v(n) \tag{1.397}$$

where the measurement noise could be taken as a normal PDF.

1.12.2 Other Tracking Cases

There are examples considering range-only tracking, [88]. In this case the measurement equation is similar to the falling body:

$$R(n) = \sqrt{x(n)^2 + y(n)^2} \tag{1.398}$$

Other examples combine angle and range tracking, [45], so the process output is a vector $\mathbf{y} = [R,\ \theta]$.

Angle-only tracking is studied in [56].

When dealing with tracking scenarios it is important to consider what happens with observability, [100].

The tracking of maneuvering targets, like an aircraft, could involve accelerations and the use of 3D coordinates.

1.12.3 Univariate Non-stationary Growth Model

The following population growth model has become a benchmark for filtering algorithms, [7, 16, 29, 41, 57, 59]:

$$x(n) = 0.5\,x(n-1) + 25\,\frac{x(n-1)}{1 + x(n-1)^2} + 8\,\cos(1.2\,(n-1)) + w(n-1) \tag{1.399}$$

$$y(n) = \frac{x(n)^2}{20} + v(n) \tag{1.400}$$

where:

$$w(n) \propto N(0,\ 1)\ ,\ \ v(n) \propto N(0,\ 1) \tag{1.401}$$

Notice the nonlinearities in the transition and the measurement equations.

1.12.4 Financial Volatility Model

This is a time series that model fluctuations in the stock market prices, exchange rates and option pricing, [52]:

$$y(n) = \sigma(n) \cdot \varepsilon(n) \tag{1.402}$$

$$\ln \sigma(n)^2 = \alpha + \beta \cdot \ln\ \sigma(n-1)^2 + \mu\, v(n) \tag{1.403}$$

where:

$$\varepsilon(n),\ v(n)\ \propto N(0,\ 1) \tag{1.404}$$

An instance of parameter values could be: $\alpha = 1.8;\ \ \beta = 0.95$.

Another equivalent expression of this mode, [101], is:

$$y(n) = \exp(x(n)/2) \cdot \varepsilon(n) \tag{1.405}$$

$$x(n) = \alpha + \beta \cdot x(n) + \kappa \eta(n) \tag{1.406}$$

1.12.5 Nonlinear Series

This is a synthetic time series that is useful for algorithm checking, [20, 66, 109]:

$$x(n) = 1 + a \cdot x(n-1) + \sin(\,w \cdot\ (n-1) \cdot \pi) + s(n) \tag{1.407}$$

where the PDF of the noise $s(n)$could be of Gamma type, or Gaussian, etc. The value of parameters could be: $a = 0.5;\ \ w = 0.04$

The measurement equation could be:

$$y(n) = \begin{cases} b\,x(n)^2 + v(n), & n \le 30 \\ c\,x(n)^2 - 2 + v(n), & n > 30 \end{cases} \tag{1.408}$$

with $b = 0.2;\ c = 0.5,\ v(n) \in N(0,\ 10^{-5})$.

1.12.6 The Pendulum

The pendulum is a natural example of nonlinear system. A state space model can be written as follows, [97]:

$$\theta(n+1) = \theta(n) + T\,\omega(n) \tag{1.409}$$

$$\omega(n+1) = \omega(n) - (T \cdot g/L)\,\sin\,\theta(n) \tag{1.410}$$

$$y(n) = \begin{bmatrix} \theta(n) \\ \omega(n) \end{bmatrix} + v(n) \tag{1.411}$$

Because of conservation of energy, there is a constraint to be satisfied:

$$-\,m\,g\,L\,\cdot\cos\,\theta(n) + m\,L^2\,\omega^2(n)\,/2 = C \tag{1.412}$$

This is an interesting example for state estimation of constrained systems, [27, 93, 103].

1.13 Resources

1.13.1 MATLAB

1.13.1.1 Toolboxes

- Kalman filter toolbox:
  http://www.cs.ubc.ca/~murphyk/Software/Kalman/kalman.html
- Kalman filtering Toolbox:
  www.l3nav.com/kalman_filtering_toolbox.htm
- KALMTOOL:
  www.iau.dtu.dk/research/control/kalmtool2.html
- EKF/UKF Toolbox:
  http://becs.aalto.fi/en/research/bayes/ekfukf/
- REBEL Toolkit:
  http://choosh.cse.ogi.edu/rebel
- NEF Nonlinear Estimation Toolbox:
  http://nft.kky.zcu.cz/nef
- PFToolbox:
  http://pftoolbox.sourceforge.net/
- MGDTnToolbox (open-loop process control):
  http://cc.oulu.fi/~iko/MGDT/MGDT.htm
- Robotics Toolbox:
  http://www.petercorke.com/Robotics_Toolbox.html

- CAS Robot Navigation Toolbox:
  http://www.cas.kth.se/toolbox/
- GPstuff Toolbox:
  http://becs.aalto.fi/en/research/bayes/gpstuff/

1.13.1.2 Matlab Code

- Dan Simon:
  http://academic.csuohio.edu/simond/
- Polypedal/kalman:
  http://www.eecs.berkeley.edu/~sburden/matlab/
  http://www.polypedal/doc/polypedal/kalman/index.html
- Tim Bailey:
  www-personal.acfr.usyd.edu.au/tbailey/
- EnKF Ensemble Kalman Filter:
  http://enkf.nersc.no/
- Hannes Nickisch:
  http://hannes.nickisch.org/
- Bethge Lab:
  http://bethgelab.org/

1.13.2 Internet

1.13.2.1 Web Sites

- The Kalman Filter:
  www.cs.unc.edu/~welch/kalman/
- Simo Särkkä:
  www.lce.hut.fi/~ssarkka/
- Van der Merwe:
  http://www.cse.ogi.edu/~rudmerwe/
- MOSAIC Group:
  http://www.mosaic.ethz.ch/Downloads/matlabtracker
- BECS AALTO Univ.:
  http://becs.aalto.fi/en/research/bayes/
- OpenDA data assimilation:
  http://www.openda.org/joomla/index.php
- Model Data Fusion (Ocean):
  http://www.cmar.csiro.au/staff/oke/MDF.htm

1.13.2.2 Link Lists

- Kernel-Machines.org:
  www.kernel-machines.org/

References

1. V.J. Aidala, Kalman filter behavior in bearings-only tracking applications. IEEE T. Aerosp. Electron. Syst. **15**(1), 29–39 (1979)
2. V.J. Aidala, S.E. Hamnel, Utilization of modified polar coordinates for bearings-only tracking. IEEE T. Autom. Control **28**(3), 283–294 (1983)
3. D.L. Alspach, H.W. Sorenson, Nonlinear Bayesian estimation using Gaussian sum approximation. IEEE T. Autom. Control **17**(4), 439–448 (1972)
4. J.L. Anderson, Ensemble Kalman filters for large geophysical applications. IEEE Control Syst. Mgz. 66–82 (2009)
5. N. Apostoloff, Vision based lane tracking using multiple cues and particle filtering. Master's thesis, Australian National Univ., 2005
6. M.S. Arulampalam, B. Ristic, N. Gordon, T. Mansell, Bearings-only tracking of manoeuvring targets using particle filters. EURASIP J. Appl. Signal Process. **2004**(15), 2351–2365 (2004)
7. S. Arulampalam, S. Maskell, N. Gordon, T. Clapp, A tutorial on particle filters for on-line non-linear/non-Gaussian Bayesian tracking. IEEE T. Signal Process. **50**(2):174–188 (2002)
8. S.P. Banks, *Control Systems Engineering* (Prentice-Hall, 1986)
9. E. Baziw, Application of Bayesian Recursive estimation for seismic signal processing. Ph.D. thesis, Univ. British Columbia, 2007
10. B.M. Bell, F.W. Cathey, The iterated Kalman filter update as a Gauss-Newton method. IEEE T. Autom. Control **38**(2):294–297 (1993)
11. J.L. Blanco, J. Gonzalez, J.A. Fernandez-Madrigal, Optimal filtering for non-parametric observation models: applications to localization and SLAM. Int. J. Robot. Res. **29**(14), 1726–1742 (2010)
12. H.A.P. Blom, Y. Bar-Shalom, The interacting multiple model algorithm for systems with Markovian switching coefficients. IEEE T. Autom. Control **33**(8), 780–783 (1988)
13. M. Briers, A. Doucet, S. Maskell, Smoothing algorithms for state-space models. Ann. Inst. Stat. Math. **62**(1), 61–89 (2010)
14. M. Briers, S.R. Maskell, R. Wright, A Rao-Blackwelised unscented Kalman filter, in *Proceedings 6thIEEE International Conference on Information Fusion*, pp. 55–61 (2003)
15. J.V. Candy, *Bayesian Signal Processing* (Wiley, IEEE, 2009)
16. O. Cappé, S.J. Godsill, E. Moulines, An overview of existing methods and recent advances in sequential Monte Carlo. Proc. IEEE **95**(5), 899–924 (2007)
17. R. Casarin, J.M. Marin, Online data processing: comparison of Bayesian regularized particle filters. Electron. J. Stat. **3**, 239–258 (2009)
18. C. Chang, R. Ansari, Kernel particle filter for visual tracking. IEEE Signal Process. Lett. **12**(3), 242–245 (2005)
19. Z. Chen, Bayesian filtering: form Kalman filters to particle filters, and beyond. Technical Report, McMaster Univ., 2003. http://www.dsi.unifi.it/users/chisci/idfric/Nonlinear_filtering_Chen.pdf
20. Q. Cheng, P. Bondon, An efficient two-stage sampling method in particle filter. IEEE T. Aerosp. Electron. Syst. **48**(3), 2666–2672 (2012)
21. Y. Cheng, J.L. Crassidis, Particle filtering for sequential spacecraft attitude estimation, in *Proccedings AIAA Guidance, Navigation, and Control Conference* (2004)
22. R. Chrobok, Theory and application of advanced traffic forecast methods. Ph.D. thesis, Univ. Duisburg-Essen, Germany (2005)

23. S. Citrin, M.R. Azimi-Sadjadi, A full-plane block Kalman filter for image restoration. IEEE T. Image Process. **1**(4):488–495 (1992)
24. J.L. Crassidis, Sigma-point Kalman filtering for integrated GPS and inertial navigation. IEEE T. Aerosp. Electron. Syst. **42**(2), 750–756 (2006)
25. N. Cui, L. Hong, J.R. Layne, A comparison of nonlinear filtering approaches with an application to ground target tracking. Signal Process. **85**, 1469–1492 (2005)
26. F. Daum, Nonlinear filters: beyond the Kalman filter. IEEE A&E Syst. Mag. **20**(8), 57–69 (2005)
27. M.P. Deisenroth, M.F. Huber, U.D. Hanebeck, Analytic moment-based Gaussian process filtering, in *Proceedings 26thACM Annual International Conference Machine Learning*, pp. 225–232 (2009)
28. A. Doucet, M. Briers, Efficient block sampling strategies for sequential Monte Carlo methods. J. Comput. Graph. Stat. **15**(3):693–711 (2006)
29. A. Doucet, S. Godsill, C. Andrieu, On sequential Monte Carlo sampling methods for Bayesian filtering. Stat. Comput. **10**, 197–208 (2000)
30. M.C. Dunn, Applying particle filter and path-stack methods to detecting anomalies in network traffic volume. Ph.D. thesis, Carnegie-Melon Univ., 2004
31. H. Durrant-Whyte, T. Bailey, Simultaneous localisation and mapping (SLAM): Part I. IEEE Robot. Autom. Mgz. 99–108 (2006)
32. S. Erturk, Real-time digital image stabilization using Kalman filters. Real-Time Imag. **8**, 317–328 (2002)
33. G. Evensen, The ensemble Kalman filter: theoretical formulation and practical implementation. Ocean Dyn. **53**, 343–367 (2003)
34. D. Fox, Adapting the sample size in particle filters through KLD-sampling. The Intl. J. Robotics Research, 22, 12, 985–1003 (2003)
35. D. Fox, J. Hightower, L. Liao, D. Schulz, G. Borriello, Bayesian filtering for location estimation. IEEE Pervasive Comput. 24–33 (2003)
36. D. Fraser, J. Potter, The optimum linear smoother as a combination of two optimum linear filters. IEEE T. Autom. Control **14**(4), 387–390 (1969)
37. G. Galanis, P. Louka, P. Katsafados, G. Kallos, I. Pytharoulis, Applications of Kalman filters based on non-linear functions to numerical weather prediction. Ann. Geophys. **24**, 1–10 (2006)
38. S. Gannot, Speech processing utilizing the Kalman filter. IEEE Instrum. Meas. Mgz. 10–14 (2012)
39. S. Gannot, D. Burshtein, E. Weinstein, Iterative and sequential Kalman filter-based speech enhancement algorithms. IEEE T. Speech Audio Process. **6**(4), 373–385 (1998)
40. G.C. Goodwin, *Adaptive Filtering Prediction and Control* (Prentice-Hall, 1984)
41. N.J. Gordon, D.J. Salmond, A.F.M. Smith, Novel approach to nonlinear/non-Gaussian Bayesian state estimation. IEE Proc.-F **140**(2), 107–113 (1993)
42. M.S. Grewal, A.P. Andrews, Applications of Kalman filtering in aerospace 1960 to the present. IEEE Control Syst. Mag. 69–78 (2010)
43. G. Grisetti, C. Stachniss, W. Burgard, Improved techniques for grid mapping with Rao-Blacwellized particle filters. IEEE T. Robot. **32**(1), 34–46 (2007)
44. F. Gustaffson, Particle filter theory and practice with positioning applications. IEEE A&E Mag. **25**(7), 53–81 (2010)
45. F. Gustafsson, F. Gunnarsson, N. Bergman, U. Forssell, J. Jansson, R. Karlsson, P.J. Nordlund, Particle filters for positioning, navigation, and tracking. IEEE T. Signal Process. **50**(2), 425–437 (2002)
46. F. Gustafsson, G. Hendeby, Some relations between extended and unscented Kalman filters. IEEE T. Signal Process. **60**(2), 545–555 (2012)
47. S. Hamis, Dynamics of oil and electricity spot prices in ensemble stochastic models. Master's thesis, Lappeenranta Univ. of Technology, 2012
48. A.C. Harvey, *Forecasting, Structural Time Series Models and the Kalman Filter* (Cambridge University Press, 1990)
49. A.M. Hasan, K. Samsudin, A.R. Ramli, R.S. Azmir, S.A. Ismael, A review of navigation systems (integration and algorithms). Australian J. Basic Appl. Sci. **3**(2), 943–959 (2009)

50. J.D. Hol, T.B. Schon, F. Gustafsson, On resampling algorithms for particle filters, in *Proceedings IEEE Nonlinear Statistical Signal Processing Workshop*, pp. 79–82 (2006)
51. K. Ito, K.Q. Xiong, Gaussian filters for nonlinear filtering problems. IEEE T. Autom. Control **45**(5), 910–927 (2000)
52. E. Jacquier, N.G. Polson, P.E. Rossi, Bayesian analysis of stochastic volatility models with fat-tails and correlated errors. J. Econ. **122**(1), 185–212 (2004)
53. S.J. Julier, J.K. Uhlmann, Unscented filtering and nonlinear estimation. IEEE Proc. **92**(3), 401–422 (2004)
54. S. Kalla, Reservoir characterization using seismic inversion data. Ph.D. thesis, Louisiana State Univ., 2008
55. R. Kalman, A new approach to linear filtering and prediction problems. ASME J. Basic Eng. **82**, 35–45 (1960)
56. R. Karlsson, Various topics on angle-only tracking using particle filters. Technical Report, Linköping University, 2002. http://www.diva-portal.org/smash/get/diva2:316636/FULLTEXT01.pdf
57. G. Kitagawa, Non-Gaussian state-space modeling of nonstationary time-series. J. Am. Stat. Assoc. **82**, 1032–1063 (1987)
58. J. Ko, D. Fox, GP-BayesFilters: Bayesian filtering using Gaussian process prediction and observation models. Auton. Robots **27**, 75–90 (2009)
59. J.H. Kotecha, P.M. Djuric, Gaussian particle filtering. IEEE T. Signal Process. **51**(10), 2592–2601 (2003)
60. J.H. Kotecha, P.M. Djuric, Gaussian sum particle filtering. IEEE T. Signal Process. **51**(10), 2602–2612 (2003)
61. B.L. Kumari, K.P. Raju, V.Y.V. Chandan, R.S. Krishna, V.M.J. Rao, Application of extended Kalman filter for a free falling body towards Earth. Int. J. Adv. Comput. Sci. Appl. **2**(4), 134–140 (2011)
62. S. Lakshmivarahan, D.J. Stensrud, Ensemble Kalman filter. IEEE Control Syst. Mag. 34–46 (2009)
63. X.R. Li, V.P. Jilkov, Survey of maneuvering target tracking. part I: Dynamic models. IEEE T. Aerosp. Electron. Syst. **39**(4), 1333–1364 (2003)
64. X.R. Li, V.P. Jilkov, Survey of maneuvering target tracking. part II: Motion models and ballistic and space targets. IEEE T. Aerosp. Electron. Syst. **46**(1), 96–119 (2004)
65. L. Liang-qun, J. Hong-bin, L. Jun-hui, The iterated extended Kalman particle filter, in *Proceedings IEEE Intlernational Symposium on Communications and Information Technology*, pp. 1213–1216 (2005)
66. L. Liang-Qun, J. Hong-Bing, L. Jun-Hui, The iterated extended Kalman particle filter, in *Proceedings IEEE International Symposium Communications and Information Technology, ISCIT 2005*, vol. 2, pp. 1213–1216 (2005)
67. F. Lindsten, Rao-Blackwellised particle methods for inference and identification. Master's thesis, Linköping University, 2011. Licenciate Thesis
68. X. Luo, I.M. Moroz, Ensemble Kalman filter with the unscented transform. Physica D: Nonlinear Phenomena **238**(5), 549–562 (2009)
69. N. Ma, Speech enhancement algorithms using Kalman filtering and masking properties of human auditory systems. Ph.D. thesis, Univ. Otawa, 2005
70. C. Maina, Approximate Bayesian inference for robust speech processing. Ph.D. thesis, Drexel Univ., 2011
71. M. Mallick, L. Mihaylova, S. Arulampalam, Y. Yan, Angle-only filtering in 3Dusing modified spherical and log spherical coordinates, in *Proceedings of 14th IEEE International Conference on Information Fusion*, pp. 1–8 (2001)
72. J.M. Mendel, *Lessons in Digital Estimation Theory* (Prentice-Hall, 1987)
73. L. Mihaylova, A. Hegyi, A. Gning, R. Boel, Parallelized particle and Gaussian sum particle filters for large scale freeway traffic systems. IEEE T. Intell. Transp. Syst. 1–13 (2012)
74. T.P. Minka, A family of algorithms for approximate Bayesian inference. Ph.D. thesis, MIT, 2001

75. A. Molina-Escobar, Filtering and parameter estimation for electricity markets. Ph.D. thesis, Univ. British Columbia, 2009
76. V.M. Moreno, A. Pigazo, *Kalman Filter Recent Advances and Applications* (Intech, 2009)
77. K.P. Murphy, Switching Kalman filters. Technical report, Univ. Berkeley, 1998
78. C. Musso, P.B. Quang, F. Le Gland, Introducing the Laplace approximation in particle filtering, in *Proceedings of 14th International Conference on Information Fusion*, pp. 290–297 (2001)
79. R. Nagayasu, N. Hosoda, N. Tanabe, H. Matsue, T. Furukawa, Restoration method for degraded images using two-dimensional block Kalman filter with colored driving source, in *Proceedings IEEE Digital Signal Processing Workshop*, pp. 151–156 (2011)
80. J.C. Neddermeyer, Sequential Monte Carlo methods for general state-space models. Master's thesis, University Ruprecht-Karls Heidelberg, 2006. Diploma dissertation
81. P-J. Nordlund, Efficient estimation and detection methods for airborne applications. Ph.D. thesis, Linköping University, 2008
82. M. Norgaard, N.K. Poulsen, O. Ravn, New developments in state estimation for nonlinear systems. Automatica **36**, 1627–1638 (2000)
83. M.A. Oliveira, An application of neural networks trained with Kalman filter variants (EKF and UKF) to heteroscedastic time series forecasting. Appl. Math. Sci. **6**(74), 3675–3686 (2012)
84. M.K. Pitt, N. Shephard, Filtering via simulation: auxiliary particle filters. J. Am. Stat. Assoc. **94**(446), 590–599 (1999)
85. J.A. Quinn, Bayesian condition monitoring in neonatal intensive care. Ph.D. thesis, University of Edinburgh, 2007
86. K. Radhakrishnan, A. Unnikrishnan, K.G. Balakrishnan, Bearing only tracking of maneuvering targets using a single coordinated turn model. Int. J. Comput. Appl. **1**(1), 25–33 (2010)
87. H.E. Rauch, F. Tung, C.T. Striebel, Maximum likelihood estimates of linear dynamic systems. J. Am. Inst. Aeronaut. Astronaut. **3**(8), 1445–1450 (1965)
88. B. Ristic, S. Arulampalam, N. Gordon, Beyond the Kalman filter. IEEE Aerosp. Electron. Syst. Mgz. **19**(7), 37–38 (2004)
89. M. Rixen et al., Improved ocean prediction skill and reduced uncertainty in the coastal region from multi-model super-ensembles. J. Marine Syst. **78**, S282–S289 (2009)
90. A.M. Sabatini, Quaternion-based extended Kalman filter for determining orientation by inertial and magnetic sensing. IEEE T. Biomed. Eng. **53**(7), 1346–1356 (2006)
91. S.I. Sadhar, A.N. Rajagopalan, Image recovery under nonlinear and non-Gaussian degradations. J. Opt. Soc. Am. **22**(4), 604–615 (2005)
92. R. Sameni, M.B. Shamsollahi, C. Jutten, G.D. Clifford, A nonlinear Bayesian filtering framework for ECG denoising. IEEE T. Biomed. Eng. **54**(12), 2172–2185 (2007)
93. S. Särkkä, *Bayesian Filtering and Smoothing*, vol. 3 (Cambridge University Press, 2013)
94. O. Sayadi, M.B. Shamsollahi, Life-threatening arrhythmia verification in ICU patients using the joint cardiovascular dynamical model and a Bayesian filter. IEEE T. Biomed. Eng. **58**(10), 2748–2757 (2011)
95. B. Sherlock, B. Herbstm *Introduction to the Kalman Filter and Applications* (University of Stellenbosch, South Africa, 2002). http://dip.sun.ac.za/~hanno/tw796/lesings/kalman.pdf
96. D. Simon, *Optimal State Estimation* (Wiley, 2006)
97. D. Simon, Kalman filtering with state constraints: a survey of linear and nonlinear algorithms. IET Control Theory Appl. **4**(8), 1303–1318 (2010)
98. L. Smith, V. Aitken, The auxiliary extended and auxiliary unscented Kalman particle filters, in *Proceedings IEEE Canadian Conference on Electrical and Computer Engineering*, pp. 1626–1630. Vancouver (2007)
99. A. Solin, Cubature integration methods in non-linear Kalman filtering and smoothing. Master's thesis, Aalto University School of Science and Technology, Finland, 2010. Bachelor's thesis
100. T.L. Song, Observability of target tracking with range-only measurements. IEEE J. Oceanic Eng. **24**(3), 383–387 (1999)
101. J.R. Stroud, N.G. Polson, P. Müller, Practical filtering for stochastic volatility models, ed. by S. Koopmans, A. Harvey, N. Shephard. *State Space and Unobserved Component Models*, pp. 236–247 (Cambridge University Press, 2004)

102. M. Tarvainen, Estimation methods for nonstationary biosignals. Ph.D. thesis, University of Kuopio, Finland, 2004
103. B.O. Teixeira, J. Chandrasekar, L.A. Tôrres, L.A. Aguirre, D.S. Bernstein, State estimation for linear and non-linear equality-constrained systems. Int. J. Control **82**(5), 918–936 (2009)
104. G.A. Terejanu, Tutorial on Monte Carlo techniques. Technical report, University of Buffalo, 2008. together with other tutorials. https://cse.sc.edu/~terejanu/files/tutorialMC.pdf
105. S. Thrun, W. Burgard, D. Fox, *Probabilistic Robotics* (MIT Press, 2005)
106. S. Tully, H. Moon, G. Kantor, H. Choset, Iterated filters for bearing-only SLAM, in *Proceedings IEEE Intl. Conf. on Robotics and Automation*, pp. 1442–1448 (2008)
107. G. Valverde, V. Terzija, Unscented Kalman filter for power system dynamic state estimation. IET Gener. Trans. Distrib. **5**(1), 29–37 (2010)
108. R. Van der Merwe, Sigma-point Kalman Filters for probabilistic inference in dynamic state-space models. Ph.D. thesis, Oregon Health & Science University, 2004
109. R. Van der Merwe, N. Freitas, A. Doucet, E. Wan, The unscented particle filter. Adv. Neural Inf. Process. Syst. **13** (2001)
110. R. Van der Merwe, E. Wan, Efficient derivative-free Kalman filters for online learning, in *Proceedings European Symposium on Artificial Neural Networks*, pp. 205–210 (Bruges, 2001)
111. R. Van der Merwe, E. Wan, Gaussian mixture sigma-point particle filters for sequential probabilistic inference in dynamic state-space models, in *Proceedings IEEE International Conference Acoustics, Speech and Signal Processing*, vol. 6, pp. 701–704 (2003)
112. M. Verhaegen, P. van Dooren, Numerical aspects of different Kalman filter implementations. IEEE T. Autom. Control **31**(10), 907–917 /(1986)
113. N. Vlassis, B. Terwijn, B. Krose, Auxiliary particle filter robot localization from high-dimensional sensor observations, in *Proceedings IEEE International Conference Robotics and Automation*, vol. 1, pp. 7–12 (2002)
114. M.R. Walter, Sparse Bayesian information filters for localization and mapping. Ph.D. thesis, MIT, 2008
115. E.A. Wan, R. Van der Merwe, The unscented Kalman filter, ed. by S. Haykin. *Kalman Filtering and Neural Networks*, Chap. 13, pp. 221–280 (Wiley, 2001)
116. D. Wang, H. Hua, H. Cao, Algorithm of modified polar coordinates UKF for bearings-only target tracking, in *Proceedings IEEE International Conference Future Computer and Communication*, vol. 3, pp 557–560 (Wuhan, 2010)
117. K. Wang, Vibration monitoring on electrical machines using Vold-Kalman filter order tracking. Master's thesis, Univ. Pretoria, 2008
118. Q. Wang, L. Xie, J. Liu, Z. Xiang, Enhancing particle swarm optimization based particle filter tracker, in *Computational Intelligence, LNCS 4114/2006*, pp. 1216–1221 (Springer Verlag, 2006)
119. G. Welch, G. Bishop, An introduction to the Kalman filter. Technical report, UNC-Chapel Hill, TR, 2006
120. N. Whiteley, A.M. Johansen, Recent developments in auxiliary particle filtering, eds. by B. Cemgil, Chiappa. *Bayesian Time Series Models* (Cambridge University Press, 2010)
121. D.H. Wilson. Assistive intelligent environments for automatic health monitoring. Ph.D. thesis, Carnegie-Mellon Univ., 2005
122. R.A. Wiltshire, G. Ledwich, P. O'Shea, A Kalman filtering approach to rapidly detecting modal changes in power systems. IEEE T. Power Syst. **22**(4), 1698–1706 (2007)
123. Y. Wu, D. Hu, M. Wu, X. Hu, A numerical-integration perspective of Gaussian filters. IEEE T. Signal Process **54**(8), 2910–2921 (2006)
124. A. Yilmaz, Object tracking: a survey. ACM Comput. Surveys **38**(4), 1–45 (2006)
125. J. Yu, Y. Tang, X. Chen, W. Liu, Choice mechanism of proposal distribution in particle filter, in *Proceedings IEEE World Congress Intelligent Control and Automation*, pp. 1051–1056 (2010)
126. A.Z. Zambom, R. Dias, *A Review of Kernel Density Estimation with Applications to Econometrics* (Universidade Estadual de Campinas, 2012). arXiv preprint arXiv:1212.2812

Part II
Sparse Representation. Compressed Sensing

Chapter 2
Sparse Representations

2.1 Introduction

Sparse representations intend to represent signals with as few as possible significant coefficients. This is important for many applications, like for instance compression.

Consider for example the signal $\cos(5\,t)$ along 7 min. For computer storage this signal could be saved as a file with hundreds of bytes, if one takes samples and save them. On the other hand, this signal can be represented with a text file such [C 5 7m] with just 4 bytes. Obviously, this last representation is more economical in terms of symbols. In many cases, it is possible to obtain this economy by choosing adequate bases combined with dictionaries.

When using wavelets it is frequently noticed that a great compression rate can be obtained, with almost unnoticeable loss of information. Supposing that the signal comes from a sensor, it would be very convenient to have the sensor yielding already compressed information. This is called '*compressed sensing*'.

Almost immediately, the topics already mentioned lead to inverse problems (related to reconstruction). In the recent times, some interesting methods have been proposed for their solution.

Due to pressing demands from digital devices and networked systems consumers, a lot of interest is being paid to finding sparsest representations. This turns out to be a not-so-standard optimization problem. It also suggests that some space should be devoted, in our book, to mathematical optimization. Since it would be too long for this chapter, an appendix has been included with selected optimization topics, leaving for our present chapter some specific aspects of optimization related to sparse representations.

In the next pages the reader will find numerous references to key publications, which may well serve for a fruitful diving into the various aspects of sparse representations. In the end, what is behind the scene is a new look to sampling [18, 34, 42, 68, 160].

J.M. Giron-Sierra, *Digital Signal Processing with Matlab Examples, Volume 3*, Signals and Communication Technology, DOI 10.1007/978-981-10-2540-2_2

2.2 Sparse Solutions

This section considers a general problem, with a well-known mathematical expression:

$$A\mathbf{x} = \mathbf{b} \tag{2.1}$$

The target is to obtain sparse solutions.

First of all, looking at the $m \times n$ matrix A and the corresponding system of linear equations, there are three cases: $m > n$, $m = n$, and $m < n$.

The case of more equations than unknowns is typical of statistics and regression problems.

On the other hand, the undetermined system is the basic scenario considered in compressive sensing (which will be studied in more detail in the next section).

Usually, in order to get sparse solutions, the question is stated in an optimization framework.

A matrix or a vector is said to be sparse if it has only a few (compared to the total size) non-zero entries. Of course, one needs to formalize this initial concept.

In the case of a vector, its $l0$ norm, $\|\mathbf{x}\|_0$, is simply the number of non-zero entries in the vector.

Continuing with vectors, the $l1$norm, $\|\mathbf{x}\|_1$, is the sum of the absolute value of the entries in a vector. The $l2$ norm of a vector is its Euclidean length:

$$\|\mathbf{x}\|_2 = \left(\sum_{i=1}^{n} |x_i|^2 \right)^{1/2}.$$

From the point of view of sparsity, $l0$ is the most appropriate norm, while $l2$ says almost nothing. On the other hand, from the point of view of mathematical analysis, finding the sparsest solution is an optimization problem and in this context, the $l2$ *norm* is a well know terrain while $l0$ is unusual.

As we shall see, our efforts will concentrate on using the $l1$norm, which is linked to sparsity and is easier for mathematical work.

The main focus of this section is put on methods for solving sparsity optimization problems. The basic literature is [3, 27, 76, 177].

2.2.1 The Central Problem

Consider the problem of finding a sparse $n \times 1$ vector $\mathbf{x}$ such that:

$$A\mathbf{x} = \mathbf{b} \tag{2.2}$$

where the $m \times 1$ vector $\mathbf{b}$ and the $m \times n$ matrix A, are given.

Recall that in the case of regression problems, with $m > n$, this is an overdetermined system that usually has no solution. But one could be interested on establishing a model, based on the following problem:

$$\text{minimize } \|A\mathbf{x} - \mathbf{b}\|_2^2$$

In general this minimization would not enforce sparsity. Some suitable terms should be added, as we shall see.

In the case of undetermined system, with $m < n$, there would be in principle infinite candidate solutions. A typical proposal is to choose the solution with the minimum $l2$ norm, so the problem is:

$$\text{minimize } \|\mathbf{x}\|_2^2, \text{ subject to } A\,\mathbf{x} = \mathbf{b}$$

Using Lagrange multipliers, and taking derivatives, it is possible to analytically get the solution:

$$\mathbf{x} = (A^T A)^{-1} A^T\, \mathbf{b} \tag{2.3}$$

Again, in general this solution would be not sparse. The problem should be stated with other terms.

But, why it would be interesting to obtain sparse solutions?

If you are trying to establish a model in a regression problem, a sparse solution will reveal the pertinent variables as those with non-zero values. This helps to identify and select the correct variables (in the specialized literature variable selection, or feature selection, is a frequently cited issue).

When dealing with signals, it is sometimes desired to transform a set of measurements $\mathbf{b}$ into a sparse vector $\mathbf{x}$. Later on, the vector $\mathbf{x}$ could be used to exactly reproduce $\mathbf{b}$, or, in certain applications it would be sufficient with an approximate recovery. You may remember from past chapters, that this was frequently the case with wavelets, where most information was contained in large entries of $\mathbf{x}$, while other entries were negligible. Then, it would make sense to consider an approximation of the form:

$$\|A\,\mathbf{x} - \mathbf{b}\| < \varepsilon \tag{2.4}$$

where some of the entries of $\mathbf{x}$ were set to zero. This is a sparse approximation.

Another case of interest is the design of digital filters with a minimum number of non-zero coefficients, [17].

2.2.1.1 Basic Approaches

Suppose that it is an undetermined system, with infinite candidate solutions. One wants the sparsest solution:

- In formal terms, the problem of sparsest representation can be written as follows:

$$\text{minimize } \|\mathbf{x}\|_0 \,, \text{ subject to } A\,\mathbf{x} = \mathbf{b}$$

- There is a method, called '*Basis Pursuit' (BP)*, which replaces $l0$ with $l1$, so now the problem is:

$$\text{minimize } \|\mathbf{x}\|_1 \,, \text{ subject to } A\,\mathbf{x} = \mathbf{b}$$

  Under certain conditions BP will find a sparse solution or even the sparsest one. Another method, called the '*Lasso'*, states the following problem:

$$\text{minimize } \|A\mathbf{x} - \mathbf{b}\|_2^2 \,, \text{ subject to } \|\mathbf{x}\|_1 < c$$

This method can be used in the regression context, and for undetermined systems to find sparse approximations. Like BP, under certain conditions it will find the sparsest solution.

2.2.1.2 Some Mathematical Concepts and Facts

Just in the phrases above, the term "norm" has been taken with somewhat excessive license. A function $\|\cdot\|$ is called a *norm* if:

(a) $\|\mathbf{x}\| = 0$ if and only if $\mathbf{x} = 0$.
(b) $\|\lambda\,\mathbf{x}\| = |\lambda|\; \|\mathbf{x}\|$ for all λ and $\mathbf{x}$
(c) $\|\mathbf{x} + \mathbf{y}\| \leq \|\mathbf{x}\| + \|\mathbf{y}\|$ for all $\mathbf{x}$ and $\mathbf{y}$ (triangle inequality)

If only (b) and (c) hold the function is called a *semi-norm.*
If only (a) and (b) hold the function is called a *quasi-norm.*

The functions already taken as measures have the form:

$$\|\mathbf{x}\|_p = (\sum_{i=1}^{n} |x_i|^p)^{1/p} \tag{2.5}$$

If $1 \leq p < \infty$ this function is a norm. However, if $0 < p < 1$ this function is a quasi-norm.

$\|\mathbf{x}\|_0$ is the limit as $p \to 0$. It is neither a norm or a quasi-norm.

As regards sparsity, a pertinent definition is the following:

A vector $\mathbf{x}$ is *s-sparse* if at most s of its entries are nonzero.

Suppose you have obtained a solution $\mathbf{x}$ for $A\,\mathbf{x} = \mathbf{b}$ with a few nonzero entries. Then you may ask yourself if it is the sparsest solution. Here, it becomes relevant to consider the following definition:

The '*spark*' of a matrix A is the smallest number of columns of A that are linearly-dependent.

According with this definition, the vectors in the null space of the matrix (i.e. $A\,\mathbf{x} = 0$) must satisfy: $\|\mathbf{x}\|_0 \geq spark(A)$.

Now, there is a theorem that establishes the following: if $A\,\mathbf{x} = \mathbf{b}$ has a solution $\mathbf{x}$ obeying $\|\mathbf{x}\|_0 < spark(A)\,/2$, this solution is necessarily the sparsest possible [76].

2.2.2 *Norms and Sparsity*

Specialists on sparse representations do speak of norms that promote sparsity. This is an interesting aspect, which can be illustrated with some plots.

Suppose that the matrix A is a 1×2 matrix. Then, the solutions of $A\,\mathbf{x} = \mathbf{b}$ form a line L on the 2D plane.

Consider the following ball of radius r:

$$B_p(r) = \{\mathbf{x} \mid \|\mathbf{x}\|_p \leq r\} \tag{2.6}$$

Then the solution that minimizes the norm lp is obtained at the intersection of the ball and the line L.

Figure 2.1 shows three cases, corresponding to the intersection of the line and the balls $B_{1/2}$, B_1, and B_2. Notice that the intersection of the line L with the vertical axis gives a sparsest solution (only one nonzero entry).

The left hand plot, (a), in Fig. 2.1 makes clear that quasi-norms with $0 < p < 1$ promote sparsity. The central plot, (b), shows that usually $l1$ promote sparsity (unless L was parallel to a border of B_1). Finally, the right hand plot shows that $l2$ do not promote sparsity.

This simple example can be generalized to problems with more dimensions.

There are other functions that promote sparsity, like for instance $1 - \exp(|\mathbf{x}|)$, or $\log(1 + |\mathbf{x}|)$, or $|\mathbf{x}|/(1 + |\mathbf{x}|)$ [76].

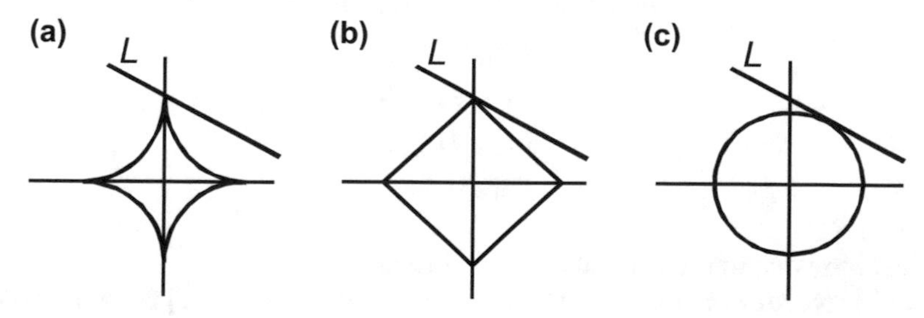

Fig. 2.1 The line L and, **a** the ball $B_{1/2}$, **b** the ball B_1, **c** the ball B_2

2.2.3 *Solving Sparsity Optimization Problems*

There are several ways for solving the sparsity optimization problems. The BP problem can be tackled with linear programming. The Lasso problem can be treated with quadratic programming, but there is a better approach. In certain scenarios it would be more appropriate to use first order methods.

Let us enter into specific details.

2.2.3.1 Basis Pursuit

The basic basis pursuit problem,

$$\text{minimize } \|\mathbf{x}\|_1, \text{ subject to } A\,\mathbf{x} = \mathbf{b}$$

can be expressed as a Linear Programming (LP) problem (see appendix on optimization), as it is detailed in [53]. The main steps are the following:

- First, notice that $\|\mathbf{x}\|_1 = |x_1| + |x_2| + \cdots + |x_n|$. One could use a set of auxiliary variables $t_1, t_2, \ldots, t_n$ to obtain an equivalent problem:

$$\text{minimize } (t_1 + t_2 + \ldots + t_n),$$

$$\text{subject to } |x_1| \leq t_1\,;\ |x_2| \leq t_2\,;\ \ldots;\ |x_n| \leq t_n \text{ and } A\,\mathbf{x} = \mathbf{b}$$

- Next, each inequality can be expressed as two inequalities; for example, $|x_1| \leq t_1$ is equivalent to: $x_1 \leq t_1$, $x_1 \geq -t_1$. Then, the problem can be expressed in the following compact form:

$$\begin{aligned} &\text{minimize } (t_1 + t_2 + \ldots + t_n), \\ &\text{subject to:} \\ &I\mathbf{x} - I\mathbf{t} \leq 0 \\ &I\mathbf{x} + I\mathbf{t} \geq 0 \\ &A\mathbf{x} = \mathbf{b} \end{aligned} \tag{2.7}$$

This last formulation corresponds to a LP format.

Another way for obtaining an LP expression is to use a +/− split, so each component x_i is represented as follows:

$$x_i = p_i - n_i \tag{2.8}$$

with:

$$p_i = \begin{cases} x_i \; if \; x_i > 0 \\ 0 \; else \end{cases} \quad \text{and } n_i = \begin{cases} -x_i \; if \; x_i < 0 \\ 0 \; else \end{cases}$$

Then: $\|\mathbf{x}\|_1 = \mathbf{p} + \mathbf{n}$

With these changes, the problem now would be:

$$\text{minimize } \mathbf{p} + \mathbf{n}, \text{ subject to } A(\mathbf{p} - \mathbf{n}) = \mathbf{b}\,; \; \mathbf{p}, \, \mathbf{n} \geq 0$$

which is LP.

The reader would find in [53] programs in MATLAB for basis pursuit, using the *linprog( )* function.

Of course, other methods, like for instance interior-point methods, can also be applied to solve this problem.

2.2.3.2 Lasso

The Lasso approach can be expressed in two equivalent forms:

- penalized form:

$$\text{minimize} \left(\|A\mathbf{x} - \mathbf{b}\|_2^2 + \lambda \; \|\mathbf{x}\|_1 \right) \tag{2.9}$$

- constraint form:

$$\text{minimize } \|A\mathbf{x} - \mathbf{b}\|_2^2, \text{ subject to } \|\mathbf{x}\|_1 < c$$

It sometimes happens that different types of problems lead to similar equations. In the case of Lasso, it appears at least in two contexts: statistics and signal processing.

In statistics, one finds that certain regression problems [156] can be treated with a Lasso approach. For example, the basic regression method (least squares) consists in minimizing $\|A\mathbf{x} - \mathbf{b}\|_2^2$ (where A and $\mathbf{b}$ are given); however, it has been noticed in practical applications, that this approach may have numerical difficulties or unsatisfactory results (like for instance, overfitting). By adding penalty terms, both regularization and suitable fitting could be obtained.

'*ridge regularization*' substitutes the minimization of $\|A\mathbf{x} - \mathbf{b}\|_2^2$ with the minimization of $\left(\|A\mathbf{x} - \mathbf{b}\|_2^2 + \lambda \; \|\mathbf{x}\|_2^2 \right)$. A *l2* penalty is added.

The ridge regularization is obviously related to the Tikhonov regularization. Eventually, the ridge regularization was introduced by a most cited article [101], in 1970; while the work of Tikhonov, which started in the 1940s, was widely know after the publication in 1977 of his first book in English.

One of the good effects of ridge regularization is a certain shrinking of the solution components. According with [166], higher values of λ force these components to be more similar to each other.

In 1996, Tibshirani [176] introduced the method *'Least Absolute Shrinkage and Selection Operator' (LASSO)*. A $l1$ penalty is added (2.9).

There are analysis of large data and data mining scenarios where it is important, as said before, to assess which variables are more important, more influential. The $l1$ penalty has the advantage of promoting sparsity. Contrary to the $l2$ penalty, the $l1$ penalty leads to solutions with many zero components, and this helps to determine the relevant variables. This is a main reason of Lasso popularity.

As we shall see in the next sections, the $l1$ pènalty (and the $l1$ norm) plays an important role both in compressed sensing and image processing. In other contexts [156], $l1$ is used for robust statistics in order to obtain results insensitive to outliers [187], and for maximum likelihood estimates in case of Laplacian residuals.

It is also shown, in an appendix of [187], that the two formulations:

$$\min_x \| \mathbf{b} - A\mathbf{x}\|_2 + c_1 \|\mathbf{x}\|_1 \text{ and } \min_x \| \mathbf{b} - A\mathbf{x}\|_2^2 + c_2 \|\mathbf{x}\|_1$$

are equivalent.

With respect to analytical solutions, recall that the least squares problem in undetermined systems has the following solution:

$$\mathbf{x} = (A^T A)^{-1} A^T \mathbf{b} \tag{2.10}$$

The solution for the ridge regularization would be:

$$\mathbf{x} = (A^T A + \lambda I)^{-1} A^T \mathbf{b} \tag{2.11}$$

In the case of Lasso, finding the solution is more involved, and many methods have been proposed. Indeed, it can be treated as a Quadratic Programming (QP) problem (see appendix on optimization). In fact, Tibshirani proposed two ways of stating the Lasso as QP [166, 176], much in the line described above for expressing the basis pursuit as LP. Concretely, one of the proposed ways uses a +/– split, and so the Lasso is expressed as follows:

$$\text{minimize } \left(\|A\,(\mathbf{p} - \mathbf{n}) - \mathbf{b}\| + \lambda(\mathbf{p} + \mathbf{n})\right), \text{ subject to } \mathbf{p}, \mathbf{n} \geq 0$$

which is a QP problem.

Nowadays there are better algorithms to solve Lasso problems. A popular one is *LARS* (*'Least Order Regression'*), which was introduced in 2004 [75]. It is a stepwise iterative algorithm that starts with $c = 0$ (constraint form), and detects the most influential variables as c increases. Figure 2.2 shows an example of the paths of the solutions when c increases: for small values of c there is only one non-zero component of $\mathbf{x}$, when c reaches a certain value a second non-zero component rises,

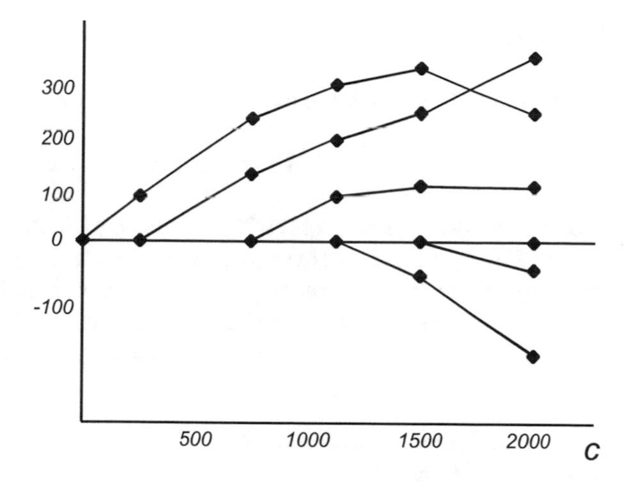

Fig. 2.2 An example of the solution paths obtained with LARS

then for larger c a third non-zero component rises, and so on. Noticc that components with negative values can also appear. The next paragraphs intend now to introduce LARS in more detail.

Let us follow the scheme of [75] in order to explain LARS. A first point is that the LARS algorithm relates to the *'Forward Selection' (FS)* algorithm described in [185] (1980). The FS algorithm applies the following steps:

- Set all variables to zero
- Find the column $\mathbf{A}_j$ of matrix A most correlated with $\mathbf{b}$. Based on $\mathbf{A}_j$ perform a simple linear regression to obtain x_j. Then, a residue is obtained:

$$\mathbf{r}_j = \mathbf{A}_j \, x_j - \mathbf{b} \tag{2.12}$$

- Find the column $\mathbf{A}_k$ most correlated to $\mathbf{r}_j$ and perform a linear regression to obtain x_k, obtain a new residue:

$$\mathbf{r}_k = \mathbf{A}_k x_k - \mathbf{r}_i \tag{2.13}$$

- Repeat until a stop criterion is reached (for instance in terms of the statistical characteristics of the residual).

(The correlation of a column $\mathbf{A}_i$ with a residual $\mathbf{r}_l$ can be computed as $|\mathbf{A}_i^T \mathbf{r}_l|$).

Once a final solution $\mathbf{x}$ was built, a model $\boldsymbol{\mu} = A\,\mathbf{x}$ is obtained that can be used for several purposes.

Along the process, a series of components, x_i, x_j, x_k, x_l ,..., were selected. At the time of implementing this kind of algorithm it is usual to maintain a set, called *the active set*, which includes the indexes of the selected components:

$$J = \{i, j, k, l...\} = \{j_i\} \tag{2.14}$$

Likewise, one can assemble the selected columns $\mathbf{A}_i$ into a matrix:

$$A_J = \{\mathbf{A}_i, \mathbf{A}_j, \mathbf{A}_k, \mathbf{A}_l...) \tag{2.15}$$

These notations will be useful when describing LARS.

The FS algorithm is not in general recommended; in [75] words: it is an aggressive technique that can be too much greedy. However, there is a cautious version, called *'Forward Stagewise'*, which can be more acceptable. It also starts with all variables set to zero and with $\mathbf{r} = \mathbf{b}$, and then it repeats many times the following:

- Find the $\mathbf{A}_j$ most correlated with $\mathbf{r}$
- $\delta = \gamma \, sign \, (\mathbf{A}_j^T \, \mathbf{r})$
- $x_j \leftarrow x_j + \delta$
- $\mathbf{r} \leftarrow \mathbf{r} - \delta \, \mathbf{A}_j$

where the learning rate γ is small.

In [75] an example was studied with 10 columns of data from 442 diabetes patients. Using forward stagewise, a solution path figure (like Fig. 2.2) was obtained, after 6000 stagewise steps. The article remarks that the cautious nature of the algorithm, with many small steps, implies considerable computational effort. One of the benefits of LARS is that it would require much less steps.

Like the others, the LARS algorithm begins with $\mathbf{x} = 0$, $\mathbf{r} = \mathbf{b}$, and a model $\boldsymbol{\mu}_0 = 0$. It finds the $\mathbf{A}_j$ most correlated with $\mathbf{r}$, and then it augments the model as follows:

$$\boldsymbol{\mu}_1 = \boldsymbol{\mu}_0 + \gamma_1 \, \mathbf{A}_j \tag{2.16}$$

(the component x_j is selected, with the value γ_1)

The forward stagewise algorithm would use a small γ_1; while the FS would use a large enough γ_1 to make $\boldsymbol{\mu}_1$ equal to the projection of $\mathbf{b}$ into $\mathbf{A}_j$. LARS takes an intermediate value.

The idea in LARS is to take the largest step possible in the direction of $\mathbf{A}_j$ until some other $\mathbf{A}_k$ has as much correlation with the residual $\mathbf{r}_1 = \mathbf{b} - \boldsymbol{\mu}_1$. The corresponding geometry is that the residual would bisect the angle between $\mathbf{A}_j$ and $\mathbf{A}_k$ (the least angle direction). Let $\mathbf{u}_1$ be the unit vector along $\mathbf{r}_1$, the next step would be:

$$\boldsymbol{\mu}_2 = \boldsymbol{\mu}_1 + \gamma_2 \, \mathbf{u}_1 \tag{2.17}$$

This step will be the largest possible until a third $\mathbf{A}_l$ is found such as $\mathbf{A}_j$, $\mathbf{A}_k$ and $\mathbf{A}_l$ had the same correlation with $\mathbf{r}_2 = \mathbf{b} - \boldsymbol{\mu}_2$. The next step will proceed equiangulary between the three already selected A columns. And so on.

According with [75] the equiangular direction can be computed as follows:

• Denote:

$$G = A_J^T A_J; \quad Q = (1_J^T G^{-1} 1_J)^{-1/2} \tag{2.18}$$

• Then:

$$\mathbf{u}_J = A_J \, \omega_J, \text{ where } \omega_J = Q \, G^{-1} \, 1_J$$

(the vector 1_J is a vector of 1's)
The equiangular direction $\mathbf{u}_J$ has the following characteristics:

$$A_J^T \, \mathbf{u}_J = Q \, 1_J, \text{ and } \| \, \mathbf{u}_J \|_2^2 = 1$$

In the algorithm steps, the step size is computed as follows:

• Denote:

$$\mathbf{a} = A^T \, \mathbf{u}_J$$

$$\bar{c} = A^T \, (\mathbf{b} - \mu_J) \text{ (current correlation)}$$

$$C = \max_i \{|c_i|\} \tag{2.19}$$

• Then:

$$\gamma = \min_i{}^+ \left\{ \frac{C - c_i}{Q - a_i}, \frac{C + c_i}{Q + a_i} \right\} \tag{2.20}$$

($\min^+$ means that the minimum is taken over only positive components).

In order to use LARS for the Lasso problem it is necessary to add a slight modification: if an active component passes through zero, it must be excluded from the active set. It may happen that later on this variable would enter again into the active set.

When that modification is included, the LARS algorithm is able to find all the solutions of the Lasso problem.

A simple, direct implementation of the LARS algorithm is offered in the Program 2.1, which deals with a table of diabetes data. The reader can edit the *lasso_f* flag (see the code) in order to run LARS, or LASSO. The program is based on the information and listings provided by [1], which has a link to the diabetes data (Stanford University). Figure 2.3 shows the solution paths obtained with LARS.

Figure 2.4 shows the solution paths obtained with LASSO, which have noticeable differences compared to the LARS solution paths.

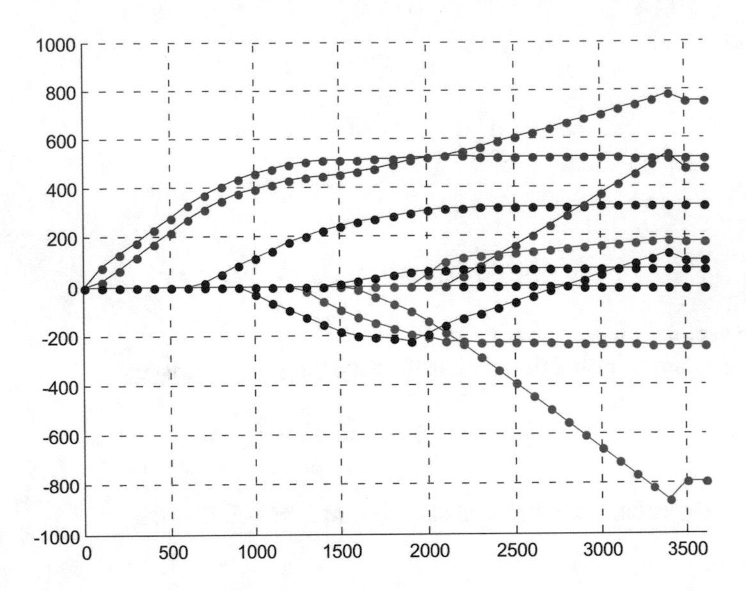

Fig. 2.3 Solution paths using LARS for diabetes set

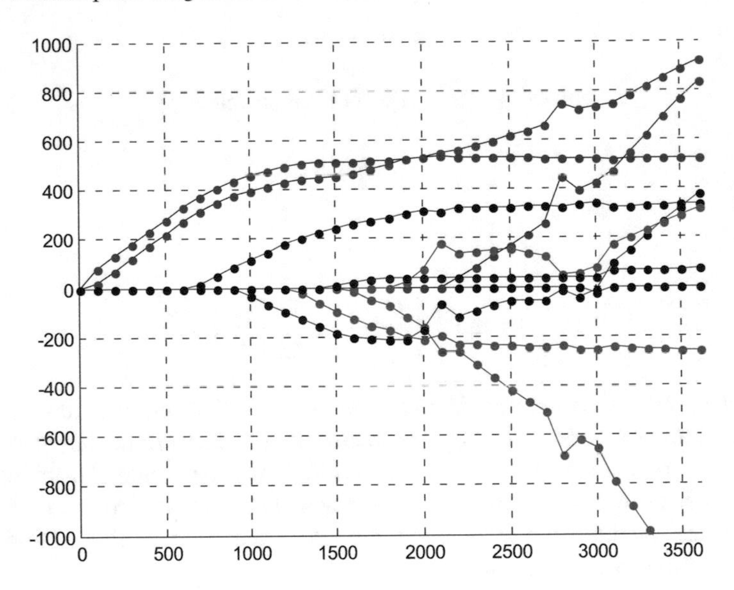

Fig. 2.4 Solution paths using LASSO for diabetes set

Program 2.1 LARS/LASSO example

```
% LARS/LASSO example
% diabetes data
% data matrix (the last column is the response)
data=load('diabetes_table.txt');
% Data preprocessing:
% zero mean, unit L2 norm, response with no bias
[ll,cc]=size(data); A=zeros(ll,cc-1); b=zeros(ll,1);
for j=1:cc-1,
  A(:,j)=data(:,j)-mean(data(:,j)); %mean removal
  aux=sqrt(A(:,j)'*A(:,j)); A(:,j)=A(:,j)/aux; %unit L2-norm
end;
b=data(:,cc)-mean(data(:,cc));
% Initialization of variables
hbeta=[];
maxnav=min(ll,cc-1); %maximum number of active variables
gamma_av=0;
%valid sign flag:
vls=1; %valid
% LARS/LASSO: ---------------------------------
lasso_f=0; % flag (0- LARS; 1-LASSO) (Edit here)
ni=1; nitmax=3600;
for t=0:100:nitmax,
  %Initialization of variables
  beta=zeros(cc-1,1);
  mu=zeros(ll,1);
  c=[]; maxc=0; maxc_ix=0;
  J=[]; %indices of the active set
  signJ=[];
  %active set variables (av):
  nav=0; %counter of active variables
  gamma_av=0;
  %valid sign flag:
  vls=1; %valid
  inz=1; %initialize flag
  bkloop=0; %for loop breaking
  norm1=0;
  oldnorm1=0;
  while nav<maxnav,
    c=A'*(b-mu); %current corr
    maxc=max(abs(c));
    %initialization
    if inz==1, [aux, maxc_ix]=max(abs(c)); inz=0; end;
    if vls==1, J=[J, maxc_ix]; nav=nav+1; end; %add index to J
    signJ=sign(c(J));
    complJ=setdiff(1:maxnav,J); %complement of J
    LcomplJ=length(complJ);
    Aj=A(:,J);
```

```
for k=1:length(J),
  Aj(:,k)=signJ(k)*Aj(:,k);
end;
G=Aj'*Aj;
Ij=ones(length(J),1);
iG=inv(G);
Q=1/sqrt(Ij'*iG*Ij);
wj=Q*iG*Ij;
uj=Aj*wj;
a=A'*uj;
if nav==maxnav, gamma_av =maxc/Q; %unusual last stage
else
  gamma=zeros(LcomplJ,2); %min of two terms
  for k=1:LcomplJ,
    n=complJ(k);
    gamma(n,:)=[((maxc-c(n))/(Q-a(n))),...
    ((maxc+c(n))/(Q+a(n)))];
  end;
  %remove complex elements, reset to Inf
  [pi,pj]=find(0~=imag(gamma));
  for nn=1:length(pi), gamma(pi(nn),pj(nn))=Inf; end;
    % find minimum
    gamma(gamma<=0) = Inf;
    mm=min(min(gamma)); gamma_av=mm;
    [maxc_ix,aux]=find(gamma==mm);
  end;
  %update coeff estimate:
  baux(J)=beta(J)+gamma_av*diag(signJ)*wj;
  bkloop=0;
  Jold=J;
  % The LASSO option -----------
  if lasso_f==1,
    vls=1; %valid sign
    dh=diag(signJ'*wj);
    gamma_c = -beta(J)./dh;
    %remove complex elements, reset to Inf
    [pi,pj]=find(0~=imag(gamma_c));
    for nn=1:length(pi), gamma_c(pi(nn),pj(nn))=Inf; end;
    % find minimum
    gamma_c(gamma_c <= 0) = Inf;
    mm=min(min(gamma_c)); gamma_w=mm;
    [gamma_w_ix,aux]=find(gamma_c==mm);
    %Lasso modification:
    if isnan(gamma_w), gamma_w=Inf; end;
    if gamma_w < gamma_av,
      gamma_av = gamma_w;
      baux(J)=beta(J)+gamma_av*diag(signJ)*wj;
      J(gamma_w_ix)=[]; %delete zero-crossing element
      nav=nav-1;
```

```
            vls=0;
          end;
        end;
        % -----------------------------
        norm1=norm(baux(Jold),1);
        if oldnorm1<=t && norm1>=t,
          baux(Jold)=beta(Jold)+Q*(t-oldnorm1)*diag(signJ)*wj;
          bkloop=1; %for loop break
        end;
        oldnorm1=norm1;
        mu=mu+ gamma_av*uj;
        beta(Jold)=baux(Jold);
        if bkloop==1, break; end %while end
    end;
    beta(ni,:)=beta';
    ni=ni+1;
end;
% display --------------------------------------
[bm,bn]=size(hbeta);
aux=[0:100:nitmax];
q=0; %color switch
figure(1)
hold on; q=0;
for np=1:bn,
  if q==0,
    plot(aux,hbeta(:,np),'r',aux,hbeta(:,np),'k.');
    axis([0 nitmax -1000 1000]);
  if lasso_f==0,
    title('LARS, diabetes set');
  else
  title('LASSO, diabetes set');
end;
end;
if q==1,
  plot(aux,hbeta(:,np),'b',aux,hbeta(:,np),'m.');
  axis([0 nitmax -1000 1000]); end
if q==2,
  plot(aux,hbeta(:,np),'k',aux,hbeta(:,np),'b.');
  axis([0 nitmax -1000 1000]); end
  q=q+1; if q==3, q=0; end;
end;
grid;
```

It is interesting to look at a special case: that A is orthonormal and so $A^T A = I$. In this case the ordinary least square (OLS) regression has the following solution, that we will denote as $\hat{x}$:

$$\hat{x} = A^T \mathbf{b} \tag{2.21}$$

- The solution for the ridge regression would be:

$$\mathbf{x} = \frac{1}{1+\lambda}\,\hat{x} \tag{2.22}$$

- And the solution for the Lasso would be:

$$x_i \;=\; S_{\lambda/2}(\hat{x}_i)\;,\; i = 1,\, 2,\, \ldots, n \tag{2.23}$$

In this last expression the soft-thresholding operator $S_{\lambda/2}$ is used, its values are:

$$S_{\lambda/2}(x) \;=\; \begin{cases} x + (\lambda/2)\,, & if\ x \leq -\lambda/2 \\ 0\,, & if\, |x| < \lambda/2 \\ x - (\lambda/2), & if\ x \geq \lambda/2 \end{cases} \tag{2.24}$$

The action of this operator is expressed in Fig. 2.5. This operator appears in other optimization contexts.

In two most cited papers, [201, 206], the oracle properties of Lasso were studied. These articles include relevant references to previous work. The desired oracle properties are that the right model is identified, and that it has optimal estimation rate. It may happen that, for a certain value of λ (the regularization parameter), a good estimation error was achieved but with an incorrect (inconsistent) model.

One of the sections of [206] shows that the Lasso variable selection could be inconsistent. In coincidence with this, [201] writes that, in general, if an irrelevant $\mathbf{A}_k$ is highly correlated with columns $\mathbf{A}_i$ of the true model, Lasso may not be able to distinguish it from the true model columns. [201] introduces an *"irrepresentable condition"*, which resembles the following constraint:

$$|(A_J^T\, A_J)^{-1}\, A_J^T\, A_I| \;<\; 1 \tag{2.25}$$

where J is the active set and I the inactive set.

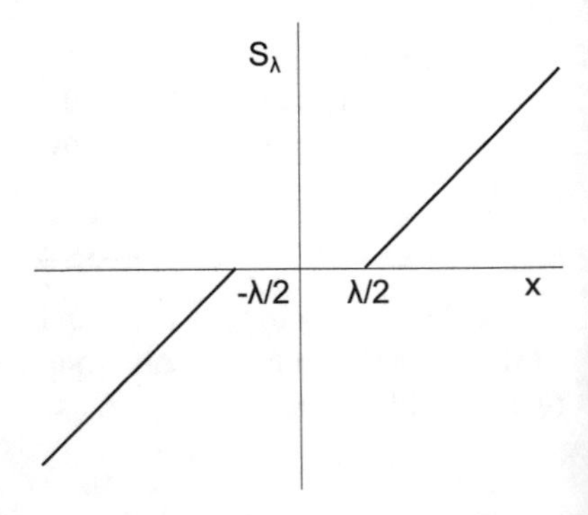

Fig. 2.5 Soft-thresholding operator

This last expression implies that relevant variables were not too highly correlated with nuisance variables.

The irrepresentable condition of [201] is almost necessary and sufficient for Lasso to select the true model. The lesson for experiment designers is to avoid spurious variables that may incur in high correlation with relevant variables.

In order to get better oracle properties, the *'adaptive Lasso'* was introduced in [206]. The original constrained Lasso is transformed to:

$$\text{minimize} \left(\| A\mathbf{x} - \mathbf{b} \|_2^2 + \lambda \sum_{i=1}^{n} \frac{|x_i|}{|\hat{x}_i|^{\nu}} \right)$$

(where $\nu > 0$)

Notice that a weighting has been introduced in the penalty term. This weighting is data-dependent. It is shown in [206] that consistency in variable selection can be achieved.

By the way, the reader is referred to [206] for an interesting comparison of the adaptive Lasso and the nonnegative *'garotte'*. The garotte finds a set of scaling factors c_j for the following problem:

$$\text{minimize} \left(\left\| \mathbf{b} - \sum_{j=1}^{n} \mathbf{A}_j \hat{x}_j c_j \right\|_2^2 + \lambda \sum_{j=1}^{n} c_j \right)$$

where $c_j \geq 0$).

In some problems it is known that the variables belong to certain groups, and it is opportune to shrink the members of a group together. The *'group Lasso'* has been introduced in [195]. Suppose the n variables are divided into L groups. Denote A_k the matrix obtained from A by selecting the columns corresponding to the *k-th* group. Likewise, $\mathbf{x}_k$ is the vector with components being the variables of the *k-th* group. The group Lasso set the problem as follows:

$$\text{minimize} \left(\| A\mathbf{x} - \mathbf{b} \|_2^2 + \lambda \sum_{k=1}^{L} \sqrt{n_k} \, \| \mathbf{x}_k \|_2 \right)$$

where n_k is the number of member of the *k-th* group.

The penalty term can be viewed as an intermediate between l_1 and l_2.

A remark from [84] is that the group Lasso does not yield sparsity within a group. A more general penalty was then proposed [84] that promotes sparsity both at the group and individual levels. It is called *'sparse group Lasso'*:

$$\text{minimize} \left(\| A\mathbf{x} - \mathbf{b} \|_2^2 + \lambda_1 \sum_{k=1}^{L} \sqrt{n_k} \, \| \mathbf{x}_k \|_2 + \lambda_2 \, \| \mathbf{x} \|_1 \right)$$

In another most cited paper, [207] introduced the *'elastic net'*, which is a two-step method. For the first step, the paper introduces the *'naïve elastic net'* as the following problem:

$$\text{minimize } \left(\| A\mathbf{x} - \mathbf{b} \|_2^2 + \lambda_1 \, \|\mathbf{x}\|_1 \, + \, \lambda_2 \, \|x\|_2^2 \right)$$

Denote as $\tilde{x}$ the estimate obtained by the first step. The second step is just the following correction:

$$\mathbf{x} = \frac{\tilde{x}}{\sqrt{1 + \lambda_2}} \tag{2.26}$$

The elastic net can be regarded as compromise between ridge regression and Lasso. Because of the quadratic term, the elastic net penalty is strictly convex.

As it was shown in Fig. 2.2, the solutions of the Lasso problem follow polygonal paths. This fact was observed in [141], in the year 2000. Based on this fact, the paper proposed a homotopy algorithm to solve the problem. It uses an active set, which is updated through the addition and removal of variables. The LARS algorithm, introduced four years later, is an approximation to the homotopy algorithm (see the brief historical account given in [71]). Later on, year 2006, it was shown, [72], that the LARS/homotopy methods, first proposed for statistical model selection, could be useful for sparsity research.

With relative frequency new variants of the LARS and homotopy algorithms appear in the literature, and different aspects of them are investigated, like for instance [123] on the number of linear segments of the solution paths. See [107] for some variants of group Lasso in the context of structured sparsity.

2.2.3.3 First Order Methods

Although most of the sparse approximation problems can be tackled with LP or QP algorithms, there are applications in which, due to the size or the characteristics of the problem, computational difficulties appear [20]. Some of these difficulties are detailed in [81]. Therefore, other solution alternatives have been explored, including the first order methods described in the appendix on optimization.

An example of gradient based method is the *'gradient projection sparse representation' (GPSR)* proposed in [81]. Positive and negative variables are separated, and the problem:

$$\text{minimize } \left(\frac{1}{2} \, \| \, \mathbf{b} - A\mathbf{x} \, \|_2^2 + \lambda \, \|\mathbf{x}\|_1 \right)$$

is rewritten as:

$$\text{minimize } Q(\mathbf{x}) \; = \left(\frac{1}{2} \, \|\mathbf{b} - \, [A, -A][\mathbf{x}_+; \; \mathbf{x}_- \,]\|_2^2 + \lambda \, 1^T \, (\mathbf{x}_+ + \mathbf{x}_- \right)$$

with: $\mathbf{x}_+ \geq 0$; $\mathbf{x}_- \geq 0$

Now, denote:

$$\mathbf{z} = [\mathbf{x}_+; \mathbf{x}_-], \ \mathbf{c} = \lambda 1 + [-A^T \mathbf{b}; A^T \mathbf{b}], \text{ and:}$$

$$B = \begin{pmatrix} A^T A & -A^T A \\ -A^T A & A^T A \end{pmatrix} \tag{2.27}$$

Then:

$$Q(\mathbf{z}) = \frac{1}{2} \mathbf{z}^T B \, \mathbf{z} + \mathbf{c}^T \mathbf{z} \tag{2.28}$$

and the gradient: $\nabla_z \, Q(\mathbf{z}) = B \, \mathbf{z} + \mathbf{c}$

Therefore, a steepest descent algorithm can be enounced:

$$\mathbf{z}_{k+1} = \mathbf{z}_k - \alpha_k \nabla_z \, Q(\mathbf{z}_k) \tag{2.29}$$

It was shown in [81] that the computation of the matrix B can be done more economically than its size suggests, and that the gradient of Q requires one multiplication each by A and A^T. No inversion of matrices is needed.

There is an important family of methods in connection with the *'iterative shrinkage/thresholding' (IST)* algorithm. Initially, IST was introduced as an EM algorithm for image deconvolution [136]. According with [81], it can be used for Lasso problems and only requires matrix-vector multiplications involving A and A^T. Let us describe the idea; suppose you have the following composite problem:

$$\text{minimize } f(\mathbf{x}) + \lambda \, g(\mathbf{x})$$

Consider the next iterative approximation of the solution:

$$\begin{aligned} \mathbf{x}_{k+1} &= \arg \min_x \{ f(\mathbf{x}_k) + (\mathbf{x} - \mathbf{x}_k)^T \nabla f(\mathbf{x}_k) + \\ &+ \tfrac{1}{2} \, \|\mathbf{x} - \mathbf{x}_k\|_2^2 \, \nabla^2 f(\mathbf{x}_k) + \lambda \, g(\mathbf{x}) \} \approx \\ &\approx \arg \min_x \{ (\mathbf{x} - \mathbf{x}_k)^T \nabla f(\mathbf{x}_k) + \tfrac{\alpha_k}{2} \, \|\mathbf{x} - \mathbf{x}_k\|_2^2 + \lambda \, g(\mathbf{x}) \} \end{aligned} \tag{2.30}$$

(the Hessian $\nabla^2 f(\mathbf{x}_k)$ is approximated by a diagonal matrix $\alpha_k I$)

Denote:

$$\mathbf{u}_k = \mathbf{x}_k - \frac{1}{\alpha_k} \nabla f(\mathbf{x}_k) \tag{2.31}$$

Then, the iterative procedure can be written as:

$$\mathbf{x}_{k+1} = \arg \min_x \{ \frac{1}{2} \, \|\mathbf{x} - \mathbf{u}_k\|_2^2 + \frac{\lambda}{\alpha_k} \, g(\mathbf{x}) \} \tag{2.32}$$

(the $\approx$ was replaced by =)

In the case of $g(\mathbf{x})$ being $\|\mathbf{x}\|_1$, which is separable, then there is a component-wise closed-form solution:

$$(\mathbf{x}_{k+1})_i = \arg\min_x \{ \frac{1}{2} ((\mathbf{x})_i - (\mathbf{u}_k)_i)^2 + \frac{\lambda}{\alpha_k} |(\mathbf{x})_i| \} = S_T((\mathbf{u}_k)_i) \qquad (2.33)$$

were we used $(\mathbf{x})_i$ to denote the *i-th* component of $\mathbf{x}$. The last, right-hand term includes a soft-thresholding operator $S_T(y)$:

$$S_T(y) = \begin{cases} y + T, & if\ y \le -T \\ 0, & if |y| < T \\ y - T, & if\ y \ge T \end{cases} \qquad (2.34)$$

with: $T = \frac{\lambda}{\alpha_k}$

Therefore, a simple iterative shrinkage/thresholding algorithm, named ISTA, has been devised [20]. There are two parameters to specify, λ and α_k. The research has proposed a number of strategies for this specification.

Another, equivalent form of expressing the soft-thresholding operator is:

$$S_T(y) = sgn(y) \cdot (|y| - T)_+ \qquad (2.35)$$

In case the composite problem was of Lasso type:

$$\text{minimize} \left(\frac{1}{2} \| \mathbf{b} - A\mathbf{x} \|_2^2 + \lambda \, \|\mathbf{x}\|_1 \right)$$

then:

$$\nabla f(\mathbf{x}_k) = -A^T (\mathbf{b} - A\mathbf{x}_k)$$

And so, the iterative algorithm would be:

$$\mathbf{x}_{k+1} = S_T \left(\mathbf{x}_k + \frac{1}{\alpha_k} A^T (\mathbf{b} - A\,\mathbf{x}_k) \right) \qquad (2.36)$$

(the operator should be component-wise applied)

The algorithm is clearly related with the *'Landweber algorithm'*, introduced in the 1950s to solve ill-posed linear inverse problems and that has the following general expression:

$$\mathbf{x}_{k+1} = \mathbf{x}_k + \lambda \, A^T M (\mathbf{b} - A\,\mathbf{x}_k) \qquad (2.37)$$

(where M is a positive symmetric matrix and λ is a parameter)

The (2.36) algorithm is called the *'thresholded Landweber iteration' (TLI)* in an interesting paper of Daubechies, et al. [63]. This paper proposes an improvement of the TLI algorithm using a projection on $l1$-balls (resulting in a variable thresholding).

Notice that TLI is a particular case of ISTA for the Lasso problem. However, it happens that some papers use the name ISTA to refer just to TLI, so loosing generality.

An illustrative example is the case of a sparse signal that we observe through a filter and that is further contaminated with noise. Program 2.2 attacks this case, applying ISTA for the recovery of the sparse signal. One of the outputs of the program is Fig. 2.6, which shows the original signal, the observed signal (contaminated with noise), and the signal being recovered by ISTA. The program is based on [168].

Another output of Program 2.2 is Fig. 2.7, which shows the evolution of the objective function along the minimization process.

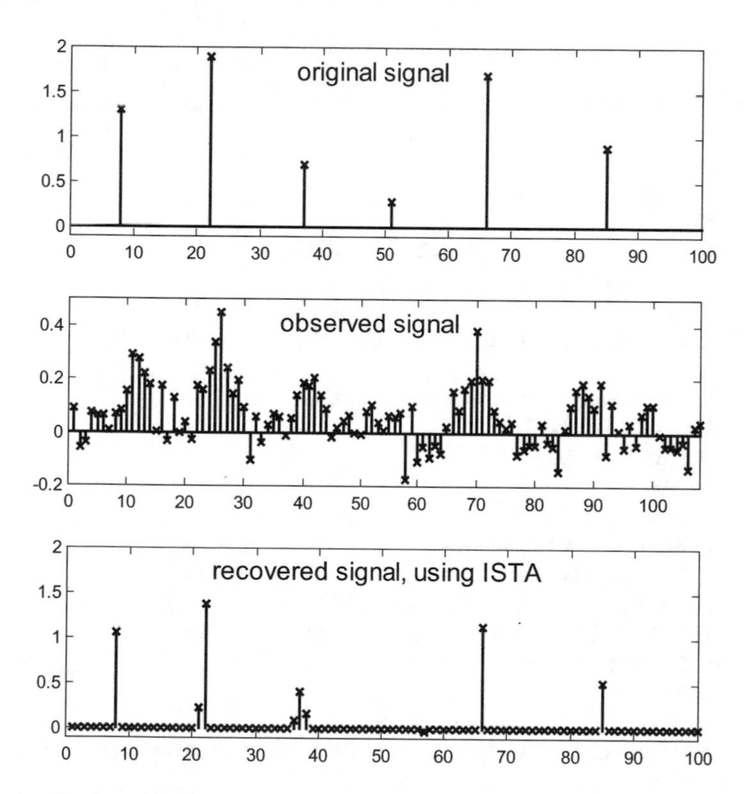

Fig. 2.6 Application of ISTA for a sparse signal recovery example

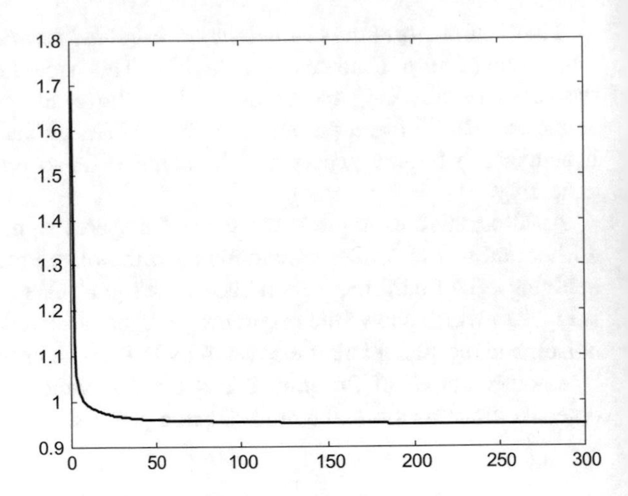

Fig. 2.7 Evolution of objective function along ISTA iterations

Program 2.2 ISTA example

```
% ISTA example
% 1D sparse signal recovery
% original sparse signal:
x=zeros(1,100);
x(8)=1.3; x(22)=1.9; x(37)=0.7; x(51)=0.3;
x(66)=1.7; x(85)=0.9;
% the signal is observed through a filter
h=[1,2,3,4,5,4,3,2,1]/25; %filter impulse response
y=conv(h,x); L=length(y);
A=convmtx(h',100); %convolution matrix
% gaussian noise is added
y=y+0.06*randn(1,L);
y=y'; %column
% ISTA algorithm
nit=300; %number of iterations
lambda=0.1; %parameter
J=zeros(1,nit); %objective function
z=0*A'*y; %initialize z, the recovered signal
T=lambda/2; %threshold
for k=1:nit,
  q=A*z;
  J(k)=sum(abs(q(:)-y(:)).^2) + lambda*sum(abs(z(:)));
  p=z+(A'*(y-q));
  %soft thresholding (component-wise)
  for n=1:100,
    if abs(p(n)) <= T,
      z(n)=0;
    else
      if p(n)>T,
        z(n)=p(n)-T;
```

```
        else
           z(n)=p(n)+T;
        end;
     end;
  end;
end;
figure(1)
subplot (3,1,1)
[i,j,v]=find(x);
stem(j,v,'kx'); hold on;
axis([0 100 -0.1 2]);
plot([0 100],[0 0],'k');
title('original signal');
subplot(3,1,2)
stem(y,'kx'); hold on;
axis([0 L -0.2 0.5]);
plot([0 L],[0 0],'k');
title('observed signal');
subplot(3,1,3)
stem(z,'kx');
axis([0 100 -0.1 2]);
title('recovered signal, using ISTA')
figure(2)
plot(J,'k')
title('Evolution of objective function')
```

Since the ISTA algorithm involves a gradient descent, it is a good idea to use the Nesterov's accelerated gradient descent (see section on gradient descent in the appendix on optimization). The result is the *'fast iterative shrinkage/thresholding algorithm' (FISTA)*, which was introduced in 2009 [20].

While ISTA can be expressed as:

$$\mathbf{x}_{k+1} = S_{\lambda/\alpha_k}(\mathbf{x}_k - \frac{1}{\alpha_k}\nabla f(\mathbf{x}_k)) \tag{2.38}$$

FISTA can be expressed as the next three steps:

$$\mathbf{x}_{k+1} = S_{\lambda/L}(\mathbf{y}_k - \frac{1}{L}\nabla f(\mathbf{y}_k)) \tag{2.39}$$

$$t_{k+1} = \frac{1+\sqrt{1+4t_k^2}}{2} \tag{2.40}$$

$$\mathbf{y}_{k+1} = \mathbf{x}_k + \left(\frac{t_k - 1}{t_{k+1}}\right)(\mathbf{x}_{k+1} - \mathbf{x}_k) \tag{2.41}$$

There are direct generalizations of the ISTA and FISTA algorithms by replacing the soft-thresholding operator with the proximal operator. This is frequent in the most recent literature. In particular, ISTA is equivalent to the proximal gradient method.

The convergence rate of subgradient methods is $O(1/\sqrt{k})$, proximal gradient methods obtain $O(1/k)$, and accelerated proximal gradient methods further reduce it to $O(1/k^2)$.

Notice that an example of projected subgradient iterations for $l1$ minimization was already included in the appendix on optimization. That example was clearly related to a basis pursuit (BP) problem.

Another idea of Nesterov was to substitute non-smooth objective functions by suitable smooth approximations [134]. One application of this idea for our problem, is to consider a smooth approximation of the $l1$ norm minimization as follows.

- First, the minimization is written as minimax:

$$\min_x \|\mathbf{x}\|_1 = \min_x \max_u \mathbf{u}_T\mathbf{x} \; ; \quad -1 \le u_i \le 1 \tag{2.42}$$

- Second, the smoothed approximation is introduced:

$$smoothed \min_x \|\mathbf{x}\|_1 = \min_x \max_u \mathbf{u}_T\mathbf{x} + \frac{\mu}{2} \|\mathbf{u}\|_2^2 \tag{2.43}$$

Based on this approach, a popular algorithm called NESTA was introduced in [21]. A MATLAB implementation is available (see the section on resources). Other relevant algorithms are TwIST [23], SALSA [5], and SpaRSA [186].

2.2.3.4 Interior-Point Methods

Since the standard sparsity optimization settings correspond to LP or QP problems, the interior-point methods described in the appendix on optimization can be chosen for solving them. Actually, some of the first algorithms found in the literature on basis pursuit or on ridge regularization adopted this approach [55, 165]. See also the page *l1-magic* on Internet.

Large scale problems may be out of reach, or require excessive computational effort, for conventional interior-point methods. In such situations other approaches, like for instance first-order methods, would be preferred. Responding to this concern, an interior-point method was introduced by [106] that is able to cope with large scale Lasso problems. It is based on a preconditioned conjugate gradient algorithm. This aspect, preconditioning of the CG in relation with sparsity, is becoming important for practical applications [105].

A matrix-free interior-point method has recently been introduced in [83]. The method performs favorably with large scale one-dimensional signals.

2.2.3.5 Alternating Directions

Let us now focus on the *'alternating direction method of multipliers' (ADMM)*. According with [192], this approach was not widely employed in thc field of image and signal processing until recently, from 2009. It is an iterative method.

Consider the problem:

$$\text{minimize } f(\mathbf{x}) + g(\mathbf{x})$$

where $f(\mathbf{x})$ and $g(\mathbf{x})$ are closed proper convex functions. Both functions can be non-smooth.

The ADMM method can be expressed as follows [145]:

$$\mathbf{x}_{k+1} = \Pr ox_{\lambda f}(\mathbf{z}_k - \mathbf{u}_k) \tag{2.44}$$

$$\mathbf{z}_{k+1} = \Pr ox_{\lambda g}(\mathbf{x}_{k+1} + \mathbf{u}_k) \tag{2.45}$$

$$\mathbf{u}_{k+1} = \mathbf{u}_k + \mathbf{x}_{k+1} - \mathbf{z}_{k+1} \tag{2.46}$$

Typically $g(\mathbf{x})$ encodes the constraints and $\mathbf{z}_k \in \; dom \; g(\mathbf{x})$ (the domain of $g(\mathbf{x})$). On the other hand, $\mathbf{x}_k \in \; dom \; f(\mathbf{x})$. Along the iterations, $\mathbf{x}_k$ and $\mathbf{z}_k$ converge to each other. The ADMM method is helpful when the proximal mappings of $f(\mathbf{x})$ and $g(\mathbf{x})$ are easy to compute but the proximal mapping of $f(\mathbf{x}) + g(\mathbf{x})$ is not so.

Equivalently, the method can be expressed in terms of an *'augmented Lagrangian'*, starting from the following problem:

$$\text{minimize } f(\mathbf{x}) + g(\mathbf{z}) \text{ subject to } \mathbf{x} - \mathbf{z} = 0$$

(the problem is expressed in *'consensus form'*)

An augmented Lagrangian can be written:

$$L(\mathbf{x}, \mathbf{y}, \mathbf{z}) = f(\mathbf{x}) + g(\mathbf{z}) + \mathbf{y}^T(\mathbf{x} - \mathbf{z}) + \frac{\rho}{2} \|\mathbf{x} - \mathbf{z}\|_2^2 \tag{2.47}$$

and the ADMM would be:

$$\mathbf{x}_{k+1} = \arg\min_x \; L(\mathbf{x}, \mathbf{z}_k, \mathbf{y}_k) \tag{2.48}$$

$$\mathbf{z}_{k+1} = \arg\min_z \; L(\mathbf{x}_{k+1}, \mathbf{z}, \mathbf{y}_k) \tag{2.49}$$

$$\mathbf{y}_{k+1} = \mathbf{y}_k + \rho\,(\mathbf{x}_{k+1} - \mathbf{z}_{k+1}) \tag{2.50}$$

It is shown in [145] that these three equations can be translated to the proximal form introduced above.

The ADMM method has connections with classical methods, like the alternating projections of von Neumann (see [9] for an interesting presentation). An extensive treatment of ADMM, with application to Lasso type problems, can be found in [26]. Also, [94] includes accelerated versions, and application examples concerning the elastic net, image restoration, and compressed sensing.

Some times, in problems of the type found in this chapter, it is convenient to consider the indicator function. Given a set C, its indicator function $i_C(\mathbf{x})$ is 0 if $\mathbf{x} \in C$, and ∞ otherwise. The corresponding proximal is $\Pr ox_{\lambda f}(\mathbf{x}) = \arg\min_{\mathbf{u}} \|\mathbf{u} - \mathbf{x}\|_2^2 = P_C(\mathbf{x})$, which is the projection over set C.

One of the applications of ADMM treated in [26] is the Basis Pursuit (BP) problem. This problem can be written as:

$$\text{minimize } f(\mathbf{x}) + \|\mathbf{z}\|_1$$
$$\text{subject to } \mathbf{x} - \mathbf{z} = 0$$

where $f(\mathbf{x})$ is the indicator function of the set $C = \{\mathbf{x} \mid A\mathbf{x} = \mathbf{b}\}$. In this setting, the ADMM iterations are the following:

$$\mathbf{x}_{k+1} = P_C(\mathbf{z}_k - \mathbf{u}_k) \tag{2.51}$$

$$\mathbf{z}_{k+1} = S_{(1/\rho)}(\mathbf{x}_{k+1} + \mathbf{u}_k) \tag{2.52}$$

$$\mathbf{u}_{k+1} = \mathbf{u}_k + \mathbf{x}_{k+1} - \mathbf{z}_{k+1} \tag{2.53}$$

---

where $S_{(1/\rho)}$ is the soft thresholding operator, and the projection P_C can be computed as follows:

$$\mathbf{x}_{k+1} = (I - A^T(A\,A^T)^{-1} - A)(\mathbf{z}_k - \mathbf{u}_k) + A^T(A\,A^T)^{-1}\,\mathbf{b} \tag{2.54}$$

Program 2.3 provides an implementation example of ADMM for the BP problem, closely related to the examples given in an ADMM web page (see the section on resources). The program assigns random values to the matrix A and to $\mathbf{b}$. Then, it finds the sparsest solution to $A\mathbf{x} = \mathbf{b}$.

Of course, each time one executes the Program 2.3 different sparsest solutions are found, since both A and $\mathbf{b}$ are different each time. Figure 2.8 shows one of these sparsest solutions.

Figure 2.9 shows the evolution of $\|x\|_1$ during a program execution.

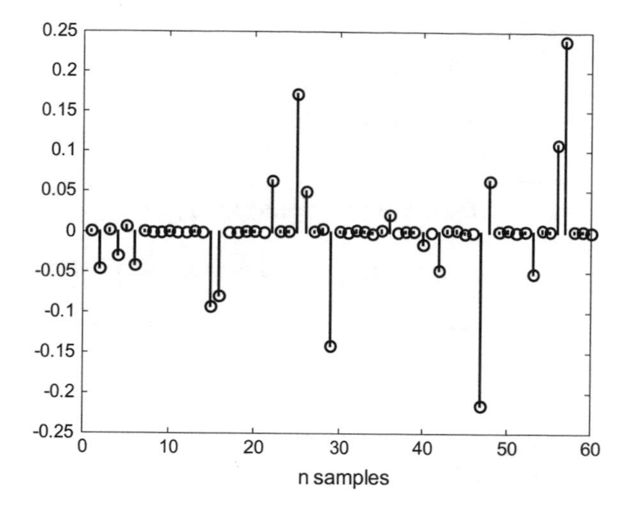

Fig. 2.8 A BP sparsest solution

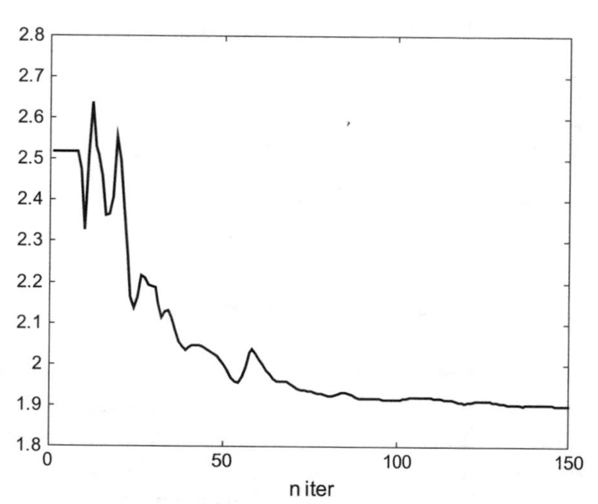

Fig. 2.9 Evolution of objective function $\|x\|_1$ along ADMM iterations

Program 2.3 Example of ADMM for Basis Pursuit

```
% Example of ADMM for Basis Pursuit
n=60; m=20;
A=randn(m,n);
b=rand(m,1);
% ADMM algorithm for finding sparsest solution
niter=150;
x=zeros(n,1); z=zeros(n,1); u=zeros(n,1);
rob=zeros(niter,1);
alpha=1.0; mu=1.0;
uA=inv(A*A');
aux1=eye(n)-(A'*uA*A);
```

```
aux2=A'*uA*b;
for nn=1:niter,
  % x update
  x=(aux1*(z-u))+aux2;
  % z update
  zo=z;
  xe=(alpha*x)+((1-alpha)*zo);
  aux=xe+u; imu=1/mu;
  z=max(0,aux-imu)-max(0,-aux-imu); %shrinkage
  % u update
  u=u+(xe-z);
  % recording
  rob(nn)=norm(x,1);
end
% display
figure(1)
stem(x,'k');
title('the sparsest solution x');
xlabel('n samples');
figure(2)
plot(rob,'k');
title('evolution of norm(x,1)');
xlabel('n iter');
```

2.2.3.6 Douglas-Ratchford Splitting

As recognized by the scientific literature, the Douglas-Rachford algorithm is one of the most successful optimization methods, able to tackle nonsmooth objective functions. The method has also been applied to nonconvex problems. An interesting series of applications is cited in [8], including image retrieval, graph coloring, protein studies, matrix completion, and even nonograms.

The optimization problem is, again:

$$\text{minimize } f(\mathbf{x}) + g(\mathbf{x})$$

Using the formulation given in [61], the algorithm is simply:

$$\mathbf{x}_k = \Pr ox_{\lambda\, g}(\mathbf{y}_k) \tag{2.55}$$

$$\mathbf{y}_{k+1} = \mathbf{y}_k + \lambda_k(\Pr ox_{\lambda\, f}(2\mathbf{x}_k - \mathbf{y}_k) - \mathbf{x}_k) \tag{2.56}$$

where $\lambda_k \in [\varepsilon\, ,\, 2 - \varepsilon]$.

See the Thesis [74] for an extensive treatment of splitting methods. The history of the Douglas-Rachford algorithm and other aspects of interest are concisely included in [61]. There is a web page focusing on Douglas-Rachford (see the section on resources).

An implementation example is provided in the next section for a compressive sensing case.

2.2.3.7 Matching Pursuits

There is a series of methods based on *'matching pursuit' (MP)*. The basic idea is simple and familiar. Imagine you have a puzzle with a certain number of pieces. To assemble the puzzle, the usual procedure would be to select pieces that match with certain empty regions of the puzzle. You select and place pieces one after one. This methodology can be applied to many types of problems; actually, some authors refer to it as a *meta-scheme*.

In the next sections we shall see the method being applied to *'dictionaries'*. If you place yourself in the Fourier context, it would be natural to combine sinusoidal harmonics. Or, perhaps, it would be better for your problem to combine arcs and line segments to analyze image shapes, etc.

For now, let us focus on the problem considered in this section: to find a sparse vector $\mathbf{x}$ such that:

$$A\,\mathbf{x} = \mathbf{b} \tag{2.57}$$

Most of the section has been centered on optimization. Matching pursuit is more related with variable selection. Like in the LARS algorithm, one uses residuals and select most correlated columns of A. After setting the initial residual as $\mathbf{r}_0 = \mathbf{b}$ and a counter $k = 0$, the basic MP would be as follows:

- Find the $\mathbf{A}_j$ most correlated with $\mathbf{r}_k$
- Update the active set J with the selected j.
- $x_j \leftarrow (\mathbf{A}_j^T\,\mathbf{r}_k)$
- $\mathbf{r}_{k+1} \leftarrow \mathbf{r}_k - \mathbf{A}_j x_j$
- $k \leftarrow k + 1$

(repeat until stop criterion)

This basic algorithm suffers from slow convergence. The *'orthogonal matching pursuit' (OMP)* has better properties. Due to the orthogonalization, once a column is selected, it is never selected again in the next iterations. The OMP algorithm starts as the MP algorithm, and then:

- Find the $\mathbf{A}_j$ most correlated with $\mathbf{r}_k$
- Update the active set J with the selected j.

- Update the matrix A_J
- $\mathbf{x} \leftarrow (A_J^{-1}\mathbf{b})$ (solve the least squares regression)
- $\mathbf{r}_{k+1} \leftarrow \mathbf{b} - \mathbf{A}\,\mathbf{x}$
- $k \leftarrow k+1$

(repeat until stop criterion)

Once the suitable columns of A were selected, a matrix B can be built with these columns. Now, there is a theorem that establishes that OMP will recover the optimum representation if:

$$\max_{A_i} \left\| B^+ A_i \right\|_1 < 1 \tag{2.58}$$

where A_i are the columns of A not in B, and B^+ is the pseudo-inverse of B, [27].

Other variants of this approach are *'stagewise orthogonal matching pursuit' (StOMP)*, *'regularized orthogonal matching pursuit' (ROMP)*, and *'compressive sampling matching pursuit' (CoSaMP)*. A concise description of them is included in [27].

An example of OMP application is given below. The case $A\,\mathbf{x} = \mathbf{b}$ is considered. Both A and $\mathbf{b}$ are created with random values. Using OMP, a sparse $\mathbf{x}$ is found. Figure 2.10 shows the result.

A record of the residual values along iteration steps was done. Figure 2.11 plots the evolution of the residual until a certain specified low value is reached. Both Figs. 2.10 and 2.11 have been generated with the Program 2.4, which includes a MATLAB implementation of OMP. Note that *pinv()* has been used for the least square regression.

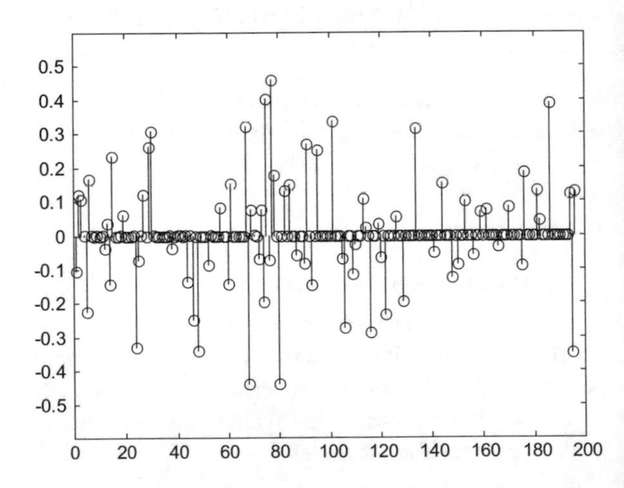

Fig. 2.10 Sparse solution obtained with OMP

Fig. 2.11 Evolution of the norm of the residual

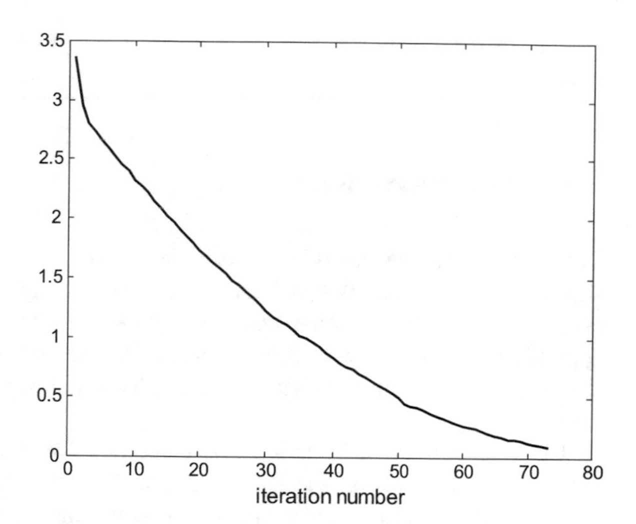

Program 2.4 Example of OMP

```
% Example of OMP
clear all;
% random A and b
N=100;
A=rand(N,2*N);
b=rand(N,1);
M=1; %measure of the residual
hM=zeros(200,1);
% OMP algorithm
Aj=[];
R=b; %residual
n=0; %counter
while M>0.1, %(edit this line)
  [v,ix]=max(abs(A'*R));
  Aj(:,ix)=A(:,ix);
  A(:,ix)=0;
  %x=Aj\b;
  x=pinv(Aj)*b;
  R=b-Aj*x;
  M=norm(R);
  n=n+1; hM(n)=M; %keep a record of residuals
end;
%display
figure(1)
stem(x,'k');
title('sparse solution');
axis([0 200 -0.6 0.6]);
figure(2)
plot(hM(1:n),'k');
```

```
title('evolution of the residual norm');
xlabel('iteration number');
```

2.3 Compressed Sensing

The compressed sensing (CS) theory claims that it is possible to recover certain signals from fewer samples than the Nyquist rate required by Shannon. This is based on the observation that many natural signals, such sound or images, can be well approximated with a sparse representation in some domain. For example, most of the energy of typical images is preserved with less than 4 % of the wavelet coefficients [153, 175].

The term 'compressed sensing' was coined by Donoho [68] in 2006. Other important contributions came, the same year, in [34, 41]. Since then, the theory has advanced and nowadays there are books on the mathematics of CS, like [62, 82]; however, there are still open questions, as we shall see.

Instead of acquiring with a sensor large amounts of data and then compress it, the new paradigm proposes to directly obtain compressed data. The recovery of the signal would be done via an optimization process.

These ideas are supported by the scenario that has appeared several times in this book, having the form of a system of linear equations $A\,\mathbf{x} = \mathbf{b}$. In particular, our interest focuses on undetermined systems.

One of the peculiar aspects of CS is that it is convenient to select a random matrix to be used as matrix A.

2.3.1 *Statement of the Approach*

The idea of CS is based on the following fact: if you have a signal $\mathbf{x}$ (n samples) and a $m \times n$ matrix A, with $m < n$, you can write the following equation:

$$A\,\mathbf{x} = \mathbf{b} \tag{2.59}$$

This is an undetermined system of linear equations: more unknowns than equations. However, if the matrix A is appropriate and the signal is sparse, then it is possible to recover $\mathbf{x}$ from $\mathbf{b}$.

Exploiting this fact, the practical use of compressed sensing can be to apply a linear operator A to a sparse signal $\mathbf{x}$; in other words, compute $\mathbf{b} = A\,\mathbf{x}$. Then, you just store $\mathbf{b}$ instead of $\mathbf{x}$. This $\mathbf{b}$ has less numbers (is denser) than $\mathbf{x}$.

The application of the linear operator could be done in real time, in your sensor. From the point of view of CS, the application of the operator A to the signal $\mathbf{x}$ is regarded as a sampling process. Therefore A is called the *sampling matrix* (or the *sensing matrix*). A $m \times n$ matrix A takes m samples from $\mathbf{x}$. In the case that A was a random matrix, CS would tell that there is a random sampling of $\mathbf{x}$ via A.

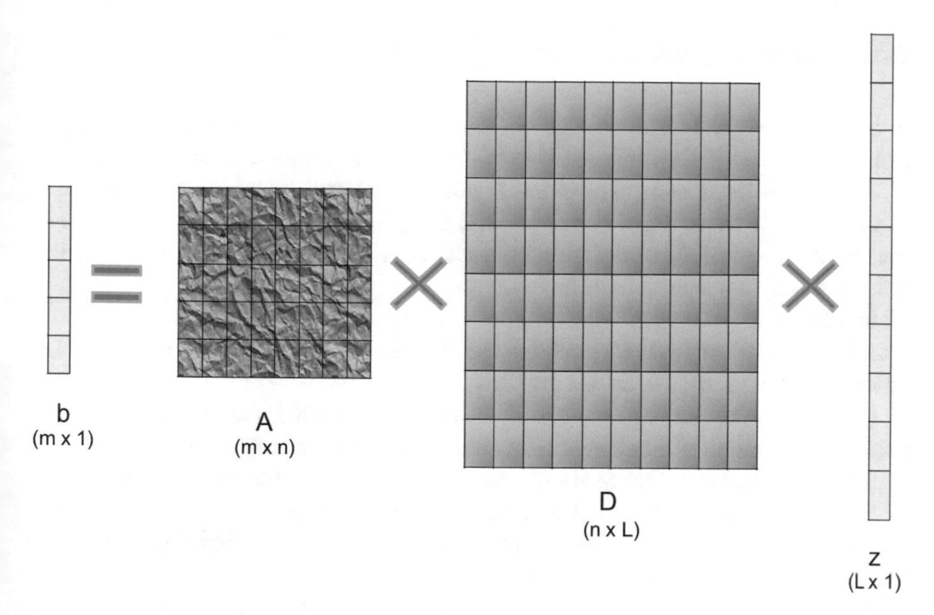

Fig. 2.12 The CS scheme

If you have a signal $\mathbf{z}$ that can be expressed in a sparse format, like for instance $\mathbf{x} = D\,\mathbf{z}$, then you can use $\mathbf{b} = A\,\mathbf{x} = A\,D\,\mathbf{z}$ for adhering to a compressed sensing. A typical example would be a signal $\mathbf{z}$ that can be expressed with a few Fourier harmonics. In general, D would represent the use of a dictionary; and the operation $D\,\mathbf{z}$ could be regarded as a *'sparsification'* of the signal $\mathbf{z}$. The diagram shown in Fig. 2.12 illustrates the approach.

An analysis of these ideas under the view of information flow, would say that you are exploiting the low-rate of information being transmitted by $\mathbf{z}$; so, in fact, you can concentrate the same information in $\mathbf{b}$.

In relation with these points, it must be said that (strictly) sparse signals are rarely encountered in real applications. What is more probable is to find signals with a small number of elements significantly non-zero (the rest could also be non-zero). Such signals are considered as *'compressible'*. One way of formalizing this concept is to look for an exponential decay of values; the signal components are rearranged in a decreasing order of magnitude $|x_{(1)}| \geq |x_{(2)}| \geq \ldots \geq |x_{(n)}|$ and then one checks that:

$$|x_{(k)}| \leq R\,k^{1/p} \tag{2.60}$$

In general, compressible signals are well approximated by s-sparse signals $\mathbf{x}_s$:

$$\|\mathbf{x} - \mathbf{x}_s\|_1 \leq C_p\,R\,s^{1-1/p} \tag{2.61}$$

$$\|\mathbf{x} - \mathbf{x}_s\|_2 \leq D_p\,R\,s^{1/2-1/p} \tag{2.62}$$

with: $C_p = (1/p - 1)^{-1}$and $D_p = (2/p - 1)^{-1/2}$ (see [133]).

2.3.2 Compression and Recovery. The Matrix A

In order to apply the CS approach, one has to guarantee that the compressed signal can be recovered. This is the question that we want to consider now.

A first and important point is that, while in regression problems you inherit a given matrix A formed from experimental data, in the case of compressed sensing you are the responsible of designing the matrix A. In consequence, it is convenient to have some guidance for building A.

In relation to this aspect –the design of A—an important step in the early development of CS theory was to establish the *'restricted isometry property' (RIP)*. Let us introduce with some care this property.

Recall that a vector $\mathbf{x}$ is *s-sparse* if at most s of its entries are non-zero. A first, natural question would be: how many samples are necessary to acquire s-sparse signals? The answer is that m should be $m \geq 2s$ (note that the dimensions of A are $m \times n$). As explained in [133], the sampling matrix must not map two different sparse signals to the same set of measurement samples. Hence, each collection of $2s$ columns from A must be nonsingular.

In this line of thinking, [35] considered that the geometry of sparse signals should be preserved under the action of the sampling matrix. The *'s-th restricted isometry constant* δ_s' was defined as the smallest number δ_s such that:

$$(1 - \delta_s)\, \|\mathbf{x}\|_2^2 \;\leq\; \|A\,\mathbf{x}\|_2^2 \leq (1 + \delta_s)\, \|\mathbf{x}\|_2^2 \tag{2.63}$$

When $\delta_s < 1$, the expression above implies that each collection of s columns of A is non-singular, so *(s/2)-sparse* signals can be acquired. In case $\delta_s << 1$, the action of A would nearly maintain the $l2$ distance between each pair of signals (the term *isometry* refers to keep the distance).

It is said that A satisfies the restricted isometry property (RIP) if it has an associated isometry constant with value $\delta_s < 1$.

Now, let us see if we will be able to recover $\mathbf{x}$ from $\mathbf{b}$. One of the alternatives considered in CS is to state and solve a BP problem:

$$\text{minimize } \|\mathbf{x}\|_1\,, \text{ subject to } A\,\mathbf{x} = \mathbf{b}$$

According with [37], having got the BP solution $\mathbf{x}^*$, this solution recovers $\mathbf{x}$ exactly if the signal is sufficiently sparse and the matrix A has the RIP property. Moreover, assume that $\delta_{2s} < \sqrt{2} - 1$, then (theorem):

$$\left\|\mathbf{x}^* - \mathbf{x}\right\|_1 \leq C_0\, \|\mathbf{x} - \mathbf{x}_s\|_1 \tag{2.64}$$

and:

$$\left\|\mathbf{x}^* - \mathbf{x}\right\|_2 \leq C_0\, s^{1/2}\, \|\mathbf{x} - \mathbf{x}_s\|_1 \tag{2.65}$$

where $\mathbf{x}_s$ is the vector $\mathbf{x}$ with all but the s-largest entries set to zero. See [37] for the constant C_0, which is rather small. In case that $\mathbf{x}$ was s-sparse, the recovery is exact.

Two consequences of the theorem are that:

- if $\delta_{2s} < 1$ the $l0$ problem has a unique s-sparse solution
- if $\delta_{2s} < \sqrt{2} - 1$ the solution of the $l1$ problem is the same as the $l0$ problem solution

Part of the research is trying to find larger values of δ_s still guaranteeing the recovery. For instance, recently (year 2013), a value of 0.5746 (instead of 0.4142) has been established in [202]. This paper includes a table, with key references, of the previous results from other specialists.

Normally, one wants a matrix A with a small δ_s. It has been found that many types of random matrices have very good δ_s. Often a value $\delta_s \leq 0.1$ can be obtained. On the contrary, it is difficult for deterministic matrices to have a good δ_s.

Random matrices can be obtained in several ways. Here is a short selection:

- **Gaussian matrices**: the entries of A are independent normal variables, with zero mean and $1/m$ variance; that is: $a_{i,j} \in N(0, 1/m)$
- **Bernouilli matrices**: the entries of A are:

$$a_{i,j} = \begin{cases} +1/\sqrt{m}, & with\ probability\ 1/2 \\ -1/\sqrt{m}, & with\ probability\ 1/2 \end{cases} \tag{2.66}$$

- **Partial Fourier matrices**: the entries of A are random samples of a Fourier matrix

See [42] and [133] for more alternatives, and details. A contemporary research topic is to deterministically generate matrices with good RIP.

Supposing that the signal recovery would be done with BP, it is important to consider also the *'null space property' (NSP)*. Let us introduce this property.

Recall that the null space $Nul(A)$ of the matrix A is the set of al solutions of $A\,\mathbf{x} = 0$.

Consider the set of indexes $L = \{1, 2, \ldots n\}$ and choose for example a subset $S \supset L$ with $|S| = s$. Then, any vector $\mathbf{x}$ can be written as the sum $\mathbf{x}_S + \mathbf{x}_{\bar{S}}$, where $\bar{S}$ is the complement of S, the non-zero elements of $\mathbf{x}_S$ are in S, and the non-zero elements of $\mathbf{x}_{\bar{S}}$ are in $\bar{S}$. For example, suppose that $\mathbf{x} = \{4, 3, 2, 1, 5, 6\}$, a possible decomposition could be: $\mathbf{x}_S = \{0, 0, 2, 1, 5, 0\}$ and $\mathbf{x}_{\bar{S}} = \{4, 3, 0, 0, 0, 6\}$.

The matrix A satisfies the null space condition of order s if for any non-zero vector $\mathbf{x} \in Nul(A)$ and any index subset S with $|S| = s$, one has:

$$\|\mathbf{x}_S\|_1 < \left\|\mathbf{x}_{\bar{S}}\right\|_1 \tag{2.67}$$

Intuitively, the NSP implies that non-zero vectors in the null space cannot be too sparse. The problem we want to avoid is having a sparse vector with $A\,\mathbf{x} = 0$, which would clearly interfere with the recovery of other vectors.

When dealing with exactly sparse vectors, the spark characterizes when recovery is possible. However, when dealing with approximately sparse signals (for instance when there is noise), a more restrictive condition about the vectors in $Nul(A)$ is needed, and this is why the NSP [64]. In relation with RIP, it can be shown that RIP implies NSP; that is: RIP is stronger than NSP.

A theorem establishes that BP will obtain exact recovery iff A satisfies the NSP [3, 157].

For the interested reader it could be opportune to consult [179] about the relationship between the irrepresentable condition, already mentioned in the Lasso context, and the RIP. The irrepresentable condition is more restrictive.

Nowadays, experience with some applications is showing that the RIP condition –which is a sufficient condition is too stringent. The theory is advancing, and new, weaker conditions are being discovered or re-discovered [4].

As an example of sparse signal recovery, suppose that one has a sparse 1D signal, then one uses a sensing matrix A to obtain samples (a vector $\mathbf{b}$), and then applies the Douglas-Rachford algorithm to recover the original signal. This example is borrowed from a contribution of G. Peyre to the MATLAB Central file exchange site (Toolbox of Sparse Optimization).

Program 2.5 handles this example and, at the same time, provides an example of implementation of the Douglas-Rachford algorithm.

Figure 2.13 shows the original signal to be sampled.

The signal recovery can be considered as the following problem:

$$\text{minimize } f(\mathbf{x}) + g(\mathbf{x})$$

where $f(\mathbf{x})$ is the indicator function and $g(\mathbf{x})$ is $\|\mathbf{x}\|_1$.

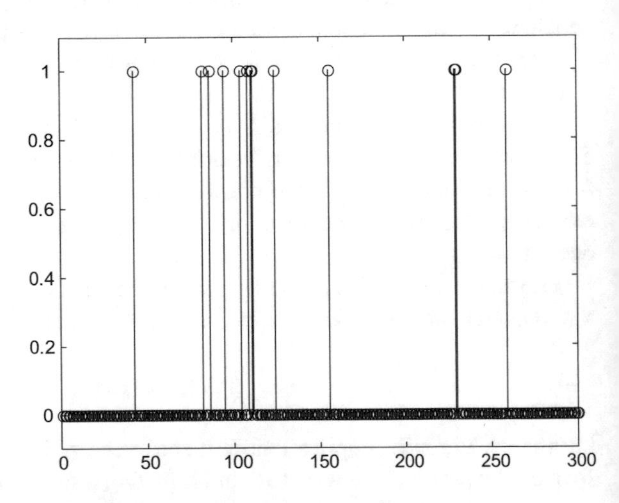

Fig. 2.13 An sparse signal being measured

The Douglas-Rachford algorithm can be written in this case as follows:

$$\mathbf{x}_k = \Pr ox_{\lambda f}(\mathbf{y}_k) = \mathbf{y}_k + A^T \, (A\, A^T)^{-1} \, (\mathbf{b} - A\, \mathbf{y}_k) \tag{2.68}$$

$$\mathbf{y}_{k+1} = \mathbf{y}_k + \lambda_k (\Pr ox_{\lambda g}(2\mathbf{x}_k - \mathbf{y}_k) - \mathbf{x}_k) \tag{2.69}$$

where $\Pr ox_{\lambda g}()$ is soft thresholding.

(notice that $f(\mathbf{x})$ and $g(\mathbf{x})$ have been exchanged with respect to the description of the algorithm given in the previous section; both are equivalent).

Figure 2.14 shows the recovered signal.

The evolution of the algorithm can be followed in several ways. For example, Fig. 2.15 shows the evolution of $\|\mathbf{x}\|_1$ along the iterations.

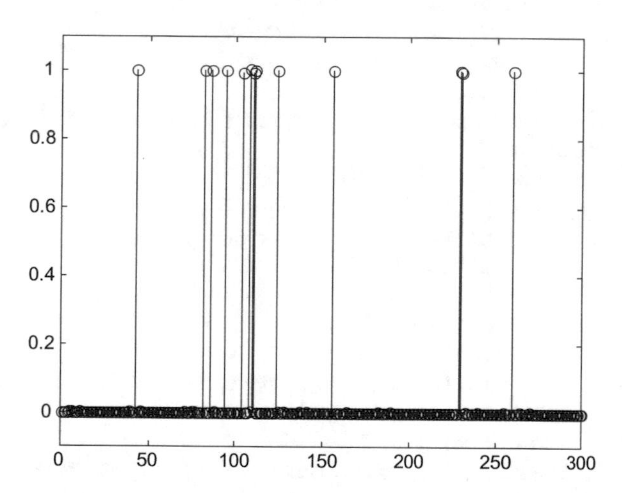

Fig. 2.14 Recovered signal

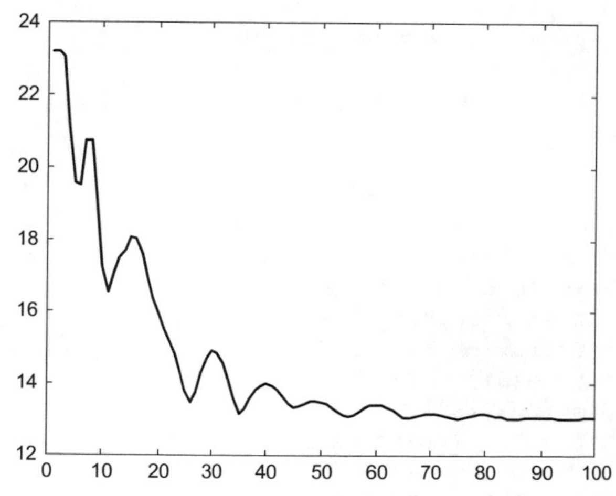

Fig. 2.15 Evolution of $\|\mathbf{x}\|_1$ *along iterations*

Program 2.5 Example of Douglas-Rachford

```
% Example of Douglas-Rachford
% for compressed sensing of 1D signal
clear all;
N=300; %problem dimension
q=N/4; %number of measurements
A=randn(q,N)/sqrt(q); %random Gaussian sensing matrix
% A s-sparse signal, with s non-zero values (=1)
s=13;
x0=zeros(N,1);
aux=randperm(N);
x0(aux(1:s))=1;
% display
% the signal to be measured
figure(1)
stem(x0,'k');
axis([0 N -0.1 1.1]);
title('the original signal')
% perform random mesurements
b=A*x0;
% Algorithm for recovering the measured signal
niter=100;
lambda=1.0; ; gamma=1.0;
x=zeros(N,1);
y=zeros(N,1);
rnx=zeros(niter,1);
uA=A'*inv(A*A');
for nn=1:niter,
  % x update
  x=y+(uA*(b-(A*y))); %proxF(indicator function)
  %proxG (soft thresholding):
  aux=(2*x)-y;
  S=max(0,1-gamma./max(0,abs(aux))).*aux; %soft th.
  % y update
  y=y+ (lambda*(S-x));
  % recording
  rnx(nn)=norm(x,1);
end
% display
% recovered signal
figure(2)
stem(x,'k');
axis([0 N -0.1 1.1]);
title('recovered signal')
% evolution of ||x||1
figure(3)
plot(rnx,'k');
title('evolution of signal norm-1')
axis([0 100 12 24]);
```

2.3.3 Incoherence and Sensing

Some of the CS authors have proposed a kind of uncertainty principle, in the vein of the time-frequency duality [42, 76]. It is illustrative to have a quick look to this aspect.

Two important examples of orthogonal matrices are the identity matrix I and the Fourier matrix F. Both matrices correspond to two orthobases, one allows for time-domain representation of signals, and the other for frequency-domain representation. More in general, given two orthobases Ψ and Φ, the signal $\mathbf{b}$ can be represented as follows:

$$\mathbf{b} = \Psi\alpha = \Phi\beta \tag{2.70}$$

Suppose that the matrix A was the concatenation of two orthogonal matrices Ψ and Φ. A particular example could be $A = [I, F]$; a sparse approximation of the signal $\mathbf{b}$ would be a superposition of spikes and sinusoids.

The *'mutual-coherence'* $\mu(A)$ is defined as follows [76]:

$$\mu(A) = proximity(\Psi, \Phi) = \max_{i,j} |\mathbf{\Psi}_i^T \cdot \mathbf{\Phi}_j| \tag{2.71}$$

where $\mathbf{\Psi}_i$ and $\mathbf{\Phi}_j$ are columns. It can be shown that $(1/\sqrt{n}) \leq \mu(A) \leq 1$. For the case $A = [I, F]$, the mutual-coherence is $\mu(A) = (1/\sqrt{n})$.

If the two orthogonal matrices were not normalized, the mutual coherence would be expressed as follows:

$$\mu(A) = \max_{i,j} \frac{|\mathbf{\Psi}_i^T \cdot \mathbf{\Phi}_j|}{\|\mathbf{\Psi}_i\|_2 \, \|\mathbf{\Phi}_j\|_2} \tag{2.72}$$

The interesting fact is that, according with a theorem [76], one has:

$$\|\alpha\|_0 + \|\beta\|_0 \geq \frac{2}{\mu(A)} \tag{2.73}$$

Therefore, if the mutual-coherence of two bases is small, then α and β cannot both be very sparse. This can be regarded as an uncertainty principle. In particular, a signal cannot be sparsely represented both in time and frequency.

There is a theorem in [42] establishing the following: suppose the signal $\mathbf{x}$ is s-sparse and take m samples of it. Then if:

$$m \geq C\,\mu^2(A) \cdot s \cdot \log n \tag{2.74}$$

for some $C > 0$, then BP will exactly recover from $\mathbf{b} = A\,\mathbf{x}$ the signal $\mathbf{x}$ with very high probability.

In consequence, the smaller the coherence, the fewer samples are needed [42]. Then, in general, incoherent sampling would be recommended.

The case $A = [I, F]$ has maximal incoherence. The coherence of *noiselets* [60] and Haar wavelets is $(\sqrt{2}/\sqrt{n})$. The coherence of random matrices and a fixed basis would be about $(\sqrt{2 \log n}/\sqrt{n})$ (see [42]).

See [108] for details on the relations between spark, NSP, mutual coherence, and RIP.

2.3.4 Stable and Robust Recovery

Usually, signals are not exactly sparse but compressible. Given a compressible signal $\mathbf{x}$, it can be approximated by a sparse signal $\mathbf{x}_s$ being the vector $\mathbf{x}$ with all but the s-largest entries set to zero, with the following error:

$$\sigma = \|\mathbf{x} - \mathbf{x}_s\|_1 \tag{2.75}$$

In addition, real life measurement is always contaminated with noise $\mathbf{e}$. Then:

$$\mathbf{b} = A\,\mathbf{x} + \mathbf{e} \tag{2.76}$$

It is assumed that $\| \mathbf{e}\|_2 < \eta$.

Returning to basis pursuit, it was found that in the presence of noisy or imperfect data, it is better to state the problem as:

$$\text{minimize } \|\mathbf{x}\|_1 \, , \text{ subject to } \|A\,\mathbf{x} - \mathbf{b}\|_2 \leq \eta$$

This problem is called *'basis pursuit denoise' (BPDN)*, and it was proposed in the famous paper of [55]. Clearly, the problem looks like a Lasso problem. In fact, BPDN became so popular that the literature sometimes use the term BPDN to refer to Lasso.

Assuming tha t$\delta_{2s} < \sqrt{2} - 1$, one of the theorems presented in [37] establishes that the recovery obtained by BPDN solution obeys:

$$\left\|\mathbf{x}^* - \mathbf{x}\right\|_2 \leq C_0\, s^{1/2}\, \sigma + C_1\, \eta \tag{2.77}$$

for constants $C_0, \; C_1 > 0$. A small C_0 would mean that the recovery is *stable* with respect to inexact sparsity; and a small C_1 would mean that the recovery is *robust* with respect to noise. It is established in the theorem that both constants are rather small. For example, with $\delta_{2s} = 1/4$ the values of the constants would be $C_0 \leq 5.5$ and $C_1 \leq 6$, [42]. In the noiseless case, the solution could be found with BP, and will have the same value of C_0.

Some literature is considering alternatives to BPDN. One of these alternatives is the *'Dantzig selector'*, which was proposed by [36] for cases where the number of variables is much larger than the number of observations. The problem is formulated as follows:

$$\text{minimize } \|\mathbf{x}\|_1 \,, \text{ subject to } \left\| A^T(\mathbf{b} - A\mathbf{x}) \right\|_\infty \leq c$$

(the norm $\|\cdot\|_\infty$ is just the maximum absolute value).

2.3.5 *Phase Transitions*

As a preliminary step for introducing phase transitions, define two parameters:

- The undersampling ratio: $\delta = m/n \;,\; \delta \in [0, 1]$
- The oversampling ratio: $\rho = s/m \;,\; \rho \in [0, 1]$

Imagine you take pairs (δ, ρ) and represent with colour pixels, on the plane ρ *vs* δ the difficulty of signal recovery using $l1$ norm. Smaller δ and larger ρ would mean more recovery difficulty.

Based on counting faces of polytopes, Donoho and Tanner found what they called '*phase transitions*' associated to $l1$ recovery. Figure 2.13 shows a phase transition curve similar to the result of that study using a Gaussian matrix A. It can be obtained in the Monte-Carlo style with many (δ, ρ) trials; the curve represented in the figure corresponds to the points where the recovery success probability crosses 50 %. The epigraph of the curve is a region of improbable solvability (Fig. 2.16).

The curve represented in the figure corresponds to the noiseless recovery case. A kind of universality of phase transitions was observed [67], so the gaussianity of A can be considerably relaxed. See [70] for a extensive treatment that includes phase transitions.

Phase transitions provide a way to compare recovery methods that has been adopted by many research publications. An important aspect is what happens in the noisy case; according with [205] several performance regions could be recognized.

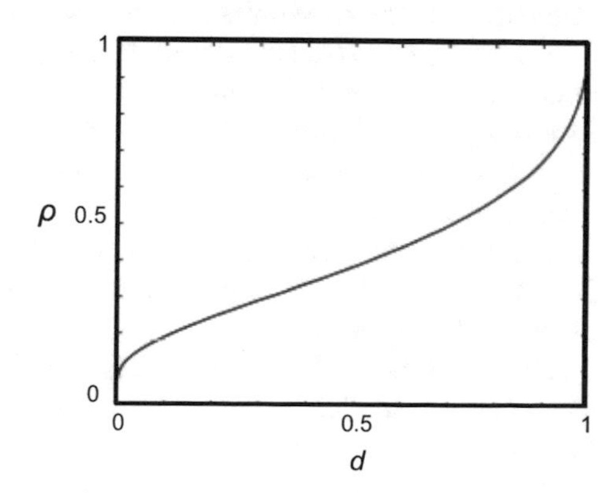

Fig. 2.16 A phase transition curve

2.3.6 Some Applications

Evidently, the potential of CS is suitable for applications in need of compression, like modern digital communication. In addition, there are applications in which something similar to sampling with random matrices occur, due to the characteristics of the sensorial system—for instance a network of sensors—or because the random nature of the process under measurement—for instance traffic-.

First examples of CS applications are being published. The intention in this part of the CS section is to show the variety and interest of what is going on in this field.

With respect to communication networks, [102] shows with some detail, and many references, that CS has a range of applications from the physical layer through the application layer, reaching in many cases performance gains of the order of ten times. In [100] a user's guide to CS for communication is offered, its first half being a good concise compendium of CS theory and algorithms. The second half of this article is devoted to CS applications for wireless channel estimation, wireless sensor networks, network tomography, cognitive radio, array signal processing, etc. Opportune references were given for each application. More specific publications are [14] on estimating sparse multipath wireless channels, and [22] on sparse channel estimation in underwater communications.

Two CS based dimensionality data reduction algorithms were introduced in [88], using for experiments a database of handwritten digits. The application of CS for speech and audio signal was considered in [57] via sparse decompositions using dictionaries of windowed complex sinusoids.

A compressed sensing based fingerprint identification for wireless systems has been introduced in [200]. This is an example of future applications for security.

A block-by-block CS was proposed by [85] for sensing of natural images. This approach has been continued by [131] and others for pictures and video.

Due to constraints of sensing devices the classes of measurement matrices are limited. Also, some kinds of signals exhibit certain structured features. So it is convenient to study algorithms that go beyond the random measurement paradigm. This subject is treated in [73] and other related papers.

There are remote control cases where the communication channel is rate-limited and so it is interesting to investigate the use of CS. This problem is discussed in [132].

A brief review of CS applied to Radar was given by [80]. It included a short realistic discussion of the practical interest of CS in this context. In [183] a CS method for SAR imaging was introduced. It outperforms the conventional SAR algorithm. There are now a significant number of papers proposing CS techniques for several types of Radar and involved issues.

There is an interesting long report on CS application in Defense sensor systems [126] that covers the single-pixel camera, SAR, Doppler Radars, sparse installations of coastal HF radars, etc.

The structural health monitoring is becoming more and more important in engineering. For example, in a detailed contribution [103] describes the application of CS for real-time accelerometer data concerning the Tianjin Yonghe cable-stayed bridge (512 m. total length). An extensive academic study of CS application in this context, focusing on vibrations, is [87]. In large constructions, the surveillance of the structure should be done with networked sensors, as discussed in [124].

A natural application field for CS is networks of sensors. The issues to be solved are described in [99], and a complete design of a large-scale wireless sensor network was presented in [119]. A centralized iterative hard thresholding algorithm was proposed in [147] for distributed CS in static or time-varying networks. The detection of events is treated by [43, 127]. The use of CS for target counting and localization in sensor networks is presented in [197].

Medical applications are abundant. For instance, [146] describes a fast compressed sensing-based cone-beam computed tomography (CBCT) reconstruction method. A number of publications, like the most cited [120], describe the application of CS for magnetic resonance imaging. An interesting contribution, concerning dynamic MRI, is [104]. The web page of J. Huang contains links to several research activities related to CS and MRI.

With respect to CS in medical ultrasound, there is an extensive academic treatment in [158]; a short review of applications is made in [112]; a description of results is given in [154].

Based on CS, [Ak] proposed a method for human activity sensing using mobile phone accelerometers.

A CS approach to urban traffic sensing with probe vehicles was introduced in [111]. This is a topic of increasing interest.

A representative publication on CS for environmental monitoring is [191], in which the methodology is demonstrated for Ozone and surface air temperature real data. An interesting CS application for ocean monitoring is described in [118].

2.4 Image Processing

Images can be treated with the Fourier transform or wavelets with a number of purposes, like compression, filtering, denoising, deblurring, etc. This has been already treated in different chapters of this book.

However, processing innovations do not stop. Several new ways of decomposing an image into components have been proposed, opening exciting possibilities. Many publications appeared, mainly in two directions: one focuses on edges and the rest, and the other on a kind of spatial dictionaries.

The objective of this section is to introduce with some detail these new approaches, indicating a number of references that can be useful for the reader to dig into the proposed methods.

2.4.1 Texture + Cartoon

In a famous work published in 2001 [128], Meyer proposed the decomposition of images into texture and cartoon parts. Different methods can be used for isolating the edges, while other can focus on texture modeling. This approach has prompted an active research on the three involved aspects: edges, textures, and decomposition.

The idea of the decomposition can be simply expressed as a decomposition of the image f into cartoon u and texture v, as follows:

$$f = u + v \tag{2.78}$$

There are authors that prefer other terms, like geometry or structure, for the u term. The v term could include constant color regions, textures that could be modeled with oscillatory functions, and other. Textiles can usually be modeled with oscillatory functions.

Of course, it would be always be possible to extract a u component from f, and then obtain v as $f - u$. However, it is more compliant with the purpose of sparse representation to try a model based approximation of v.

Notice that image denoising considers that there is an original image p and a noise w, so one has a noisy image $q = p + w$. The target of the denoising effort is to extract p. Compared with the texture + cartoon decomposition, there are similarities that could be more or less strong depending on the particular problem to be tackled. This should be taken into account when looking at the literature.

It has been pointed out that a decomposition $f = u + v$ seems to be analogous to the high-pass and low-pass decompositions described in previous chapters. However, this will not work as we want: both cartoon and texture contain high frequencies, so a linear filtering cannot separate u and v. Other alternatives should be explored, like for instance variational approaches.

In general, variational approaches try to obtain u as a function that minimizes a certain criterion, which is expressed as an image model.

Looking at the recent history of variational decompositions, as summarized in [29], an important proposal was done in 1989 by Mumford and Shah [130]. It is worthwhile to consider in some detail this contribution.

2.4.1.1 Mumford-Shah, and Image Segmentation

It may well happen that in a certain photograph one wants to separate objects of interest. For instance, cells (recall the example about thresholding in Chap. 11), roads, faces, etc. This is the type of applications addressed by image segmentation. It leads typically to the detection of edges, and so it has many things in common with $f = u + v$ decomposition.

Suppose a rectangular picture with multiple objects $O_1, O_2, \ldots, O_n$. There would be a set of regions $\Omega_1, \Omega_2, \ldots, \Omega_n$ in the image corresponding to these objects. Denote as Γ the set of smooth arcs that make up boundaries for the regions Ω_i. In total one has: $\Omega = \Omega_1 \cup \Omega_2 \cup \ldots \cup \Omega_n \cup \Gamma$. Our main goal is to capture the boundaries while the texture does not vary much inside each object. The image f could in this case be approximated by a piecewise-smooth function u.

Information is related to changes. With a certain analogy, one could speak of energies in the sense of information content. Using the Sobolev's H^1 norm, the energy of a region Ω_i would be:

$$E(\Omega_i) = \int_{\Omega_i} \left|\nabla u(\vec{x})\right|^2 d\mathbf{x} \tag{2.79}$$

(the function u must belong to the space H^1 of functions whose derivative is square-integrable).

It is known that functions belonging to H^1 cannot present discontinuities across lines, such as boundaries. Therefore, u alone cannot model the boundaries.

It is opportune to consider the energy of the boundaries, as follows:

$$E(\Gamma) = Length\,(\Gamma) \tag{2.80}$$

(more formally, this would be the Hausdorff measure of Γ)

In many applications, there is noise and blurring to be taken into account. A *'fidelity term'* can be used for this purpose:

$$E(error) = \int_{\Omega} (f(\mathbf{x}) - u(\mathbf{x}))^2\, d\mathbf{x} \tag{2.81}$$

In a most cited, seminal article [130], Mumford and Shah approximated the image f by a piecewise-smooth function u and a set Γ that solves the following problem:

$$\arg\min_{u,\Gamma} \left\{ \mu\, Length(\Gamma) + \int_{\Omega-\Gamma} |\nabla u(\mathbf{x})|^2\, d\mathbf{x} + \lambda \int_{\Omega} (f(\mathbf{x}) - u(\mathbf{x}))^2 d\mathbf{x} \right\} \tag{2.82}$$

Clearly, it is an energy minimization. It can be shown that removing any of the three terms gives a trivial, not suitable solution.

The observed result of the minimization is usually a simplified version of the original image, similar to a cartoon. It is not a perfect tool; for instance, shadows, reflections, gross textures, may cause difficulties.

Mainly because the term with Γ, it is not easy to solve the minimization problem. Many methods and approximations have been proposed, see [33, 66, 152] and references therein. An Octave implementation is included in [178]. There is a popular related segmentation method based on the Chan-Vese model; this model is described in [89] with an implementation in C available from the web.

An example of image segmentation using the Chan-Vese method will be given below. In preparation for the example, let us include a brief summary of this method. An important simplification is that $u(\mathbf{x})$ can take only two values: c_{in} for $\mathbf{x}$ inside Γ, and c_{out} for $\mathbf{x}$ outside Γ. These constants can be regarded as average gray values:

$$c_{in} = \frac{\int_{in\,\Gamma} f\, d\mathbf{x}}{\int_{in\,\Gamma} d\mathbf{x}} \qquad c_{out} = \frac{\int_{out\,\Gamma} f\, d\mathbf{x}}{\int_{out\,\Gamma} d\mathbf{x}} \tag{2.83}$$

Instead of a direct manipulation of Γ, it is represented as zero-crossings of a *'level set'* function φ (this idea was introduced in [143], a heavily cited article). This function φ is positive inside Γ and negative outside.

The energy to be minimized would be written as follows:

$$E = \mu \int_{\Omega} \delta(\varphi(\mathbf{x}))\ |\nabla\varphi(\mathbf{x})|\, d\mathbf{x} + \lambda_{in} \int_{in\Gamma} (f(\mathbf{x}) - c_{in})^2 d\mathbf{x} + \\ + \lambda_{out} \int_{out\Gamma} (f(\mathbf{x}) - c_{out})^2 d\mathbf{x} \tag{2.84}$$

The first integral in the previous equation corresponds to the length of Γ. In some cases it would be convenient to add a term with the area enclosed by Γ, to control its size, but it has been not considered in our example.

A semi-implicit gradient descent could be applied for the minimization. This is done with an iterative evolution of:

$$Q = \frac{\varphi_{xx}\,\varphi_y^2 - 2\,\varphi_x\varphi_y\varphi_{xy} + \varphi_{yy}\varphi_x^2}{(\varphi_x^2 + \varphi_y^2)^{3/2}}$$

$$\frac{\partial\varphi}{\partial t} = \delta(\varphi) \cdot \left[Q - \lambda_{in}(f - c_{in})^2 + \lambda_{out}(f - c_{out})^2\right] \tag{2.85}$$

(where sub-indexes represent partial derivatives).

The $\delta(\varphi)$ can be approximated with:

$$\delta(\varphi) = \frac{\varepsilon}{\pi(\varepsilon^2 + t^2)} \tag{2.86}$$

In each iteration, the values of c_{in} and c_{out} were updated.

Now, let us introduce the example. Figure 2.17 shows the original image. After running Program 2.6, which is a slightly modified version of code from Fields Institute (see the section on resources), an interesting image segmentation was obtained. The results are shown in Fig. 2.18.

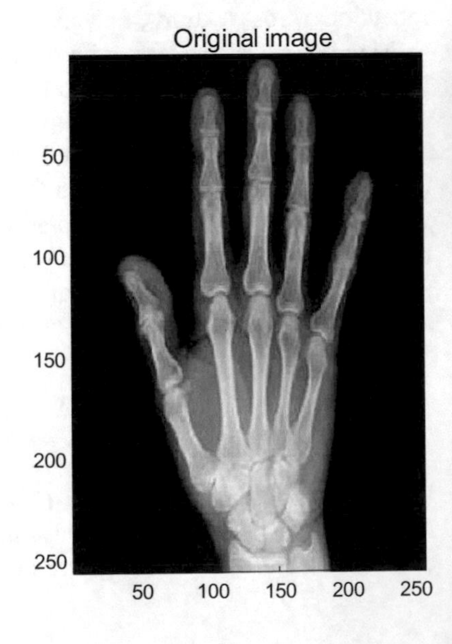

Fig. 2.17 Original image

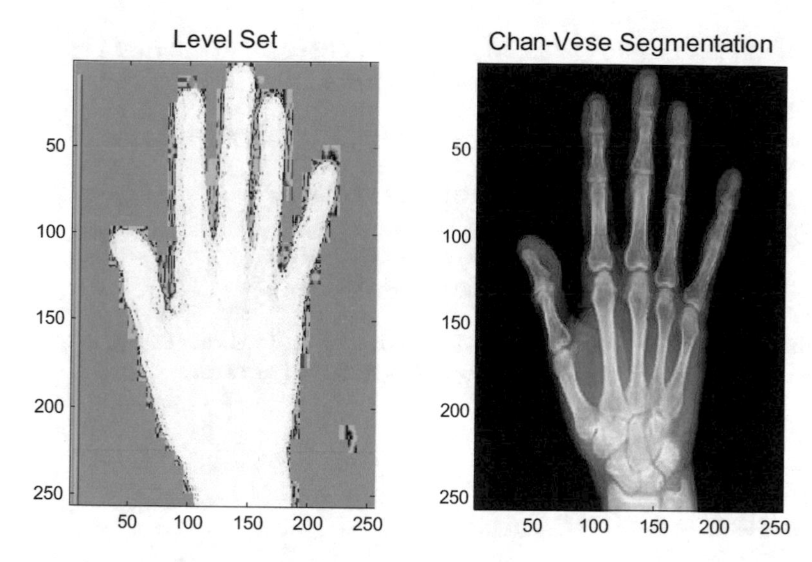

Fig. 2.18 (*right*) Chan-Vese segmentation, (*left*) level set

Program 2.6 Example of Chan-Vese algorithm

```
% Example of Chan-Vese algorithm
%(active contours segmentation, using level sets)
I=imread('hand1.jpg'); %read b&w image
I = double(I);
[m,n] = size(I);
lambda=0.1; %parameter
%Set initial box=1, with border=-1
Phi=ones(m,n);
Phi(1,1:n)=-1; Phi(m,1:n)=-1;
Phi(1:m,1)=-1; Phi(1:m,n)=-1;
%Prepare loop
disp('working...');
a = 0.01; %to avoid division by zero.
epl = 0.1; %small value
T = 10;
dt = 0.2;
for t = 0:dt:T
  ax2=2*Phi;
  P1=Phi(:,[2:n,n]); P2=Phi([2:m,m],:);
  P3=Phi(:,[1,1:n-1]); P4=Phi([1,1:m-1],:);
  %partial derivatives (approx.)
  Phi_x = (P1 - P3)/2;
  Phi_y = (P2 - P4)/2;
  Phi_xx = P1 - ax2 + P3;
  Phi_yy = P2 - ax2 + P4;
  Q1=Phi([2:m,m],[2:n,n]); Q2=Phi([1,1:m-1],[1,1:n-1]);
```

```
  Q3=Phi([1,1:m-1],[2:n,n]); Q4=Phi([2:m,m],[1,1:n-1]);
  Phi_xy = (Q1 + Q2 - Q3 - Q4)/4;
  %TV term
  Num = (Phi_xx.*Phi_y.^2 )- (2*Phi_x.*Phi_y.*Phi_xy) +
  (Phi_yy.*Phi_x.^2);
  Den = (Phi_x.^2 + Phi_y.^2).^(3/2) + a;
  %Compute averages
  c_in = sum([Phi>0].*I)/(a+sum([Phi>0]));
  c_out = sum([Phi<0].*I)/(a+sum([Phi<0]));
  %Update
  aux=( Num./Den - lambda*(I-c_in).^2 + lambda*(I-c_out).^2);
  Phi = Phi + dt*epl./(pi*(epl^2+Phi.^2)).*aux;
end;
% display of results
figure(1)
imagesc(I);
title('Original image');
colormap gray;
figure(2)
subplot(121);
imagesc(Phi);
title('Level Set');
subplot(122);
imagesc(I); hold on;
title('Chan-Vese Segmentation');
contour(Phi,[0,0],'m');
colormap gray;
```

2.4.1.2 Total Variation, and BV Functions

Further steps in the direction suggested by the Mumford-Shah model were taken by considering a total variation (TV) term. Already, a brief introduction to TV has been done in the previous book, in the context of image restoring. It is now opportune to include a more extended consideration.

There are excellent publications, with a mathematical formal orientation like [47, 164], that use test functions ϕ belonging to the set $C_0^1(\Omega\,,\,\Re^2)$ of continuously differentiable vector functions of compact support contained in Ω, and such $\|\phi\|_\infty \leq 1$. These functions are employed for the definition of TV as follows:

$$TV(u) = \sup\{ \int_\Omega u\, div\, \phi\, d\mathbf{x}\ ,\ \ \forall \mathbf{x} \in \Omega \} \tag{2.87}$$

Given a differentiable function u defined on a bounded open set Ω, its total variation is:

$$TV(u) = \int_{\Omega} |\nabla u(\mathbf{x})| \, d\mathbf{x} \tag{2.88}$$

If the TV of the differentiable function u is $TV(u) < +\infty$, then $u \in BV$, where BV is the space of *bounded variation functions*.

The set of bounded variation functions is a Banach space with the norm:

$$\|u\|_{BV} = \|u\|_{L^1} + TV(u) \tag{2.89}$$

Bounded variation functions can have sharp edges. Actually, the norm $\|\cdot\|_{BV}$ takes into account the number of edges.

If a function $u \in BV$ belongs also to the smaller Sobolev space, then the norm is just $TV(u)$. See [59] for an interesting study of BV functions, including wavelets.

The BV functions play an important role in the cartoon + texture decomposition.

2.4.1.3 The Rudin, Osher and Fatemi (ROF) Model

In 1992 Rudin, Osher and Fatemi [163] proposed to apply TV for image denoising, in a variational framework with the following expression:

$$\inf_{u} \left\{ \int_{\Omega} |\nabla u(\mathbf{x})| \, d\mathbf{x} + \lambda \int_{\Omega} (f(\mathbf{x}) - u(\mathbf{x}))^2 d\mathbf{x} \right\} \tag{2.90}$$

where $u \in BV$.

Evidently, the ROF model (enclosed in braces) is composed of a TV term and a fidelity term. This model has been cited in more than six thousand papers.

The TV term removes noise or small details, while preserving edges. The authors of the ROF model give in [163] an algorithm for computing the adequate value of λ if the noise level was known. In other cases this value has to be chosen, considering that it determines in some sense the smallest image feature to be kept.

From the point of view of optimization, the good news is that the ROF denoising problem is convex, so the solution exists in BV and is unique [164]. A detailed study of image recovery via TV minimization is [47]. The field of TV in imaging, including algorithms, is reviewed by [45].

There are a number of observed problems when adhering to the ROF approach, as described in [48]: loss of contrast, loss of geometry, staircasing, and loss of texture. Part of the recent developments cited in [48] are oriented to solve these problems.

The ROF variational method can be adapted to different types of noise, as introduced in [90]. This paper is plenty of practical numerical and analytical details, and contains a link to one implementation in C code.

Next section includes an example of ROF-TV image denoising accompanied with a MATLAB program.

2.4.1.4 Meyer's Approach

Meyer suggested to replace the $l2$ norm in the fidelity term with a weaker norm more adequate for modeling textures or oscillatory patterns. Hence, he proposed the following minimization problem:

$$\inf_u \left\{ \int_\Omega |\nabla u(\mathbf{x})| \, d\mathbf{x} + \lambda \, \| f(\mathbf{x}) - u(\mathbf{x}) \|_* \right\} \tag{2.91}$$

where $u \in BV$.

Continuing with his approach, Meyer defined the space G, which is the Banach space of all generalized functions v that can be written as $v = div(\mathbf{g})$, where $\mathbf{g} = (g_1, g_2)$ and $g_1, g_2 \in l_\infty(\Omega)$. The space G is endowed with the G-norm, which is defined as the lower bound of all $l_\infty(\Omega)$ norms of the functions $|\mathbf{g}|$, with the infimum being computed over all decompositions of v.

The space G is the dual of the closed subspace BV of BV. When applying the G-norm in the minimization problem, the second term is $\lambda \; \| f(\mathbf{x}) - u(\mathbf{x}) \|_G$. Meyer also defined two more spaces: E and F, see [128]. The spaces are related as follows: $BV \subset G \subset E \subset F$.

G-functions may have large oscillations and nevertheless small norms, which is suitable for the minimization to preserve textures.

Meyer did not propose any numerical procedure for the decomposition. A first algorithmic contribution to this aim was made in [180] (the Vese-Osher model), which soon was followed by other proposals, like [10, 193]. In [29] a simple conversion of a linear filter pair into a nonlinear filter pair was proposed, obtaining a fast separation of cartoon and texture.

2.4.1.5 TV-l1 Approach

It was suggested in [135] to replace the $l2$ term in the ROF model by a $l1$ term. This article, year 2004, contains interesting references from the 90s about the fidelity term and how to avoid outliers. According with this approach, the functional to minimize is:

$$\inf_u \left\{ \int_\Omega |\nabla u(\mathbf{x})| \, d\mathbf{x} + \lambda \int_{|f(\mathbf{x}) - u(\mathbf{x})|} d\mathbf{x} \right\} \tag{2.92}$$

As shown in [135], the $l1$ norm is well suited to remove salt and pepper noise. Further analysis by [49, 193], shows that the model enjoys interesting properties of morphological invariance and texture extraction by scale, so geometrical features are better preserved. A first, fast algorithm for solving the optimization problem was presented in [12].

All three models, Meyer, Vese-Osher, and TV-$l1$, were compared by [194] using a uniform computing approach. The three models were solved as second-order cone programs (SCOP). The comparison refers to 1D signals and 2D images. Also, [194] contains a detailed history, with references, of alternatives for solving the three models.

See [109] for a practical treatment of the TV-$l1$ decomposition, including various examples and a link to software from a web site.

2.4.1.6 Other Approaches for the Texture Term

The Report [11] proposed the following generalization:

$$\inf_u \left\{ \int_\Omega |\nabla u(\mathbf{x})| \, d\mathbf{x} + \lambda \, \| f(\mathbf{x}) - u(\mathbf{x}) \|_H^2 \right\} \tag{2.93}$$

where H is some Hilbert space. This generalization can include a number of different models, and in particular the ROF model. One of the contributions of [12] is a Hilbert space of Gabor wavelets.

From time ago there was interest on texture modeling and analysis for different purposes. A brief review of invariant texture analysis methods is offered in [198]. With the advent of the cartoon-texture approach, more research has been devoted to texture models favoring better image decompositions; see [125] for a modern perspective involving a decomposition of the functional into three terms: the fidelity term, a cartoon term, and a texture term.

2.4.1.7 Other Approaches for the Cartoon Term

As said before, it has been observed that the TV term induces image staircasing effects. A remedy for this problem was introduced in [47], by including higher order derivatives in the energy.

Several versions of the cartoon-texture decompositions were proposed in [50], by combining the improved cartoon term of [47] and three alternatives for the texture term (the third alternative considers texture + noise). This article is particularly interesting in several ways: discussion, formulae, and experimental results.

See [110] for an interesting work on second order TV, and [121] for combining TV and a fourth-order partial derivative filter.

2.4.1.8 Some Related Aspects

Let us briefly collect a number of references that deal with important aspects of the methods already introduced.

- A correlation tool was introduced in [12] to properly select the parameter λ.
- Relevant analysis results concerning BV functions are presented in [2, 58]. By the way, in [96] it was provocatively questioned if natural images are BV.
- The decomposition into cartoon + texture + noise $(u + v + w)$ was introduced by [92].
- Almost every cited paper includes a review of the decomposition topic. In addition, the reader could consult [29, 93] for more extensive reviews.

2.4.2 Patches

Science has surprising connections between seemingly disparate fields. For instance, it happens that the term *"sparse coding"* also belongs, from the 90s, to Neuroscience research.

As it will become clear soon, this biological aspect deserves some attention now. A convenient guide is offered by the review done in [171], in special connection with the work of Olshausen. Many references in [171] quote observations and conjectures made by Barlow in the 60s.

2.4.2.1 A Bit of Neural Processing

Images are captured by the retina, transmitted to the LGN, and then to the area V1 of the visual cortex; subsequent areas are V2, V4, MT, and MST. It has been reported that nobody has been able to reconstruct the input image from the recordings of neurons in V1. Cells in the visual cortex were classified as simple, complex, and hyper-complex cells. Simple and complex cells are sensitive to specific stimuli orientations.

One of the Barlow's hypotheses is that the role of early sensory neurons is to remove statistical redundancy of the input. Hence, it is not strange that the first section of [171] was devoted to PCA and ICA. This also corresponds to an important direction of the research, which assumes that sensory neurons are specially adapted to the statistical properties of those signals that occur more frequently. So, it is important to investigate the relationship between natural signals and neural processing. In general, natural images are not Gaussian and there are significant spatial correlations.

Another observation of Barlow was that neurons at later stages of processing are generally less active than those at earlier stages. It seems that we may model what happens in visual areas using multiple stages of *efficient coding*. This is, indeed, an ambitious objective. Most initial steps of the research have focused on retina and the first visual cortex areas.

Concerning the retina, it has been shown that the single-cell physiology and contrast sensitivity functions are consistent with the product of a whitening filter and a Wiener filter for noise removal and adaptation to mean luminance level (see [171] and references therein). Because non-Gaussianity, a whitened natural image still has lines, edges, contours, etc.

It seems that there is efficient coding at the retina level. The next question is if we have this in V1.

In a famous letter to *Nature* in 1996, [138] proposed to represent an image as a linear superposition of 2D basis functions, which are *image patches extracted from natural scenes*. The representation has a conventional form:

$$I(x, y) = \sum_i a_i \phi_i(x, y) \tag{2.94}$$

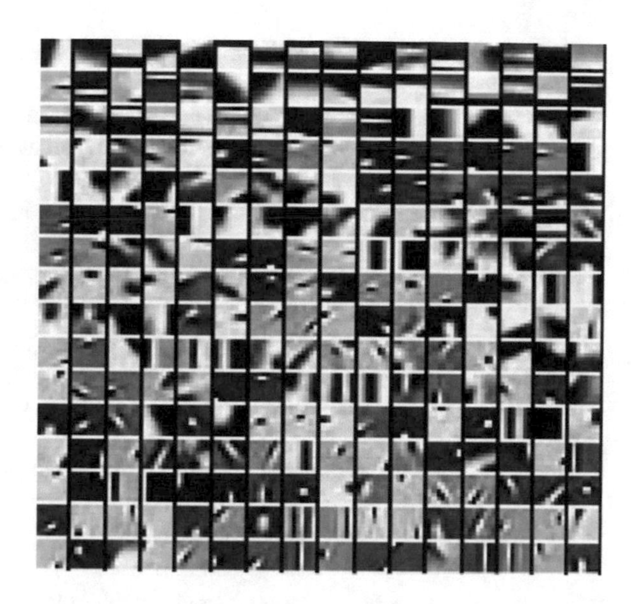

Fig. 2.19 A patch dictionary

The set of basis functions emerged after training on many (in the order of 10^5) image patches randomly extracted from natural scenes. The training tried to maximize the sparsity of the representation, searching for components that are both sparse and statistically independent (in the ICA sense). Actually, [138] shows an example of 192 basis functions, which are 16×16 pixel patches, extracted from 512×512 natural images. These functions resemble the spatial receptive field properties of simple cells, they are spatially localized, oriented, and band-pass.

Figure 2.19 shows an example of a patch dictionary, with 256 patches of 8×8 pixel each.

In 1997, [139] made the hypothesis that V1 employs sparse coding with an over-complete basis set.

The principle behind this kind of coding is that it tries to represent each image in terms of a small set of functions, chosen from an overcomplete set. Only a few neurons need to be active and expend energy. Actually, it has been estimated that at any given moment only 1/50th of the cortical neurons could afford to be active, due to energy constraints (see [140] and references therein).

It is being found that sparse coding is also employed by other senses and other neural functions [140]. Part of current research is considering space-time statistics, using natural environment movies as inputs [31, 171].

2.4.2.2 Dictionaries for Sparse Representation

Back to our signal processing atmosphere, the lessons learned from neural processing are summarized in [162] in the context of a history of transform design that includes Fourier, wavelets, etc. This summary put in contrast analytic versus trained dictionaries. Analytic dictionaries are linked to harmonic analysis, use pre-defined classes of functions, and are usually too simplistic for the nature complexity. Machine learning assumes, instead, that the structure of complex phenomena can be more accurately extracted directly from the data.

Of course, if one adheres to the use of learned dictionaries, then a training method should be devised. In his review, [161] identifies five available methods. Perhaps the most popular of them is K-SVD, which will be described in more detail below. The other methods are: generalized PCA, union of orthobases, the method of optimal directions (MOD), and parametric training methods. References are provided for all methods. The advantage of parametric training is that structured dictionaries were obtained.

Given a set of training column data vectors, $\mathbf{y}_1, \mathbf{y}_2, \cdots, \mathbf{y}_N$, the problem is to find a $(m \times K)$ dictionary D, with $K << N$, such every $\mathbf{y}_i$ is a sparse combination of elements of D. Note that the data vectors $\mathbf{y}_i$ could contain 1D signals or vectorized image patches.

Figure 2.20 shows a diagram corresponding to the problem. The representation matrix X should be sparse (with sparsity s).

In mathematical terms, the problem to solve is:

$$\min_{D, X} \sum_i \|\mathbf{y}_i - D\,\mathbf{x}_i\|^2, \text{ subject to } \|\mathbf{x}_i\|_0 \leq s$$

Notice that the optimization is over both D and X.

The problem could be relaxed from $l0$ to $l1$ norm, and be stated in Lagrangian form adding two terms.

Some authors prefer to use a simpler expression for the term $\sum_i \|\mathbf{y}_i - D\,\mathbf{x}_i\|^2$, as follows:

$$\|Y - DX\|_F^2 \tag{2.95}$$

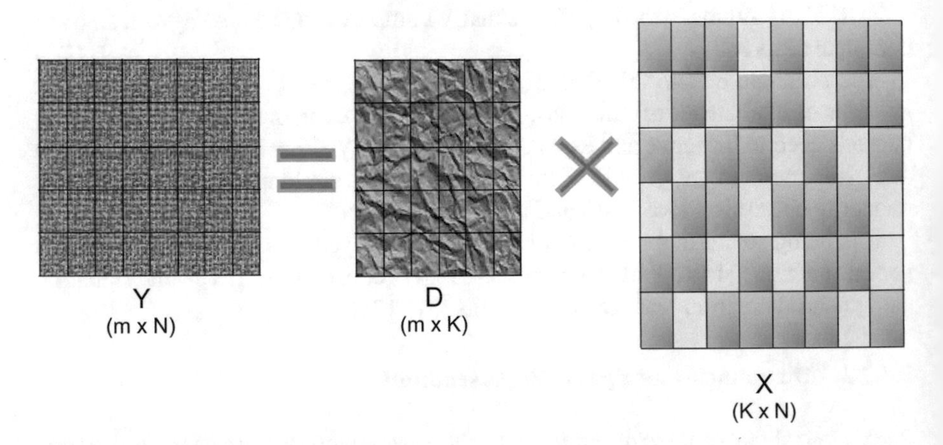

Fig. 2.20 The dictionary problem

There is a general iterative scheme that can be used for solving the problem [44]. Each iteration includes two steps:

- Sparse coding, for finding a minimizing X(for a fixed D)
- Dictionary update, finding a minimizing D (for a fixed X)

Usually this scheme would lead to a local-minimum.

The sparse coding part could be done with methods already introduced in this chapter, like for instance orthogonal matching pursuit (OMP).

The dictionary update could be done by directly solving the least square problem, using the Moore-Penrose pseudo-inverse. This is the alternative chosen by the MOD method; but it tends to lead the iteration towards the local minimum nearest to the initial guess. Other alternative could be to use a gradient descent on D.

2.4.2.3 The K-SVD Training Method

The K-SVD algorithm is described in detail in [6], which is a most cited article. It can be considered as a generalization of the K-means clustering algorithm, already seen in the chapter on data classification.

As MOD and other methods, K-SVD iterates two steps: sparse coding and dictionary update. Sparse coding can be done by any suitable method; K-SVD focuses on the dictionary update. For this update, assume that both D and X are fixed and select one column $\mathbf{d}_k$ of D; then, select the *k-th* row of X, which will be denoted as $\mathbf{x}_T^k$. Now, following [6], let us derive a convenient expression:

$$\|Y - DX\|_F^2 = \left\| Y - \sum_j \mathbf{d}_j\, \mathbf{x}_T^j \right\|_F^2 = \left\| \left(Y - \sum_{j \neq k} \mathbf{d}_j\, \mathbf{x}_T^j \right) - \mathbf{d}_k\, \mathbf{x}_T^k \right\|_F^2 = \\ = \left\| E_k - \mathbf{d}_k\, \mathbf{x}_T^k \right\|_F^2 \tag{2.96}$$

where E_k is an error matrix.

Denote as ω_k the group of indices pointing to vectors $\mathbf{y}_i$ that use $\mathbf{d}_k$ (those where $\mathbf{x}_T^k(i)$ is nonzero). The group has a certain number L of indices.

Also, denote as Ω_k a matrix with ones on the $(\omega_k(i), i)$ entries, and zeros elsewhere. This matrix will be used for shrinking purposes. For instance, the result of $\mathbf{x}_R^k = \mathbf{x}_T^k\, \Omega_k$ is a row vector $\mathbf{x}_R^k$ of length L, which is obtained from $\mathbf{x}_T^k$ by discarding the zero entries. Similarly, $E_k^R = E_k\, \Omega_k$ is the set of error columns corresponding to examples that use $\mathbf{d}_k$.

Then, for the dictionary update one has to minimize:

$$\left\| E_k \Omega_k - \mathbf{d}_k\, \mathbf{x}_T^K\, \Omega_k \right\|_F^2 = \left\| E_k^R - \mathbf{d}_k\, \mathbf{x}_R^K \right\|_F^2 \tag{2.97}$$

Here, one can use SVD (singular value decomposition) for getting the desired solution. The matrix E_k^R is decomposed as $E_k^R = U\, \Delta\, V^T$. The solution for $\mathbf{d}_k$ is the first column of U, and for the vector $\mathbf{x}_R^k$, the first column of V multiplied by the singular value $\Delta\,(1, 1)$.

Therefore, the algorithm takes the following steps:

- Set an initial dictionary with $l2$ normalized columns
- $j = 1$
- (*) repeat until convergence
- *Sparse coding step*
- *Dictionary update step*: for each column $k = 1, 2, \ldots$ in D
  - Find ω_k, build Ω_k
  - Compute E_k, and E_k^R
  - Decompose E_k^R using SVD
  - Update D with the column $\overleftarrow{d}_k$ and X with the vector $\mathbf{x}_R^k$
- $j = j + 1$, go to (*)

See [52] for a longer explanation of the K-SVD method.

It is convenient to note that part of the specialized literature use the term *atom* to refer to columns, $\mathbf{d}_k$, of the dictionary D.

A main contribution of K-SVD is that the dictionary update does not require matrix inversion; the update is made atom-by-atom, sequentially, in an efficient manner. Another difference with other algorithms is that during dictionary update, the content of X is also modified. The algorithm reduces, or at worst maintains, the error of the representation at each iteration.

A more efficient implementation of K-SVD was introduced in [161], with links to MATLAB toolboxes.

Next three figures correspond to an example of image denoising using K-SVD. Figure 2.21 depicts the original image and the same image contaminated with Gaussian noise. Figure 2.22 shows the patch dictionary obtained with the K-SVD method. Then, using this dictionary the image is denoised, as described in [78], and the result is shown in Fig. 2.23. These figures have been obtained with a program based on the software available from the Elad's web site. Although this program is a simplified version, it is relatively long and so it has been included in the appendix of long programs. The program is not only long, but also relatively slow: it would probably take around 5 min. The reader is invited to add more K-SVD iterations, or to specify less error (the constant C at the beginning of the program), or to increase the number of blocks.

The program uses two times a sparse codification function that has also been included in that appendix. This function uses sparse matrices.

2.4.2.4 Further Developments

The methods already described, like MOD or K-SVD, learn a dictionary D to sparsely represent the patches of an image, rather than the whole image itself [189]. In an important article, [78] a proposal was made in the context of image denoising.

Fig. 2.21 Original picture, and image with added Gaussian noise

Fig. 2.22 Patch dictionary obtained with K-SVD

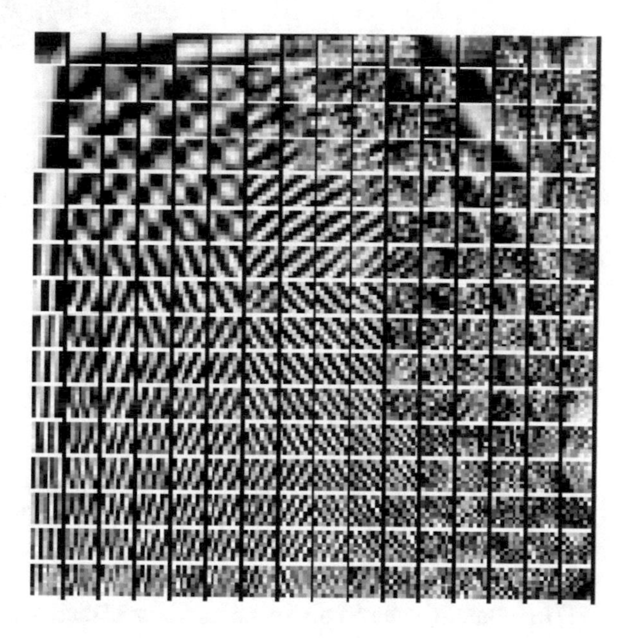

Overlapping patches were used. The idea was to denoise each patch via sparse coding, and then estimate the total image as the average of the patches together with the observed noisy image. Actually, [78] adopted a Bayesian perspective, by defining a global image prior that forces sparsity over all patches. Further elaboration in this line was presented in [77, 79].

In the case of denoising and other applications (impainting, deblurring, etc.) it is natural to consider a Bayesian treatment. If a see a noisy or corrupted image I would say: this image is not clean. So, I expected something cleaner. In consequence, I have a kind of *model* (a prior) of what an image should be.

See [189] for a fast method for whole image recovery using patch-dictionary. It has an associated web page with MATLAB code.

Fig. 2.23 Denoised image

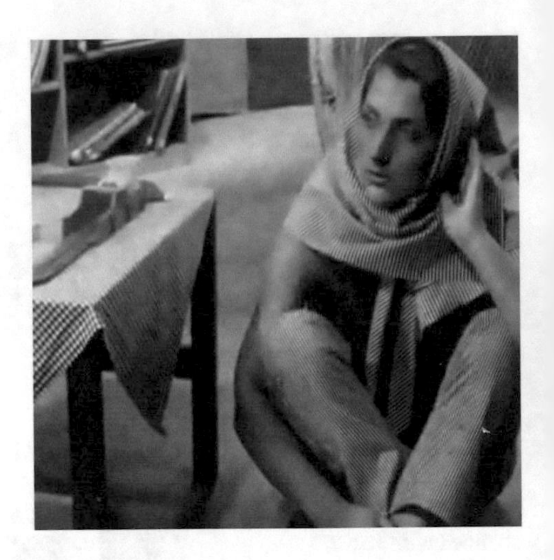

Responding to the current consumer market needs, it is important to deal with colour images. This was the target of [122], which extends the approach of [78] to colour images. The proposed scheme was also able to properly handle non-homogeneous noise; this is valuable in cases of missing data, such in image demosaicing or impainting (colloquially: filling holes).

According with [155], the use of patch dictionaries has become very popular, showing good performances. The research is trying to exploit interrelations between patches, using for instance clustering into disjoint sets to treat them adequately, etc. ([155] includes a short review of this issue).

In his interesting discussion on K-SVD, [16] remarks that a good quality dictionary for sparse coding should have a small mutual coherence constant μ. However, the μ of dictionaries obtained by K-SVD (which are highly redundant) is usually not small. Another problem is that in very redundant dictionaries, some atoms might be highly similar to others, or play an insignificant role. These arguments are used by [16] to promote, as better approach, orthogonal dictionaries with small μ. This paper also includes a convenient review.

2.4.3 Morphological Components

Consider the example depicted in Fig. 2.24, which is a synthetic image that may be similar to an astronomical picture (stars and filaments).

The stars could be well represented using wavelets, while for the lines ridgelets are more suitable. Then, it seems appropriate to use two dictionaries, wavelets and ridgelets, for representing the image. A decomposition of the image into two components would be possible, one component with the stars and the other with the lines.

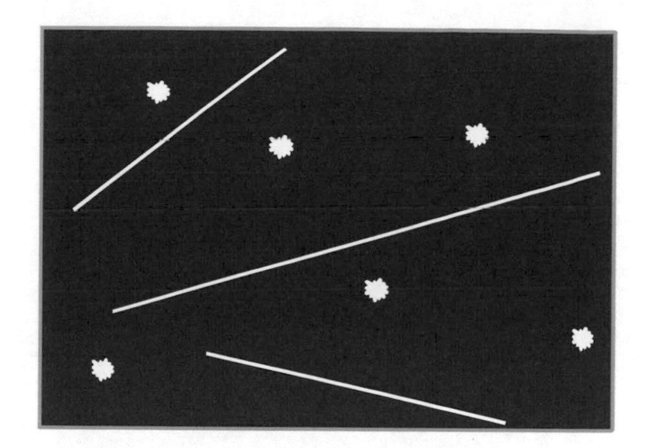

Fig. 2.24 A synthetic image

Although simplistic, this example illustrates well the idea of *morphological component analysis* (MCA), [172, 173]. As summarized in [24], the MCA method relies on an iterative thresholding algorithm, with a threshold that decreases linearly along iterations. Let us describe the algorithm, based on [24].

The case considered is a signal consisting of a sum of K signals $\mathbf{y}_i$, having different morphologies. A dictionary of bases $\{\Phi_1, \Phi_2, \ldots \Phi_K\}$ is assumed to exist. Signal $\mathbf{y}_1$ is sparse in Φ_1; signal $\mathbf{y}_2$ is sparse in Φ_2; and so on. Denote $\alpha_i = \Phi_i^T \mathbf{y}_i$.

Suppose, for simplicity, that one has only two signals ($K = 2$), so $\mathbf{y} = \mathbf{y}_1 + \mathbf{y}_2$ (the results can be easily generalized to more components). It is proposed, in order to estimate the components of $\mathbf{y}$, to solve the following minimization:

$$\min_{y_1,\, y_2} \left(\left\| \Phi_1^T \mathbf{y}_1 \right\|_1 + \left\| \Phi_2^T \mathbf{y}_2 \right\|_1 \right), \text{ subject to } \|\mathbf{y} - \mathbf{y}_1 - \mathbf{y}_2\|_2 \le \sigma$$

where σ is the noise standard deviation. Continuing with a simplified view, assume for now that $\sigma = 0$.

The first step of the MCA algorithm sets the number of iterations I_{max}, the minimum threshold λ_{min}, initial estimated values of $\mathbf{y}_1$ and $\mathbf{y}_2$, and the thresholds $\lambda_1^{(1)}$ and $\lambda_2^{(1)}$.

Then, iterations begin:

While the two thresholds are higher than a lower bound λ_{min}, do:

- $k = 2$
- Compute residuals, for $j = 1, 2$, using current estimates $\tilde{y}_i^{(k-1)}$ of the components:

$$r_j^{(k)} = \mathbf{y} - \tilde{y}_{i \neq j}^{(k-1)} \tag{2.98}$$

- Estimate, for $j = 1, 2$, current coefficients by hard thresholding:

$$\tilde{\alpha}_j^{(k)} = \delta_T\,(\Phi_j^T\, r_j^{(k)})\ ;\ T\ =\ \lambda_j^{(k)} \tag{2.99}$$

- Get, for $j = 1, 2$, new estimates $\tilde{y}_j^{(k)}$:

$$\tilde{y}_j^{(k)}\ =\ \Phi_j\, \tilde{\alpha}_j^{(k)} \tag{2.100}$$

- Decrease the thresholds

$$\lambda_j^{(k+1)}\ =\ \lambda_j^{(1)}\ -\ k\,\frac{\lambda_j^{(1)} - \lambda_{\min}}{I_{\max} - 1} \tag{2.101}$$

If there is no noise, $\lambda_{\min}$ should be set to zero. On the other hand, when there is noise, $\lambda_{\min}$ should be set to a few times σ (see [24]).

The MCA algorithm provides a good components separation when the Φ_i, the members of the dictionary, are mutually incoherent enough. In the examples provided by [173], textures are treated with DCT (discrete cosine transform) since DCT is suitable for the representation of natural periodicity (in case of non-homogeneous textures, local DCT could be used). As said before, lines are well represented with ridgelets. In addition, the curvelet transform represents well edges in images.

One of the contributions of [24] with respect to the original MCA, is a method called *'mean-of-max'* (MOM) for the decrease of thresholds. Denote:

$$m_j\ =\ \left\| \Phi_j^T\ r^{(k)} \right\|,$$

with: $r^{(k)}\ = \mathbf{y}\ -\ \mathbf{y}_1^{(k-1)}\ \ -\ \mathbf{y}_2^{(k-1)}$

Then, thresholds should be chosen as:

$$\lambda_j^{(k)}\ =\ \frac{1}{2}\,(m_1 + m_2) \tag{2.102}$$

It is also shown in [24] that MCA/MOM is clearly faster than BP (basis pursuit), being at least as efficient as BP in achieving sparse decompositions in redundant dictionaries.

See the book [174] for an extensive treatment of sparse representation and processing, including chapters on wavelets, ridgelets, curvelets, etc. In particular, the chapter 18 of that book focuses on MCA, using the MCALab package (in MATLAB) by the same authors. Several interesting application examples were presented.

A simple 1D example of MCA processing is the case of a signal composed of two sine signals with close frequencies, and three spikes at random positions. The target of MCA is to separate the signal morphological components. We chose this example and prepared a simplified version of MCALab for this case. Only two dictionaries were considered, one based on discrete cosine transform (DCT), and the other for

Dirac pulses. The simplified version is the Program 2.7. In this program, the initial value of σ was estimated using the finest details obtained by Daubechies wavelet.

During the execution of the program, the selection process is visualized with an animated figure. Figure 2.25 shows an example. When the program stops, another figure is shown with the original signal and its components (Fig. 2.26). Then, the user can see how good was the MCA work, by comparing the original and the separation result.

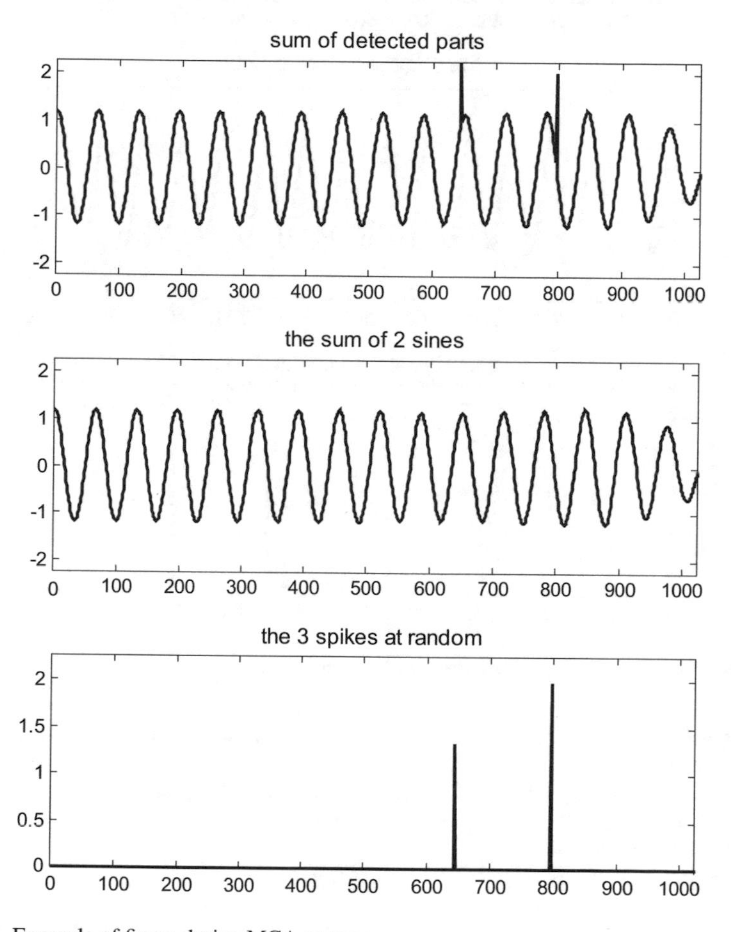

Fig. 2.25 Example of figure during MCA process

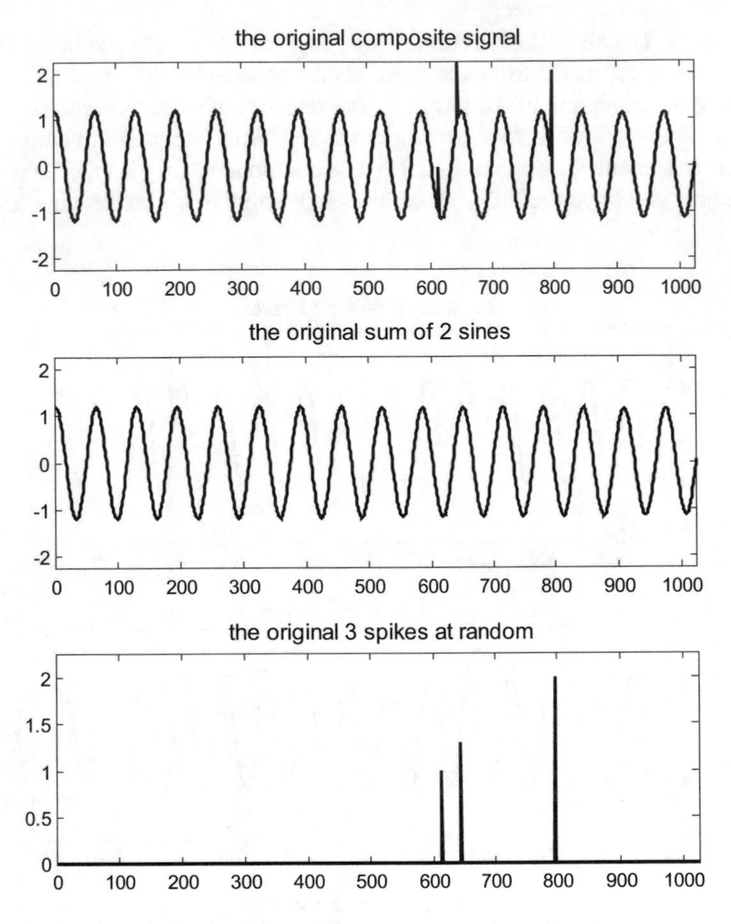

Fig. 2.26 The original composite signal and its components

Program 2.7 MCA example

```
% MCA example
% The signal is a mix of 2 sines and 3 random spikes
% signal synthesis ---------------------------
N=1024; %length
% sines have almost equal frequencies:
A=0.6; %sine amplitude
itv=1;
t=0:itv:N-itv;
s1=A*cos(pi*0.0307*t);
s2=A*cos(pi*0.0309*t);
ys=s1+s2; %sum of sines
% 3 spikes at random positions:
k=3;
v=[randn(k,1); zeros(N-k,1)]; p=randperm(N);
yk=v(p); pos=find(yk);
miny=min(yk); maxy=max(yk);
yk(pos) = (yk(pos)-miny)/(maxy-miny)+1; %normalize
```

```
% the composite signal
y = ys + yk';
signal=y(:); %column format
maxp=max(signal); %to set vertical limit in figures
nfg=0; %figure interval counter
% MCA parameters
qq=4; %to specify how fine is the approx
itermax = 200;
ndict=2; %number of dictionaries
%-----------------------------------------
%Initial variables
part= zeros(N,ndict);
%Initial sigma-----------------------------------
[Wc,Wl]=wavedec(signal,10,'db4'); %wavelet toolbox
L=sum(Wl(1:10)); %to select details
Dv=-Wc(L+1:end);
% MAD method
aux=Dv(find(~isnan(Dv)));
sigma = median(abs(aux - median(aux)))./0.6745;
%Initial coeffs (using two dictionaries: DCT and Dirac)
coeff={};
% DCT coeffs:
nv=N*qq; fq=(1:(nv-1))';
Ko=[N; 0.5*(N+sin(2*pi*fq/qq)./(2*sin(pi*fq/nv)))];
Ko=Ko.^0.5;
x=zeros(4*nv,1); L=2*N;
x(2:2:L)=signal(:);
z=fft(x);
coeff{1}= [struct('coeff',[]) struct('coeff',real(z(1:nv))./Ko)];
% Dirac coeffs:
coeff{2}= [struct('coeff',[]) struct('coeff', signal(:))];
%Initial threshold (minimum of maximal coeffs in each dictionary)
for nn=1:ndict
  aux = []; cfs = coeff{nn};
  sinx = length(cfs);
  for j = 2:sinx
    aux = [aux;cfs(j).coeff(:)];
  end
  buf(nn)=max(abs(aux(:)));
end
buf=flipud(sort(buf(:),1))';
deltamax=buf(2);
delta=deltamax;
lambda=delta/(itermax-1); %Linear decrease: slope
%Start the algorithm---------------------------
for iter=0:itermax-1
  %residual computation
  residual=signal-sum(part,2);
  % DCT part-------------
  Ra=part(:,1)+residual;
```

```
 %analysis:
 x=zeros(4*nv,1); L=2*N;
 x(2:2:L)=Ra(:); z=fft(x);
 Ca= [struct('coeff',[]) struct('coeff',real(z(1:nv))./Ko)];
 %thresholding (not the low frequency components):
 cf = Ca; ay=cf(2).coeff;
 cf(2).coeff=ay.*(abs(ay)>delta);
 Ca = cf;
 %synthesis:
 c=Ca(2).coeff; lc=length(c); M=lc/qq;
 fu=(1:(lc-1))';
 Ku=[M; 0.5*(M+sin(2*pi*fu/qq)./(2*sin(pi*fu/lc)))];
 Ku=Ku.^0.5;
 c=c./Ku;
 x=zeros(4*lc,1); L=2*M;
 x(1:lc)=c; z=fft(x);
 y=real(z(2:2:L));
 part(:,1)=y(:)/qq; %output
 % Dirac part--------------
 Ra=part(:,2)+residual;
 %analysis:
 Ca= [struct('coeff',[]) struct('coeff', Ra(:))];
 %thresholding (not the low frequency components):
 cf = Ca; ay=cf(2).coeff;
 cf(2).coeff=ay.*(abs(ay)>delta);
 Ca = cf;
 %synthesis:
 part(:,2)=Ca(2).coeff(:); %output
 % Update parameters---------------
 delta=delta-lambda; %linear decrease
 % Display along the process
 nfg=nfg+1;
 if nfg==4,
   nfg=0; %restart counter
   figure(1)
   subplot(3,1,1)
   plot(sum(part(1:N,:),2));axis tight;drawnow;
   title('sum of detected parts')
   axis([0 N -maxp maxp]);
   subplot(3,1,2)
   plot(part(1:N,1));axis tight;drawnow;
   title('the sum of 2 sines')
   axis([0 N -maxp maxp]);
   subplot(3,1,3)
   plot(part(1:N,2));axis tight;drawnow;
   title('the 3 spikes at random')
   axis([0 N 0 maxp]);
 end
end
```

```
part = part(1:N,:);
% Final display---------------------------------
%Original signals
figure(2)
subplot(3,1,1)
plot(signal);axis tight;
title('the original composite signal')
axis([0 N -maxp maxp]);
subplot(3,1,2)
plot(ys);axis tight;
title('the original sum of 2 sines')
axis([0 N -maxp maxp]);
subplot(3,1,3)
plot(yk);axis tight;
title('the original 3 spikes at random')
axis([0 N 0 maxp]);
```

2.5 An Additional Repertory of Applicable Concepts and Tools

Some of the topics to be treated in this chapter require new concepts and tools. For instance, the functions provided by MATLAB for sparse calculus, and concepts and solution techniques related to certain optimization or signal processing problems. In particular, it has been found that the Bregman algorithms are very suitable for optimal compressed sensing and other applications.

2.5.1 *Sparse Representation in MATLAB*

There are two matrix storage modes in MATLAB. *Full* storage is the default and stores the value of each element. *Sparse* storage –that should be explicitly invoked-stores only the values of nonzero elements.

The function *sparse( )* can be used to create a sparse matrix. For example, if you write:

$$S = sparse([], [], [], 1000, 1000];$$

An empty sparse matrix S is created. Then, you can specify some nonzero elements, for instance:

$$S(5, 3) = 15;\ S(312, 1) = 120.5;\ S(25, 25) = 1;$$

The function *nnz(B)* returns the number of nonzero elements of the matrix B, which can be in full or in sparse format. The function *find(B)* returns all (i, j) indices of nonzero elements. The function *nonzeros(B)* returns all the nonzero elements.

As with full matrices, with sparse matrices you also can use the expression:

$$x = A \setminus b$$

In many applications is more convenient the use of sparse matrices than full matrices. For instance, in case of having a 10,000 × 10,000 matrix with 20,000 nonzero elements. Solving $Ax = b$ will be much faster with A sparse matrix than with A full matrix.

- A full matrix B can be converted to sparse with *S=sparse(B)*. A sparse matrix S can be converted to full with *B=full(S)*.
- The function *spdiags( )* can be used to create sparse banded matrices, which have a few nonzero diagonals.
- The identity matrix is created with *speye( )*. Given a sparse matrix S, the function *spones( )* replaces the nonzero elements with ones.
- A simple visualization of the nonzero elements localization in the matrix S can be obtained with *spy(S)*.

Many more details of the use of sparse matrices in MATLAB can be found in [91].

As a first, simple example, the Program 2.8 creates a tri-banded sparse matrix, that you can list using *full(A)* without semicolon, and then applies *spy( )* to display (Fig. 2.27) the matrix structure (non-zero entries).

Program 2.8 Create a tri-banded sparse matrix

```
%Create a tri-banded sparse matrix
%
A=spdiags([2 1 0; 5 4 3; 8 7 6; 11 10 9; 0 13 12],-1:1,5,5);
full(A); %list the matrix in usual format
figure(1)
spy(A); %visualize the matrix structure (non-zero entries)
%
title('spy diagram');
```

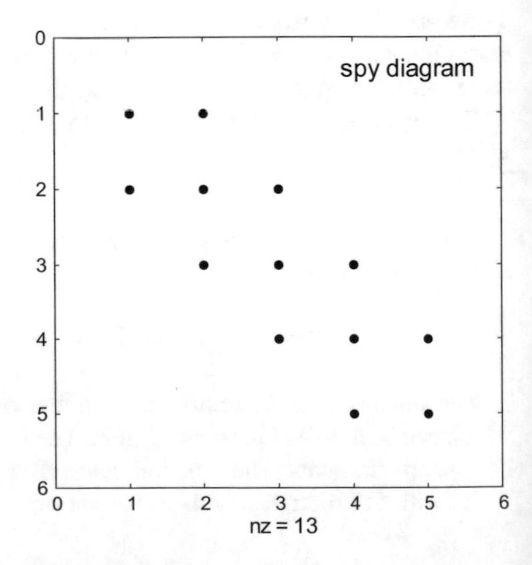

Fig. 2.27 Visualization of banded matrix using spy()

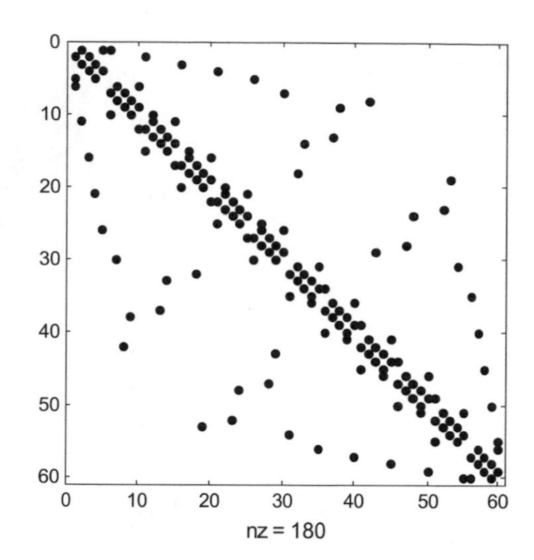

Fig. 2.28 Visualization of the Bucky ball matrix structure

Since it is a popular example, MATLAB includes data for the Bucky ball, which is composed of 60 points distributed as in a soccer ball. Each point has three neighbours. The Bucky ball models the geodesic dome made popular by Buckminster Fuller, and also the C60 molecule (a carbon molecule with 60 atoms). The Bucky ball adjacency matrix is a 60×60 symmetric matrix.

Figure 2.28 shows the *spy( )* visualization of the Bucky ball matrix.

From the data given by MATLAB it is also possible to get coordinates, and to plot with *gplot( )* the Bucky ball. Both Figs. 2.28 and 2.29 were generated with the Program 2.9.

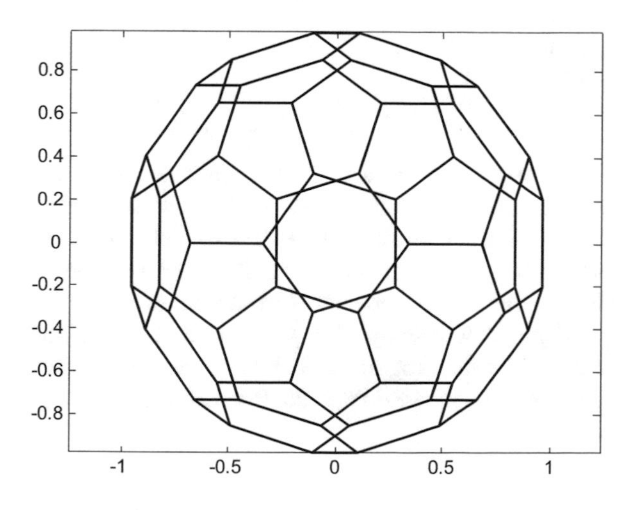

Fig. 2.29 Visualization of the bucky ball graph

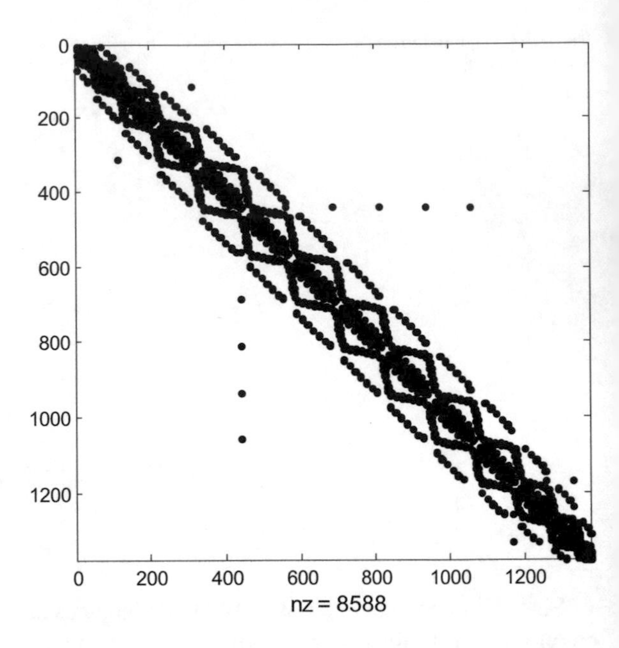

Fig. 2.30 Visualization of HB/nnc1374 matrix using spy()

Program 2.9 Load and visualize the bucky matrix

```
%Load and visualize the bucky matrix
%
A=bucky; %rapid loading of the 60x60 bucky matrix
figure(1)
spy(A,'k');
title('the bucky ball matrix');
%
[A,v]=bucky; %load the bucky matrix and the coordinates
figure(2)
gplot(A,v,'k');
axis equal;
title('the bucky ball graph');
```

Among the sources of data on Internet, there are two important collections of sparse matrices, one in the 'Matrix Market' and the other in the University of Florida Sparse Matrix Collection. From this second source, we chose the matrix HB/nnc1374 as an example. It is a matrix with *1374* × *1374* entries, and only 8588 non-zero entries. Notice that earlier versions of MATLAB cannot handle such a large matrix.

Figure 2.30 displays the *spy( )* diagram of this matrix. The figure has been generated with the Program 2.10, which shows how to extract the matrix from the data structure downloaded from the Internet repository.

Program 2.10 Load and visualize HB nnc1374

```
%Load and visualize HB nnc1374
%(note: earlier MATLAB versions cannot handle the large matrix)
%
S=load('nnc1374.mat'); %file (structure) loading
fieldnames(S) %to see structure field names
getfield(S,'Problem') %to see inside the field 'Problem'
M=getfield(S,'Problem','A'); %get the sparse matrix from 'A'
figure(1)
spy(M,'k');
title('the HB nnc1374 matrix');
```

2.5.2 *Diffusion in 2D*

Let us begin with a kind of strange example. Suppose you want to paint your inclined roof. You could come to the top edge and just drop the paint, letting the colour flow down. After some evolution time, you could expect some uniformity. Of course, this is not a recommended method, but it gives you the flavour of what will be introduced next for denoising, impainting and other applications. Notice that the colour flow will be mostly unidirectional as governed by the slope.

Another example would be the following: you take a thin aluminium plate, put a flame below and near the centre during some time, and observe how the heat flows from the centre toward the borders of the plate. The heat diffusion will be omnidirectional.

As will be seen next, the heat diffusion can be related with 2D Gaussian filtering, considering the image intensity as analogous to energy. The image filtering would take some time (some iterations), to let the diffusion evolve.

In the case of denoising, one wants to eliminate the noise using diffusion, but, at the same time, one wants to preserve edges. Is this possible? This also will be treated next.

Since diffusion directions would become important, the proper mathematical tool should be partial differential equations (PDE). Let us advance some important basic equations concerning diffusion and heat.

Denote $\phi(x, y, t)$ the density of the diffusing substance at a given position and time. The diffusion equation is:

$$\frac{\partial \phi}{\partial t} = \nabla \left[D(\phi, x, y, t) \cdot \nabla \phi \right] \tag{2.103}$$

where $D()$ is the *diffusion coefficient* (or diffusivity). This coefficient could be a scalar, a scalar function of coordinates (non-homogeneous diffusion), or a tensor (which could correspond to anisotropic diffusion). In 2D this tensor is a symmetric positive definite matrix. The equation becomes non-linear if the coefficient depends on ϕ.

The heat equation is obtained with a $D()$ being a constant:

$$\frac{\partial \phi}{\partial t} = D \cdot \nabla^2 \phi \tag{2.104}$$

In order to discretize the equation for computing the heat diffusion on a grid, one could approximate the second derivatives as follows:

$$\frac{\partial^2 \phi}{\partial x^2} = \frac{\phi(x_{i+1}, y_j) - 2\phi(x_i, y_j) + \phi(x_{i-1}, y_j)}{(dx)^2} \tag{2.105}$$

$$\frac{\partial^2 \phi}{\partial y^2} = \frac{\phi(x_i, y_{j+1}) - 2\phi(x_i, y_j) + \phi(x_i, y_{j-1})}{(dy)^2} \tag{2.106}$$

These expressions are to be introduced in: $\nabla^2 \phi = \frac{\partial^2 \phi}{\partial x^2} + \frac{\partial^2 \phi}{\partial y^2}$

For the first derivative, one could use a first-order approximation:

$$\frac{\partial \phi}{\partial t} = \frac{\phi(t + dt) - \phi(t)}{dt} \tag{2.107}$$

A simple example has been devised, in which a central region of a plate is heated, and then the heat diffusion takes place along time. The computation of the diffusion on a grid has been done with the Program 2.11. For a simpler notation, the heat has been denoted as u. Figure 2.31 shows a sequence of plots, from left to right and top to bottom, corresponding to the process evolution. The program execution may take several minutes.

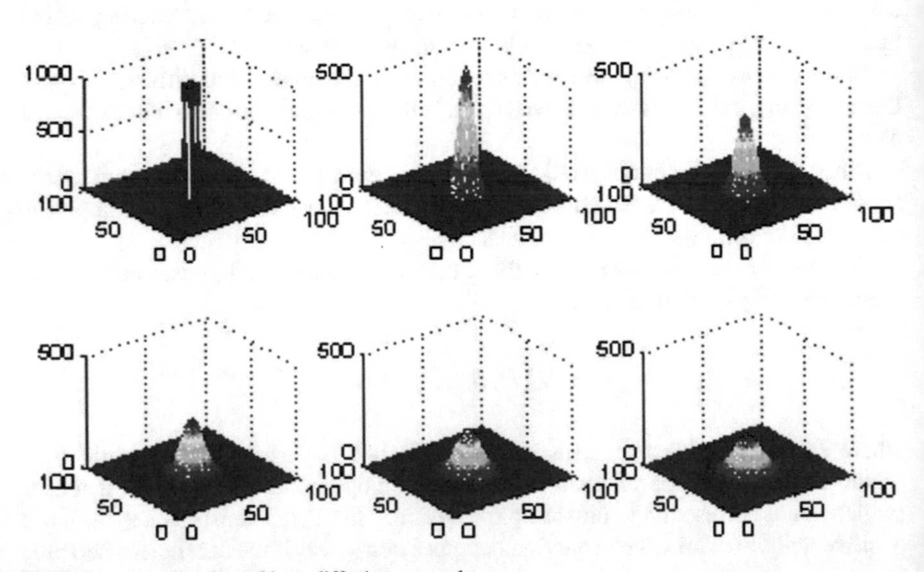

Fig. 2.31 Visualization of heat diffusion example

Program 2.11 Example of heat diffusion

```
%Example of heat diffusion
%
%Initialization of variables
D=0.25; %diffusion constant
dx=1; dy=dx; %we will use a regular grid, with dy=dx
dt=1;
x=0:dx:99; y=0:dy:99; %the grid
lx=length(x); ly=length(y);
u=zeros(lx,ly); un=u;
u(46:54,46:54)=1000; %central region is heated
%first display
figure(1)
subplot(2,3,1);
mesh(u);
axis([0 100 0 100 0 1000]);
title('heat diffusion');
for nn=2:6,
  for T=1:dt:30, %time
    for i=2:ly-1,
      for j=2:lx-1,
        aux=u(i+1,j)+u(i-1,j)+u(i,j+1)+u(i,j-1)-(4*u(i,j));
        un(i,j)=u(i,j)+((D*dt*aux)/dx^2);
      end;
    end;
    u=un;
  end
  subplot(2,3,nn); %the other plots
  mesh(un);
  axis([0 100 0 100 0 500]);
end;
```

When using the approximations already described, it is important to keep:

$$\frac{2\,D \cdot dt}{\min((dx)^2, (dy)^2)} \leq 1 \tag{2.108}$$

otherwise, the numerical scheme becomes unstable (as you may want to check).

2.5.2.1 Gaussian Diffusion

In the case of heat diffusion with homogeneous Neumann boundary conditions and $D = 1$, one has the following problem:

$$\frac{\partial \phi}{\partial t} = \nabla^2 \phi \tag{2.109}$$

$$\phi(x, y, 0) = \phi_0 \tag{2.110}$$

$$\frac{\partial \phi}{\partial \mathbf{n}} = 0\,, \ (x, y) \in b(\Omega) \tag{2.111}$$

where $b(\Omega)$ is the boundary of the region of interest Ω (usually a rectangle in the case of a picture), and $\mathbf{n}$ is the normal to this boundary.

It has been established that the solution of this problem is given by the convolution of ϕ_0 and the Gaussian function with $\sigma = \sqrt{2t}$. In the Fourier domain, the solution is:

$$\Phi(\omega) = \exp\left(\frac{-\,|\omega|^2}{2/\,\sigma^2}\right)\,\Phi_0(\omega) \tag{2.112}$$

Therefore, the Gaussian diffusion is equivalent to a special low-pass filter. In the case of a picture, the effect would be image blurring. This blurring does not respect edges, so the structure is lost.

Figure 2.32 shows the effect of Gaussian diffusion on a picture. On top, the original picture; in the middle, the image after 10 s of diffusion; in the bottom, the image

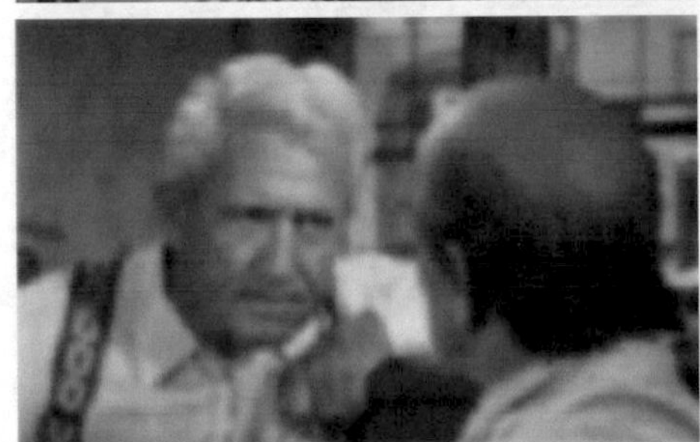

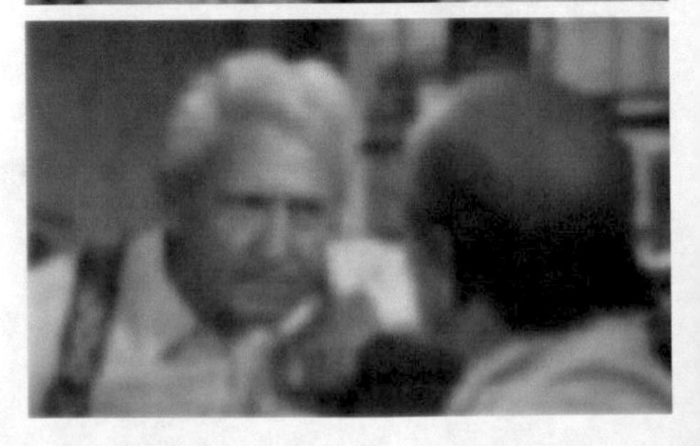

Fig. 2.32 Effect of Gaussian diffusion, original on top

after another 10 s of diffusion. The processing has been done with the Program 2.12, which is similar to the previous program for heat diffusion.

Program 2.12 Example of picture diffusion blurring

```
%Example of picture diffusion blurring
% (like heat diffusion)
%
%Initialization of variables
D=0.25; %diffusion constant
dx=1; %we will use a regular grid, with dy=dx
dt=1;
x=0:dx:399; y=0:dy:249; %the grid (image size)
lx=length(x); ly=length(y);
P=imread('spencer.jpg'); %read image
u=double(P);
un=u;
%first display
figure(1)
subplot(3,1,1);
imshow(uint8(u));
title('Gaussian diffusion');
for nn=2:3,
  for T=1:dt:10, %time
    for i=2:ly-1,
      for j=2:lx-1,
        aux=u(i+1,j)+u(i-1,j)+u(i,j+1)+u(i,j-1)-(4*u(i,j));
        un(i,j)=u(i,j)+((D*dt*aux)/dx^2);
      end;
    end;
    u=un;
  end
  subplot(3,1,nn); %the other plots
  imshow(uint8(un));
end;
```

Although it may then easily become unstable, the diffusion process could be reversed in order to sharpen (or deblur) the image. Figure 2.33 shows an example, with the original picture on top and the sharpened image below. Notice in Program 2.13 that the diffusion constant and the diffusion time have been decreased.

Program 2.13 Example of picture anti-diffusion sharpening

```
%Example of picture anti-diffusion sharpening
% (reverse heat diffusion)
%
%Initialization of variables
D=0.05; %diffusion constant
dx=1; %we will use a regular grid, with dy=dx
dt=1;
```

```
x=0:dx:399; y=0:dy:249; %the grid (image size)
lx=length(x); ly=length(y);
P=imread('spencer.jpg'); %read image
u=double(P);
un=u;
%first display
figure(1)
subplot(2,1,1);
imshow(uint8(u));
title('Gaussian anti-diffusion');
for T=1:dt:5, %time
  for i=2:ly-1,
    for j=2:lx-1,
      aux=u(i+1,j)+u(i-1,j)+u(i,j+1)+u(i,j-1)-(4*u(i,j));
      un(i,j)=u(i,j)-((D*dt*aux)/dx^2);
    end;
  end;
  u=un;
end
subplot(2,1,2); %the other plot
imshow(uint8(un));
```

Fig. 2.33 Effect of Gaussian anti diffusion, original on top

2.5.2.2 The Perona-Malik Diffusion

In 1987 Perona and Malik introduced a celebrated model [149], see also [150], that has been cited by more than eight thousand papers. The target was to protect, and even improve, the edges while smoothing more homogeneous regions of the picture.

The diffusion coefficient $D()$ will depend on the local image gradient. For small gradients, corresponding to homogeneous regions, large values of $D()$ are allowed, promoting stronger smoothing. On the other hand, for large gradients, corresponding to edges, smaller values of $D()$ are used to slow down the diffusion or even force the diffusion to go backwards.

Using now the notation commonly employed for images ($u()$ instead of $\phi()$), the Perona-Malik formulation of the diffusion becomes:

$$\frac{\partial u}{\partial t} = \nabla\,[D(|\nabla u|)\nabla\, u] \tag{2.113}$$

$$u(x, y, 0) = u_0 \tag{2.114}$$

$$\frac{\partial u}{\partial \mathbf{n}} = 0\,,\;\; (x, y) \in b(\Omega) \tag{2.115}$$

Two different choices of $D(s)$ where suggested:

$$D(s) = \frac{1}{1 + s^2/\lambda^2} \tag{2.116}$$

$$D(s) = \exp\,(-\,s^2/\lambda^2) \tag{2.117}$$

(substitute s with ∇ where appropriate)

The *flux function* is defined as $\varphi(\nabla u) = [D(|\nabla u|)\nabla\, u]$. The derivative of this function tells you if there is forward diffusion, or backward diffusion. If you selected (2.116) for $D(s)$, then:

$$\varphi'(\nabla u) \begin{cases} < 0 \; if \; |s| > \lambda \\ > 0 \; if \; |s| < \lambda \end{cases} \tag{2.118}$$

Therefore, near edges (large gradient) there is backward diffusion that will enhance these edges.

In the simple one-dimensional diffusion case, one has:

$$\frac{\partial u}{\partial t} = \varphi'(\nabla u)\, \Delta\, u \tag{2.119}$$

Figure 2.34 shows the one-dimensional diffusion coefficient and the flux function for $D(s)$ given by (2.116), with $\lambda = 3$.

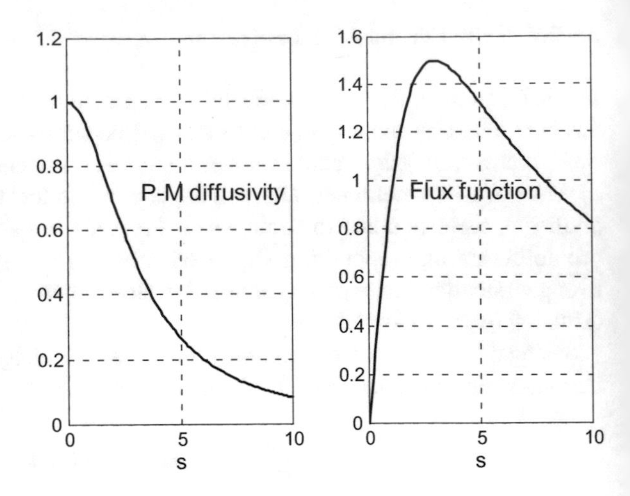

Fig. 2.34 The diffusion coefficient and the corresponding flux function

Program 2.14 Perona-Malik diffusivity function and flux function

```
%Perona-Malik diffusivity function and flux function
% (first alternative)
%
lambda=3;
s=0:0.1:10; %variable values
g=1./(1+(s.^2/lambda^2));
f=s.*g;
figure(1)
subplot(1,2,1);
plot(s,g,'k');
xlabel('s'); grid;
title('P-M diffusivity');
axis([0 10 0 1.2]);
subplot(1,2,2);
plot(s,f,'k');
xlabel('s'); grid;
title('Flux function');
axis([0 10 0 1.6]);
```

An extensive research has been devoted to the proposal contained in the Perona-Malik paper. See [184] for a detailed treatment of the topic with abundant references. Also in [184] an anisotropic diffusion was introduced using a tensorial diffusion coefficient.

The Perona-Malik method can be implemented in several discretization ways. Program 2.15 represents an example of implementation that has fast execution time. Figure 2.35 shows the result of the method for the denoising of a picture having salt & pepper noise (this noise has been added using *imnoise( )*. The noise has been softened while keeping image features.

Fig. 2.35 Denoising of image with salt & pepper noise, using P-M method

Program 2.15 Example of picture denoising using Perona-Malik diffusion

```
%Example of picture denoising using Perona-Malik diffusion
lambda=160; %constant for P-M diffusivity
D=0.1; %general diffusion constant
P=imread('face1.jpg'); %read image
A=imnoise(P,'salt&pepper',0.01); %add salt&pepper noise
u=double(A); un=u;
[ly,lx]=size(u);
for nn=1:8,
  %zero padding around image
  udif=zeros(ly+2,lx+2);
  udif(2:ly+1,2:lx+1)=u;
  %differences: north, south, east, west
  difN=udif(1:ly,2:lx+1)-u;
  difS=udif(3:ly+2,2:lx+1)-u;
  difE=udif(2:ly+1,3:lx+2)-u;
  difW=udif(2:ly+1,1:lx)-u;
  %Diffusivities
  DN=1./(1+(difN/lambda).^2);
  DS=1./(1+(difS/lambda).^2);
  DE=1./(1+(difE/lambda).^2);
  DW=1./(1+(difW/lambda).^2);
  %diffusion
  un=u+D*(DN.*difN + DS.*difS + DE.*difE + DW.*difW);
  u=un;
end;
%display:
figure(1)
subplot(1,2,1)
imshow(A);
xlabel('original');
```

```
title('Salt&pepper denoising');
subplot(1,2,2)
imshow(uint8(un));
xlabel('denoised image');
```

2.5.3 *Bregman-Related Algorithms*

A convenient iterative method for finding extrema of convex functions was proposed by Bregman in 1967, [28]. Later on, in 2005, it was shown by Osher et al. [142] that this method was very appropriate for image processing (in particular for total variation applications).

The method is based on the Bregman divergence, and it can be employed in its basic iterative version, or as split Bregman iteration [95].

2.5.3.1 Bregman Divergence

Aspects concerning divergence, distances, similarity, etc. have already been considered in the chapter on data analysis and classification. They are crucial for important applications, like for instance face recognition (and distinction between faces).

An expression of the Bregman divergence between two points $\mathbf{x}$ and $\mathbf{x}_0$ would be the following:

$$D(\mathbf{x},\,\mathbf{x}_0) = f(\mathbf{x}) - f(\mathbf{x}_0) - \nabla f(\mathbf{x}_0)^T \cdot (\mathbf{x} - \mathbf{x}_0) \tag{2.120}$$

This concept was introduced by Bregman for differentiable convex functions, and was given the name '*Bregman distance*' by Censor and Lent [46] in 1981.

Figure 2.36 gives a graphical interpretation of this distance D, sitting above the tangent L.

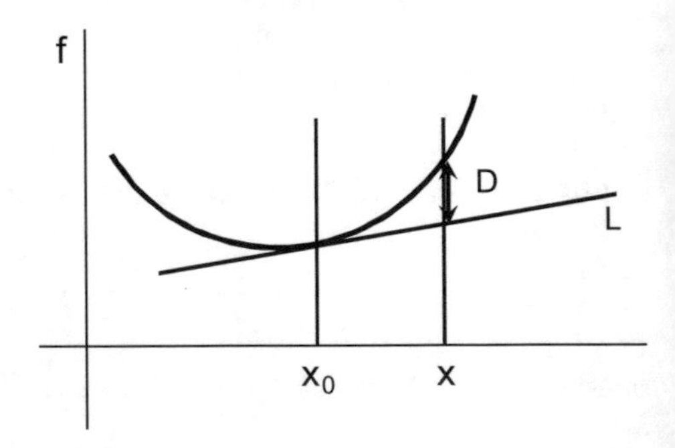

Fig. 2.36 Example of Bregman distance

Notice that a first-order Taylor expansion to approximate $f(\mathbf{x})$, would be $f_a(\mathbf{x}) = f(\mathbf{x}_0) + \nabla f(\mathbf{x}_0)^T(\mathbf{x} - \mathbf{x}_0)$, and therefore the distance D is the difference: $D(\mathbf{x} - \mathbf{x}_0) = f(\mathbf{x}) - f_a(\mathbf{x})$.

Distances are non-negative functions. A metric distance has symmetry, identity and triangle inequality properties. Bregman distances are not necessarily symmetric, nor supporting triangle inequality.

Particular cases of Bregman distance are the Euclidean distance, with the choice $f(\mathbf{x}) = \|\mathbf{x}\|^2$, the Kullback-Leibler divergence, with $f(\mathbf{x}) = \sum -x_i \log x_i$, and the Itakura-Saito distance, with $f(\mathbf{x}) = \sum -\log x_i$. The Mahalanobis distance is also a particular case.

Other equivalent expression of the Bregman distance is the following:

$$D(\mathbf{x},\, \mathbf{x}_0) = f(\mathbf{x}) - f(\mathbf{x}_0) - \ < \nabla f(\mathbf{x}_0),\, \mathbf{x} - \mathbf{x}_0 > \tag{2.121}$$

Recently, some important authors, [30, 95], used the following expression:

$$D(\mathbf{x},\, \mathbf{x}_0) = f(\mathbf{x}) - f(\mathbf{x}_0) - \ < p,\, \mathbf{x} - \mathbf{x}_0 > \tag{2.122}$$

where p is the subgradient at $\mathbf{x}_0$.

There are interesting connections of Bregman divergences with clustering, exponential functions, and information theory [15]; or in particular with Voronoi diagrams [25]. It is also worthwhile to mention [7] for a more complete view on the concept and applications of divergence.

2.5.3.2 Bregman Iteration

Consider the following constrained minimization problem:

$$\min_u J(\mathbf{u}),\ \text{subject to } H(\mathbf{u}) = 0$$

where J and H are defined in $\Re^n$ and convex. The associated unconstrained problem is:

$$\min_u J(\mathbf{u}) + \lambda H(\mathbf{u}) \tag{2.123}$$

The Bregman distance of J between $\mathbf{u}$ and $\mathbf{v}$ would be:

$$D(\mathbf{u},\, \mathbf{v}) = J(\mathbf{u}) - J(\mathbf{v}) - \ < p, \cdot\, \mathbf{u} - \mathbf{v} > \tag{2.124}$$

The *Bregman iteration* for solving the minimization problem is:

$$\mathbf{u}^{(k+1)} = \arg\min_u D^{(k)}(\mathbf{u},\, \mathbf{u}^{(k)}) + \lambda H(\mathbf{u}) \tag{2.125}$$

Suppose that H is differentiable. In this case, the Bregman iteration would be:

$$\mathbf{u}^{(k+1)} = \arg\min_{u} D^{(k)}(\mathbf{u}, \mathbf{u}^{(k)}) + \lambda H(\mathbf{u}) \tag{2.126}$$

$$p^{(k+1)} = p^{(k)} - \lambda \nabla H(\mathbf{u}^{(k+1)}) \tag{2.127}$$

For the particular problem of TV based image denoising [142], one could take:

$$H(\mathbf{u}, \mathbf{f}) = \frac{1}{2} \|A\,\mathbf{u} - \mathbf{f}\|_2^2 \tag{2.128}$$

then, the iteration can be expressed in the following simplified form:

$$\mathbf{u}^{(k+1)} = \arg\min_{u} J(\mathbf{u}) + \lambda H(\mathbf{u}, \mathbf{f}^{(k)}) \tag{2.129}$$

$$\mathbf{f}^{(k+1)} = \mathbf{f}^{(k)} + (\mathbf{f} - A\,\mathbf{u}^{(k+1)}) \tag{2.130}$$

It has been verified that these iterations converge very quickly.

2.5.3.3 Linearized Bregman Iteration

In appropriate cases, it would be possible to use the following approximation:

$$H(\mathbf{u}) \approx H(\mathbf{u}^{(k)}) + \nabla H(\mathbf{u}^{(k)}) \cdot (\mathbf{u} - \mathbf{u}^{(k)}) \tag{2.131}$$

For a better approximation a penalty quadratic term could be added, and then:

$$\mathbf{u}^{(k+1)} = \arg\min_{u} J(\mathbf{u}) + \left\langle \lambda H(\mathbf{u}^{(k)}) - p^{(k)}, \mathbf{u} \right\rangle + \frac{1}{2\delta} \left\| \mathbf{u} - \mathbf{u}^{(k)} \right\|_2^2 \tag{2.132}$$

2.5.3.4 Split Bregman Iteration

A characteristic aspect of split Bregman method is that it could separate the typical $l1$ and $l2$ portions of important image processing approaches. For instance, suppose that the minimization problem is:

$$\min_{u} \|\mathbf{d}\|_1 + H(\mathbf{u}) + \frac{\lambda}{2} \|\mathbf{d} - \phi(\mathbf{u})\|_2^2 \tag{2.133}$$

The split Bregman iteration for solving this problem is:

$$\mathbf{u}^{(k+1)} = \arg\min_{u} \; H(\mathbf{u}) + \frac{\lambda}{2} \parallel \mathbf{d}^{(k)} - \phi(\mathbf{u}) - \mathbf{b}^{(k)} \parallel_2^2 \tag{2.134}$$

$$\mathbf{d}^{(k+1)} = \arg\min_{d} \; \parallel \mathbf{d} \parallel_1 + \frac{\lambda}{2} \parallel \mathbf{d} - \phi(\mathbf{u}^{(k+1)}) - \mathbf{b}^{(k)} \parallel_2^2 \tag{2.135}$$

$$\mathbf{b}^{(k+1)} = \mathbf{b}^{(k)} + (\phi(\mathbf{u}^{(k+1)}) - \mathbf{d}^{(k+1)}) \tag{2.136}$$

As you can see, the method involves two optimization steps and a simple update of **b**. The first optimization step is differentiable and so it can be solved by a variety of methods, like Gauss-Seidel or conjugate gradient, or even in the Fourier domain. The second optimization step can be computed by shrinkage, that is:

$$\mathbf{d}_j^{(k+1)} = shrink(\phi(\mathbf{u})_j + \mathbf{b}_j^{(k)}, \; 1/\lambda) \tag{2.137}$$

where the shrinkage would be:

$$shrink(\mathbf{x}, \lambda) = \frac{\mathbf{x}}{|\mathbf{x}|} * \max(|\mathbf{x}| - \lambda), 0) \tag{2.138}$$

2.5.3.5 Using Split Bregman for ROF-TV Image Denoising

It has been shown in [95] how to apply the split-Bregman method for total variation (TV) image denoising, based on the ROF model. The method is simple and efficient. It handles two-dimensional variables.

There are two denoising formulations: the anisotropic problem, and the isotropic problem. In the *anisotropic* problem, one has to solve:

$$\min_{u} \; |\nabla_x u| + |\nabla_y u| + \frac{\mu}{2} \parallel u - f \parallel_2^2 \tag{2.139}$$

Let us replace $\nabla_x u$ by d_x and $\nabla_y u$ by d_y. In order to strengthen the constraints, two penalty terms were added:

$$\min_{u} \; |d_x| + |d_y| + \frac{\mu}{2} \parallel u - f \parallel_2^2 + \frac{\lambda}{2} \parallel d_x - \nabla_x u \parallel + \frac{\lambda}{2} \parallel d_y - \nabla_y u \parallel \tag{2.140}$$

Coming now to the split Bregman algorithm, the first optimization step would be:

$$\begin{aligned} u^{(k+1)} = \arg\min_{u} \; & \tfrac{\mu}{2} \parallel u - f \parallel_2^2 + \tfrac{\lambda}{2} \parallel d_x^{(k)} - \nabla_x u - b_x^{(k)} \parallel + \\ & + \tfrac{\lambda}{2} \parallel d_y - \nabla_y u - b_y^{(k)} \parallel \end{aligned} \tag{2.141}$$

Because the system is strictly diagonal, it is recommended in [95] to get the solution with the Gauss-Seidel method:

$$u_{i,j}^{(k+1)} = G_{i,j}^{(k)} = \frac{\mu}{\mu+4\lambda} f_{i,j} + \frac{\lambda}{\mu+4\lambda} \{u_{i+1,j}^{(k)} + u_{i-1,j}^{(k)} + u_{i,j+1}^{(k)} + u_{i,j-1}^{(k)} + \\ + d_{x,i-1,j}^{(k)} - d_{x,i,j}^{(k)} + d_{y,i,j-1}^{(k)} - d_{y,i,j}^{(k)} - b_{x,i-1,j}^{(k)} + b_{x,i,j}^{(k)} - b_{y,i,j-1}^{(k)} + b_{y,i,j}^{(k)} \}$$

Then, the complete split-Bregman algorithm, to be iterated, is:

$$u^{(k+1)} = G^{(k)} \tag{2.142}$$

$$d_x^{(k+1)} = shrink(\nabla_x u^{(k+1)} + b_x^{(k)},\ 1/\lambda) \tag{2.143}$$

$$d_y^{(k+1)} = shrink(\nabla_y u^{(k+1)} + b_y^{(k)},\ 1/\lambda) \tag{2.144}$$

$$b_x^{(k+1)} = b_x^{(k)} + (\nabla_x u^{(k+1)} - d_x^{k+1}) \tag{2.145}$$

$$b_y^{(k+1)} = b_y^{(k)} + (\nabla_y u^{(k+1)} - d_y^{k+1}) \tag{2.146}$$

In the case of *isotropic* denoising, the minimization problem is:

$$\min_u \sum \sqrt{(\nabla_x u)_i^2 + (\nabla_{\dot{x}} u)_i^2} + \frac{\mu}{2} \| u - f \|_2^2 \tag{2.147}$$

Like before, the problem is transformed to:

$$\min_u \sum_{i,j} \sqrt{d_{x,i,j}^2 + d_{y,i,j}^2} + \frac{\mu}{2} \| u - f \|_2^2 + \frac{\lambda}{2} \| d_x - \nabla_x u \| + \frac{\lambda}{2} \| d_y - \nabla_y u \| \tag{2.148}$$

Due to the coupling between d_x and d_y, the second optimization step is decomposed into two shrinkages. Hence, the split-Bregman algorithm, to be iterated, is the following:

$$u^{(k+1)} = G^{(k)} \tag{2.149}$$

$$d_x^{(k+1)} = \max(s^k - 1/\lambda, 0) \frac{\nabla_x u^{(k)} + b_x^{(k)}}{s^k} \tag{2.150}$$

$$d_y^{(k+1)} = \max(s^k - 1/\lambda, 0) \frac{\nabla_y u^{(k)} + b_y^{(k)}}{s^k} \tag{2.151}$$

$$b_x^{(k+1)} = b_x^{(k)} + (\nabla_x u^{(k+1)} - d_x^{k+1}) \tag{2.152}$$

Fig. 2.37 ROF total variation denoising using split Bregman

$$b_y^{(k+1)} = b_y^{(k)} + (\nabla_y u^{(k+1)} - d_y^{k+1}) \tag{2.153}$$

where:

$$s^k = \sqrt{\left|\nabla_x u^{(k)} + b_x^{(k)}\right|^2 + \left|\nabla_y u^{(k)} + b_y^{(k)}\right|^2} \tag{2.154}$$

According with [30], the shrinking steps could be approximated as follows:

$$d_x^{(k+1)} = \frac{s^k \lambda (\nabla_x u^{(k)} + b_x^{(k)})}{s^k \lambda + 1} \tag{2.155}$$

$$d_y^{(k+1)} = \frac{s^k \lambda (\nabla_y u^{(k)} + b_y^{(k)})}{s^k \lambda + 1} \tag{2.156}$$

Figure 2.37 shows an example of ROF-TV anisotropic denoising. It is the same example as before, with salt and pepper noise. The program is based on the implementation in [30]. It has been included in the appendix on long programs. Actually, the program is not complicated, but the discretization implies many loops; surely the algorithm could be implemented in more compact form, using matrices.

2.6 Matrix Completion and Related Problems

A matrix could be regarded as a set of signal samples. Therefore, it can be treated from the compressed sensing point of view. For instance, it may correspond to the measurements taken at a given time by a network of spatially distributed sensors, in which some sensors could fail.

The consideration of matrices has originated a fruitful new research area, which is very active and expansive. This section introduces some fundamental aspects of this area.

2.6.1 Matrix Completion

In a most cited article, [40], Candés and Recht introduced a topic of considerable interest: the recovery of a data matrix from a sampling of its entries. A number m of entries are chosen uniformly at random from a matrix M, and the question is whether it is possible to entirely recover the matrix M from these m entries.

The topic was naturally introduced by extension of compressed sensing, in which a sparse signal is recovered from some samples taken at random. In the case of matrices, if the matrix M has low rank or approximately low rank, then accurate and even exact recovery from random sampling is possible by nuclear norm minimization [40].

By the way, it is now convenient to quote a certain set of norms, denoted as *Schatten-p norms*, for $p = 1, 2$ or ∞. In particular:

- Spectral norm:

$$\text{The largest singular value: } \|X\|_S = \max(\sigma_i(X) = \|\sigma(X)\|_\infty$$

- Nuclear norm:

$$\text{The sum of singular values: } \|X\|_* = \sum_{i=1}^{N} \sigma_i(X) = \|\sigma(X)\|_1$$

- Frobenius norm:

$$\|X\|_F = \left(\sum_{i=1}^{N} \sum_{j=1}^{M} X_{ij}^2 \right)^{1/2} = \left(\sum_{i=1}^{t} \sigma_i^2(X) \right)^{1/2} = \|\sigma(X)\|_2$$

Using a simple example of a $n \times n$ matrix M that has all entries equal to zero except for the first row, it is noted in [40] that this matrix cannot be recovered from a subset of its entries. There are also more pathological cases as well. In general, we need the singular vectors of M to be spread across all coordinates. If this happens, the recovery could be done by solving a rank minimization problem.

$$\text{minimize}\, rank(X), \text{ subject to } X_{ij} = M_{ij},\ (i, j) \in \Omega$$

where X is the recovered matrix.

Like in the case of compressed sensing, where a $l1$ norm is used instead of a $l0$ norm for easiest treatment, it is preferred to consider the minimization of the nuclear norm:

$$\text{minimize } \|X\|_*, \text{ subject to } X_{ij} = M_{ij},\ (i, j) \in \Omega$$

It is shown in [40] that most matrices can be recovered provided that the number of samples m obeys:

$$m \geq C\, n^{6/5}\, r\, \log n \tag{2.157}$$

where C is some positive constant, and r is the rank.

The nuclear norm minimization problem can be solved in many ways. A special mention should be done of the singular value thresholding algorithm introduced in [32]. This algorithm is a de facto reference for comparison with other algorithms (see the web page on low-rank matrix completion cited in the section on resources).

Actually, the proximal operator corresponding to the minimization of $\|X\|_*$ is the singular value shrinkage operator that shrinks towards zero the singular values of X, by subtracting a certain threshold value [145].

An example of matrix sampling and recovery is given below. The Douglas-Rachford splitting is used [151], based on two proximal operators: one corresponds to the indicator function, and the other to the nuclear norm. In the case of the indicator function, the proximal operator is a projection that accumulates repeated samples of the same matrix entry. The algorithm is implemented in the Program 2.16, which generates two figures.

Figure 2.38 shows the evolution of the nuclear norm as the iterations go on. It converges in less than 100 iterations.

In order to show how accurate is the recovery, a simple plot has been devised. One takes a row of the original matrix and represents its entry values as 'x' points. Then, one plots on top the recovered values for this row, as a series of segments. Figure 2.39 shows the result for the 10th row. It can be seen that the recovery is fairly precise.

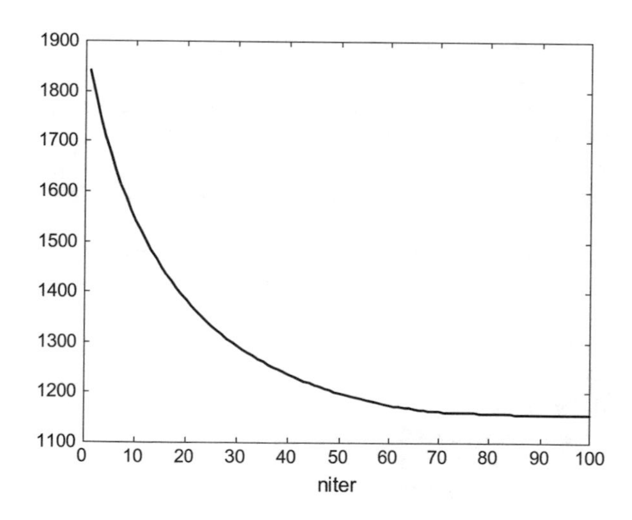

Fig. 2.38 Evolution of nuclear norm

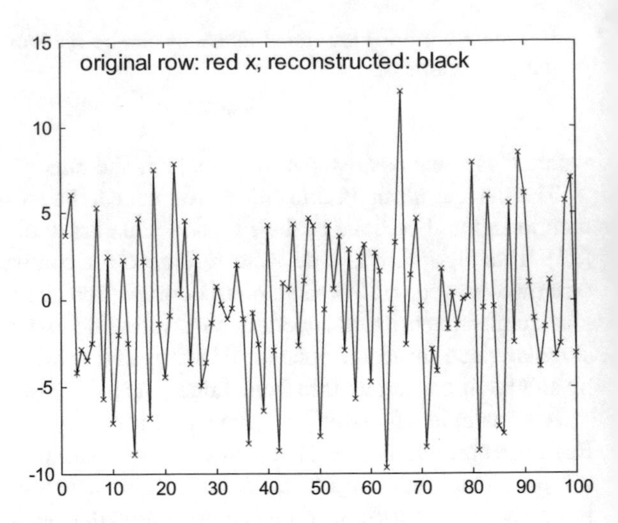

Fig. 2.39 A test of reconstruction quality

Program 2.16 Example of Matrix Completion

```
% Example of Matrix Completion
% (using Douglas-Rachford)
clear all;
disp('working...')
% a random matrix of rank r
n=100;
r=12; %rank
A=randn(n,r)*randn(r,n); % Original random matrix
disp('nuclear norm of original matrix:');
[u,d,v]=svd(A);
nuc_N = sum(diag(d))
% a subset of A is measured by
% sampling at random a number p of entries of A
p=round(n*log(n)*r); % see theory
aux=randperm(n*n);
ix=aux(1:p)'; %extract p integer random numbers
% b is a column vector:
b=A(ix); %retain a subset of entries of A
% start the algorithm --------------------------
X=zeros(n,n);
Y=zeros(n,n);
L=n*n;
niter=100;
lambda=1; gamma=2;
rnx=zeros(niter,1);
for nn=1:niter,
  % X update
  % R(b-M(Y)) term
  O=zeros(L,1);
```

```
    % accumulating repeated entries
    for j=1:p;
      ox=ix(j);
      O(ox)=O(ox)+(b(j)-Y(ox));
    end;
    Q=reshape(O,[n n]);
    % proxF (indicator function)
    X=Y+Q;
    % Y update
    %proxG (soft thresholding of singular values):
    P=(2*X)-Y;
    [U,D,V]=svd(P);
    for j=1:n,
      aux=D(j,j);
      if abs(aux)<=gamma,
        D(j,j)=0;
      else
      if aux>gamma, D(j,j)=aux-gamma; end
      if aux<-gamma, D(j,j)=aux+gamma; end;
    end;
end;
S=U*D*V'; % result of thresholding
Y=Y+ (lambda*(S-X));
% recording
[u,d,v]=svd(X);
rnx(nn)=sum(diag(d)); %nuclear norm
end
%display ------------------------------
% evolution of nuclear norm
figure(1)
plot(rnx,'k');
title('evolution of nuclear norm')
xlabel('niter');
% see a matrix row: original and reconstructed
figure(2)
Nrow=10; %(edit this number)
plot(A(Nrow,:),'r-x'); hold on;
plot(X(Nrow,:),'k');
title('original row: red x; reconstructed: black');
% a measure of error
er=A(Nrow,:)-X(Nrow,:);
E=sum(er.^2) %to be printed
```

The issues related to matrix completion have attracted a lot of research activity. For instance, what assumptions are needed to guarantee the matrix recovery? A recent article on this aspect is [98], which includes important references. See [69] for another perspective that connects phase transitions and matrix denoising. An extensive work on fundamental limits and efficient algorithms is [137]. It happens that the SVD decomposition applied in the first proposed algorithms can imply excessive computational effort for large matrices, and so many improvements or alternatives have been explored [129, 182]. Some authors have proposed the use of other norms, instead of the nuclear norm (see [117] and references therein).

2.6.2 Decomposition of Matrices

It was said in [39] that in real world applications the measured entries would be corrupted by noise (perhaps outliers, [190]). This observation originated a new type of problem, in which one has to find a decomposition of the observed matrix M into a low-rank matrix L and a sparse matrix S. The problem was recognized as a robust PCA analysis in [38] (a most cited paper), being stated as follows:

$$\text{minimize } (\|L\|_* + \lambda \|S\|_1), \text{ subject to } M = L + S$$

Again, the new problem lead to a broad range of research efforts, which have found many interesting applications.

Let us build a simple example by adding a low-rank random matrix and a sparse matrix (just a diagonal matrix). Figure 2.40 shows images corresponding to these matrices.

The result of adding the previous two matrices is shown in Fig. 2.41.

Now, the problem is to recover L and S from M. One of the methods that can be used is an adaptation of the Douglas-Rachford algorithm for this case. According with [86], it can be formulated as follows:

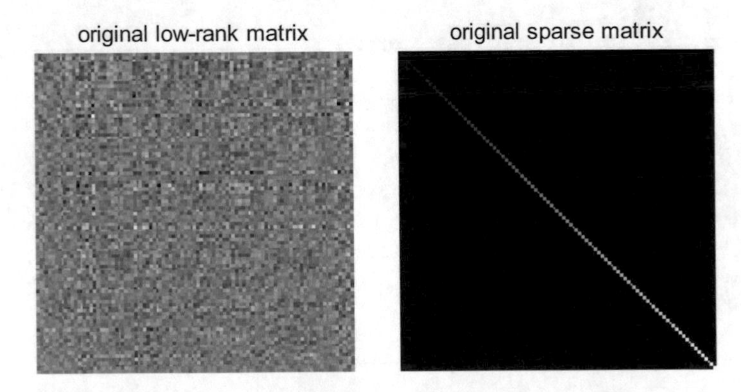

Fig. 2.40 A low-rank matrix and a sparse matrix

Fig. 2.41 The observed matrix M = L+S

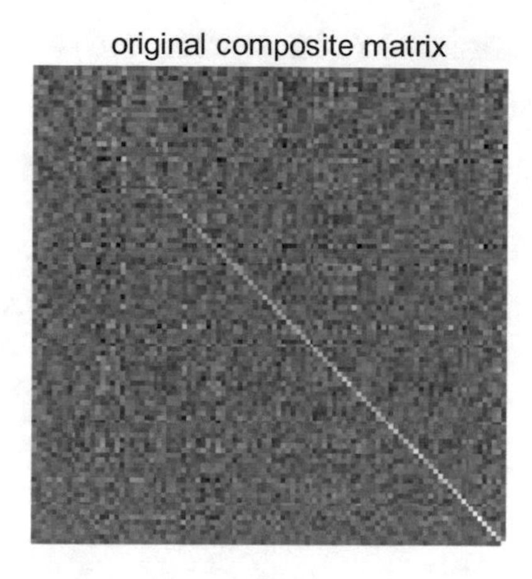

---

repeat:

$$Le = (M + L^{(k)} - S^{(k)})/2 \tag{2.158}$$

$$Se = (M - L^{(k)} + S^{(k)})/2 \tag{2.159}$$

$$L^{(k+1)} = L^{(k)} + t_k(shrink(2Le - L^{(k)}) - Le) \tag{2.160}$$

$$S^{(k+1)} = S^{(k)} + t_k(soft\_threshold(2Se - S^{(k)}) - Se) \tag{2.161}$$

until convergence.

The shrinkage corresponds to the proximity operator for the nuclear norm; and the soft-thresholding corresponds to the proximity operator for the $l1$ norm.

An implementation of this algorithm is provided by Program 2.17. The result is satisfactory as shown in Fig. 2.42.

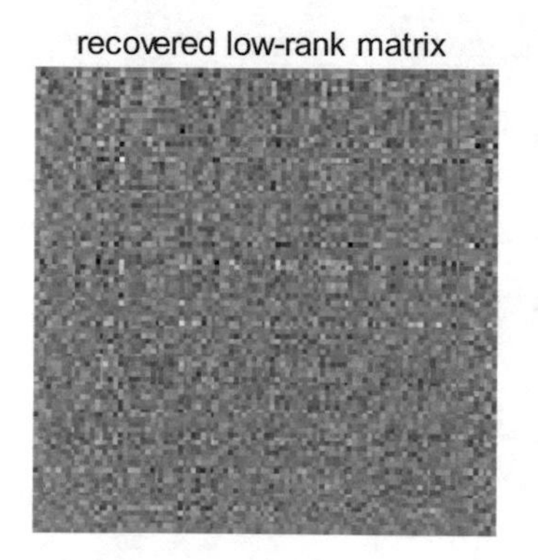

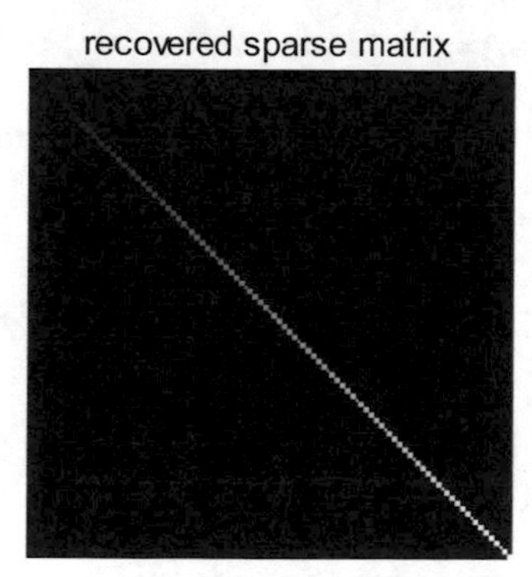

Fig. 2.42 The recovered matrices L and S

Program 2.17 Decomposition into low-rank (L) and sparse (S) matrices

```
% decomposition into low-rank (L) and sparse (S) matrices
% Douglas-Rachford
% a random matrix of rank r
n=100;
r=8; %rank
L0=randn(n,r)*randn(r,n); % a low-rank matrix
% a sparse (diagonal) matrix
nn=1:100;
S0=diag(0.2*nn,0);
% composite original matrix
M=L0+S0;
% parameter settings
lambda=1;
tk=1;
Th=3; %Threshold
nnL=30; % number of loops
rnx=zeros(nnL,1);
L=zeros(n,n); S=zeros(n,n);
% start the algorithm --------------------------------
for nn=1:nnL,
  Le=0.5*(M+L-S); Se=0.5*(M-L+S);
  % shrinking---------
  aux1=(2*Le)-L;
  [U D V]=svd(aux1);
  for j=1:n,
    D(j,j)=max(D(j,j)-Th,0);
  end;
  aux=U*D*V';
  L=L+(tk*(aux-Le));
  % soft_threshold---------
  aux1=(2*Se)-S;
  aux=sign(aux1).*max(0, abs(aux1)-lambda);
  S=S+(tk*(aux-Se));
  [u,d,v]=svd(L);
  rnx(nn)=sum(diag(d)); %nuclear norm, record
end;
% display ------------------------------
figure(1)
subplot(1,2,1)
imshow(L0,[]);
title('original low-rank matrix')
subplot(1,2,2)
imshow(S0,[]);
title('original sparse matrix')
figure(2)
imshow(M,[]);
title('original composite matrix');
```

```
figure(3)
subplot(1,2,1)
imshow(L,[]);
title('recovered low-rank matrix')
subplot(1,2,2)
imshow(S,[]);
title('recovered sparse matrix')
```

A number of methods for matrix decomposition has been proposed. Soon after the publication of [39], the *'principal component pursuit'* (PCP) was introduced. In [204] a study of PCP was presented, with mentions to robust PCA and a reference to [38] as preprint. In general, the preferred methods are based on alternating minimization schemes, which are reviewed in the introduction of [170]. One of the factors that promote the popularity of certain methods is the public availability of code [115, 116, 196]. Theoretical aspects on conditions for the recovery of matrices are treated in [51].

Some illustrative application examples are [148] for alignment of images, [199] for low-rank image textures, [54] for face recognition based on robust PCA, [113] on cognitive radio networks, [203] for medical image analysis, [13] on target tracking (for example a TV camera focusing on a basketball player during the game), [65] on computer vision, [56] on genotype imputation, or [181] for movie colorization.

2.7 Experiments

This section includes three experiments, the first is an example of 1D signal denoising, while the other two examples are applications of matrix completion and decomposition to images.

2.7.1 Signal Denoising Based on Total Variation (TV)

Let us consider an example of signal denoising based on TV, using an algorithm proposed in [169]. Given a signal $x(n)$ composed of N samples, its total variation would be:

$$TV(x) = \sum_{i=1}^{N} (x(n) - x(n-1)) \tag{2.162}$$

An equivalent expression is the following:

$$TV(x) = \|Dx\|_1 \tag{2.163}$$

with:

$$D = \begin{bmatrix} -1 & 1 & & & \\ & -1 & 1 & & \\ & & & & \\ & & & & \\ & & & -1 & 1 \end{bmatrix} \tag{2.164}$$

Suppose a noise is added to the signal, and then:

$$y = x + n \tag{2.165}$$

In order to denoise the signal y, the following criterion can be defined:

$$J(x) = \|y - x\|_2^2 + \lambda \, \|Dx\|_1 \tag{2.166}$$

and a solution x must be found for the minimization of $J(x)$.

The minimization algorithm proposed in [169] is the following:

$$x_{k+1} = y - D^T z_k \tag{2.167}$$

$$z_{k+1} = clip(z_k + \frac{1}{\alpha} Dx_{k+1} \, ; \, \lambda/2) \tag{2.168}$$

with $z_0 = 0$ and $\alpha \geq \max \, eig(D \, D^T)$.

The algorithm uses a clipping function, which is defined as follows:

$$clip(y; \, T) = \begin{cases} y \, , \; |y| < T \\ T \, sign(y), \; |y| > T \end{cases} \tag{2.169}$$

An implementation of this algorithm is provided by the Program 2.18, which is a modification of the code given in [169]. The case considered is a signal with some abrupt changes. The original signal, and the same signal with added noise, are shown in Fig. 2.43.

Figure 2.44 shows the evolution of the minimization of $J(x)$, and the denoised signal as it was obtained at the end of the iterations. Note that the abrupt changes of the signal have been preserved.

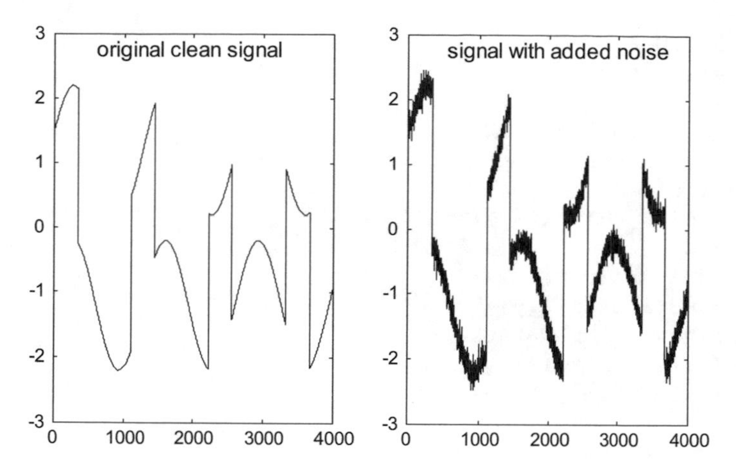

Fig. 2.43 Signal to be denoised

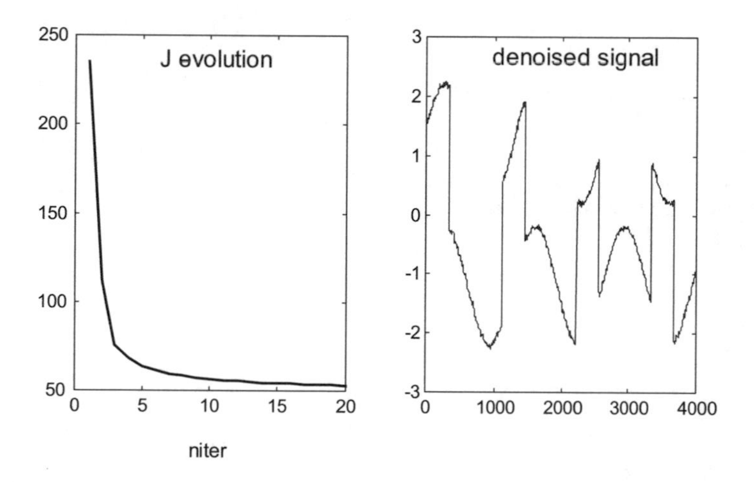

Fig. 2.44 TV denoising: (*left*) convergence, (*right*) denoised signal

Program 2.18 Example of TV denoising

```
% Example of TV denoising
clear all;
% build a test signal
t=0:0.1:400;
N=length(t);
u= sin(0.3+(0.015*pi*t));
v= 1.2*square(0.018*pi*t,30);
a=u+v; %signal with no noise
x=a+(0.1*randn(1,N)); %signal with noise
% Prepare for computations
niter=20;
```

```
z=zeros(1,N-1);
J=zeros(1,niter);
alpha=3;
lambda=0.5; th=lambda/2;
%start the algorithm -----------------------
for nn=1:niter,
  aux=[-z(1) -diff(z) z(end)];
  y=x-aux;
  aux1=sum(abs(y-x).^2);
  aux2=sum(abs(diff(y)));
  J(nn)=aux1+(lambda*aux2);
  z=z+((1/alpha)*diff(y));
  z=max(min(z,th),-th);
end
% display ---------------------
figure(1)
subplot (1,2,1)
plot(a,'k')
axis([0 4000 -3 3]);
title('original clean signal')
subplot(1,2,2)
plot(x,'k')
axis([0 4000 -3 3]);
title('signal with added noise')
figure(2)
subplot(1,2,1)
plot(J,'k')
xlabel('niter')
title('J evolution')
subplot(1,2,2)
plot(y,'k')
axis([0 4000 -3 3]);
title('denoised signal')
```

Another way of computing the total variation denoising of unidimensional signals was proposed in [167].

2.7.2 Picture Reconstruction Based on Matrix Completion

A direct example of matrix completion is the case of taking at random some pixels of a picture, and see if it is possible to recover the picture from these samples.

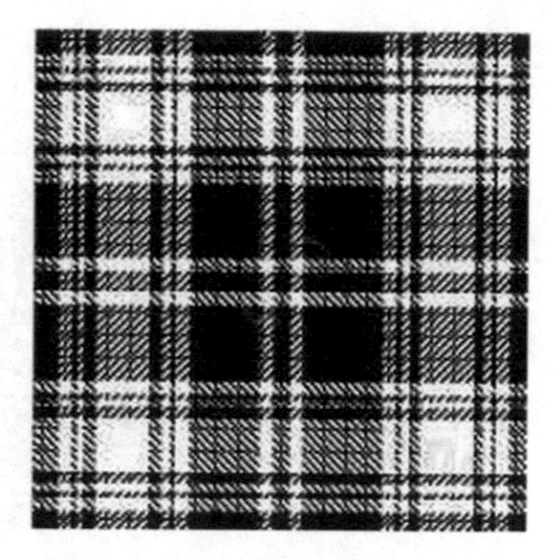

Fig. 2.45 Original image

An image with evident redundancies has been chosen, Fig. 2.45. It could be expected that this image has approximately low rank. Anyway, a certain value of p was tried.

The sampling and recovery experiment was done with the Program 2.19, which is just an adaptation of the Program 2.16.

Figure 2.46 shows the evolution of the nuclear norm along iterations.

And Fig. 2.47 shows the good result of the image recovery.

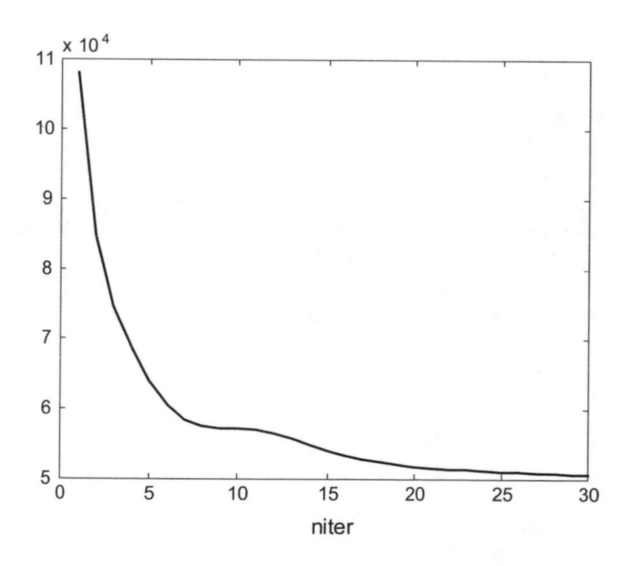

Fig. 2.46 Evolution of nuclear norm

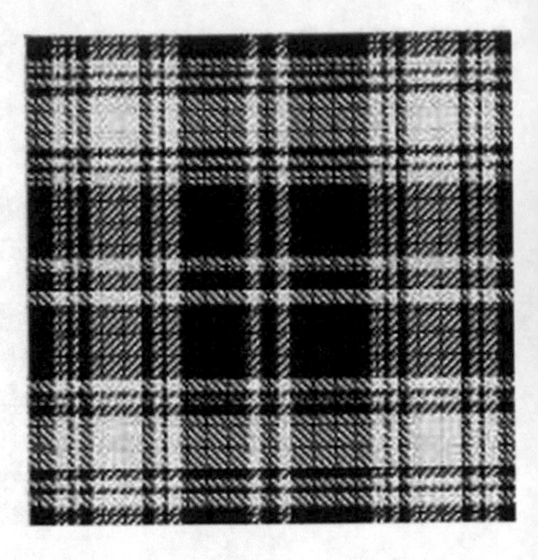

Fig. 2.47 Image reconstruction by matrix completion

Program 2.19 Example of Picture Completion

```
% Example of Picture Completion
% (using Douglas-Rachford)
clear all;
disp('working...')
% original picture
figu=imread('tartan.jpg'); %read picture
F=double(figu);
A=F(1:200,1:200); %crop;
aux=mean(mean(A));
A=A-aux;
n=200;
disp('nuclear norm of original matrix:');
[u,d,v]=svd(A);
nuc_N = sum(diag(d))
% a subset of A is measured by
% sampling at random a number p of entries of A
p=10000;
aux=randperm(n*n);
ix=aux(1:p)'; %extract p integer random numbers
% b is a column vector:
b=A(ix); %retain a subset of entries of A
% start the algorithm----------------------------
X=zeros(n,n);
Y=zeros(n,n);
L=n*n;
niter=30;
lambda=1; gamma=1000; %(notice the value of gamma)
rnx=zeros(niter,1);
for nn=1:niter,
```

```
  % X update
  % R(b-M(Y)) term
  O=zeros(L,1);
  % accumulating repeated entries
  for j=1:p;
    ox=ix(j);
    O(ox)=O(ox)+(b(j)-Y(ox));
  end;
Q=reshape(O,[n n]);
% proxF (indicator function)
X=Y+Q;
% Y update
%proxG (soft thresholding of singular values):
P=(2*X)-Y;
[U,D,V]=svd(P);
for j=1:n,
aux=D(j,j);
if abs(aux)<=gamma,
  D(j,j)=0;
  else
  if aux>gamma, D(j,j)=aux-gamma; end
  if aux<-gamma, D(j,j)=aux+gamma; end;
end;
end;
S=U*D*V'; % result of thresholding
Y=Y+ (lambda*(S-X));
% recording
[u,d,v]=svd(X);
rnx(nn)=sum(diag(d)); %nuclear norm
end
%display ----------------------------
% evolution of nuclear norm
figure(1)
plot(rnx,'k');
title('evolution of nuclear norm')
xlabel('niter');
figure(2)
imshow(A,[]);
title('original picture')
figure(3)
imshow(X,[]);
title('reconstructed picture')
```

2.7.3 Text Removal

Suppose you have a photograph and someone has written some text on it. It would be good to remove that text. Given this situation, matrix decomposition could be helpful, as far as the picture has low rank, and the text corresponds to a sparse matrix of pixels.

In order to explore this application, a synthetic problem has been fabricated: some text has been added to a wall of bricks. It is a crude simulation of a graffiti. Figure 2.48 shows the original image.

Since we wanted to explore alternating minimization schemes, a fast scheme proposed by [159]. In this paper, the matrix decomposition problem is treated as:

$$\text{minimize } (\frac{1}{2} \|L + S - M\|_F + \lambda \|S\|_1), \text{ subject to } rank(L) = t$$

The problem is solved with the following alternating minimization:

$$L^{(k+1)} = \arg\min_L (\left\| L + S^{(k)} - M \right\|_F), \text{ subject to } rank(L) = t$$

$$S^{(k+1)} = \arg\min_S (\left\| L^{(k+1)} + S - M \right\|_F + \lambda \|S\|_1) \tag{2.170}$$

These two sub-problems are solved as follows:

- The first sub-problem is solved with a partial SVD of $S^{(k)} - M$. Only t components of the SVD are selected (corresponding to the t largest singular values). This is done with the *lansvd( )* routine included in the PROPACK library [114], which is commonly found in the public available codes.
- The second sub-problem is a element-wise shrinkage of $M - L^{(k)}$.

Fig. 2.48 Simulated graffiti image

Fig. 2.49 Image decomposition into low-rank and sparse matrices

Program 2.20 presents a simple implementation of the algorithm. Figure 2.49 shows the result, which is not perfect but gives an idea of the approach. Of course, the image on the left hand would require better inpainting, and the image on the right hand would thank a better background extraction.

Program 2.20 Decomposition into low-rank (L) and sparse (S) matrices

```
% decomposition into low-rank (L) and sparse (S) matrices
% Alternating Minimization
% load image
figu=imread('wall.jpg'); %read picture
F=double(figu);
n=380;
M=F(1:n,1:n); %crop;
aux=mean(mean(M));
M=M-aux;
% parameter settings
lambda=0.5;
lambF=1.01;
Th=0.01; %Threshold
rank0=1; %intial rank guess
irk=1; %for rank increments
nnL=50; % number of loops
rank=rank0; %current rank
% start the algorithm -------------------------------
%
[UL SL VL] = lansvd(M, rank, 'L'); %partial SVD
L1=UL*SL*VL'; %initial low-rank approximation
aux=M-L1;
S1= sign(aux).*max(0,abs(aux)-lambda); %shrinkage
for nn=2:nnL,
  if irk==1,
  lambda = lambda * lambF; % lambda is modified in each iteration
  rank = rank + irk; % rank is increased " " "
```

```
end;
[UL SL VL] = lansvd(M-S1, rank, 'L'); %partial SVD
L1=UL*SL*VL'; %current low-rank approximation
aux=M-L1;
S1= sign(aux).*max(0,abs(aux)-lambda); %shrinkage
% change rank increment when appropriate
vv=diag(SL);
rho=vv(end)/sum(vv(1:end-1));
if rho<Th,
    irk=0;
  else
    irk=1;
  end;
end;
% display ------------------------------
figure(1)
imshow(M,[]);
title('original picture')
figure(2)
subplot(1,2,1)
imshow(L1,[]);
title('low-rank component')
subplot(1,2,2)
imshow(S1,[]);
title('sparse component')
```

Background extraction or subtraction is a subject of considerable interest. The web site of T. Bouwmans offer several surveys of this topic. Fast techniques are needed in the case of video, [144, 188]. One application example is video surveillance [19, 97].

2.8 Resources

The file exchange web site of Mathworks has several MATLAB programs of interest. Also, some of the methods introduced in this chapter are implemented in some routines of the MATLAB Statistics Toolbox.

2.8.1 MATLAB

2.8.1.1 Toolboxes

- l1-MAGIC:
  http://users.ece.gatech.edu/~justin/l1magic/
- SparseLab (Stanford University):
  http://sparselab.stanford.edu/

- SpaSM Toolbox (includes LARS):
  http://www.imm.dtu.dk/projects/spasm/
- SPAMS (C++ interfaced to MATLAB):
  http://spams-devel.gforge.inria.fr/doc-R/html/index.html
- NESTA (Stanford University):
  http://statweb.stanford.edu/~candes/nesta/
- Toolbox Sparse Optimization (G. Peyre):
  http://www.mathworks.com/matlabcentral/fileexchange/16204-toolbox-sparse-optmization
- Model-based Compressive Sensing Toolbox (Rice University):
  http://dsp.rice.edu/software/model-based-compressive-sensing-toolbox-v11
- Toolbox Sparsity (G. Peyre):
  https://www.ceremade.dauphine.fr/~peyre/matlab/sparsity/content.html
- UNLocBoX (convex optimization toolbox):
  http://unlocbox.sourceforge.net/

2.8.1.2 Matlab Code and Scripts

- SALSA:
  http://cascais.lx.it.pt/~mafonso/salsa.html
- TwIST:
  http://www.lx.it.pt/~bioucas/TwIST/TwIST.htm
- SpaRSA:
  http://www.lx.it.pt/~mtf/SpaRSA/
- MATLAB scripts for ADMM (Stanford University):
  http://www.web.stanford.edu/~boyd/papers/admm!
- YALL1 (Rice University):
  http://yall1.blogs.rice.edu/
- Beginners code for CS (A. Weinstein):
  http://control.mines.edu/mediawiki/upload/f/f4/Beginners_code.pdf
- MATLAB script for Chan-Vese segmentation (Fields Institute):
  http://www.math.ucla.edu/~wittman/Fields/cv.m
- Low-Rank Matrix Recovery and Completion via Convex Optimization:
  http://perception.csl.illinois.edu/matrix-rank/home.html

2.8.2 *Internet*

2.8.2.1 Web Sites

- Dave Donoho:
  http://www-stat.stanford.edu/~donoho
- Emmanuel Candès (software):
  http://statweb.stanford.edu/~candes/software.html
- Michael Elad:
  www.cs.technion.ac.il/~elad/index.html
- A. Rakotomamonjy:
  http://asi.insa-rouen.fr/enseignants/~arakoto/
- Mark Schmidt:
  http://www.cs.ubc.ca/~schmidtm/
- Mark A. Davenport (CoSaMP):
  http://users.ece.gatech.edu/~mdavenport/software/
- SeDuMi (Lehigh University):
  http://sedumi.ie.lehigh.edu/
- Tiany Zhou (CS recontruction algorithms):
  https://tianyizhou.wordpress.com/2010/08/23/compressed-sensing-review-1-reconstruction-algorithms/
- CS Audio Demonstration:
  http://sunbeam.ece.wisc.edu/csaudio/
- EPFL Signal Processing Lab (Lausanne):
  http://lts2www.epfl.ch/people/gilles/softwares
- Bio Imaging & Signal Processing Lab.:
  http://bispl.weebly.com/software.html
- J. Huang:
  http://ranger.uta.edu/~huang/index.html
- Douglas-Rachford and projection methods:
  http://carma.newcastle.edu.au/DRmethods/
- Bamdev Mishra (Riemannian matrix completion):
  https://sites.google.com/site/bamdevm/codes/qgeommc
- Principal Component Pursuit papers and code:
  http://investigacion.pucp.edu.pe/grupos/gpsdi/publicaciones-2/
- Thierry Bouwmans (surveys):
  https://sites.google.com/site/thierrybouwmans/recherche---background-subtraction---survey

2.8.2.2 Link Lists

- Numerical-tours:
  http://www.numerical-tours.com/links/
- Fast l1 Minimization Algorithms:
  http://www.eecs.berkeley.edu/~yang/software/l1benchmark/
- Compressive sensing:
  https://sites.google.com/site/igorcarron2/compressivesensing2.0
- The advanced matrix factorization jungle:
  https://sites.google.com/site/igorcarron2/matrixfactorizations
- Compressive Sensing: The Big Picture:
  https://sites.google.com/site/igorcarron2/cs
- Matrix completion solvers:
  http://www.ugcs.caltech.edu/~srbecker/wiki/Category:Matrix{_}Completion{_}Solvers
- Research in Computational Science:
  http://www.csee.wvu.edu/~xinl/source.html

References

1. S. Abeyruwan, *Least Angle Regression* (University of Miami, 2012). https://sites.google.com/site/samindaa/lars
2. R. Acar, C.R. Vogel, Analysis of bounded variation penalty methods for ill-posed problems. Inverse Prob. **10**(6), 1217 (1994)
3. B. Adcock, *An Introduction to Compressed Sensing* (Department of Mathematics, Simon Fraser University, 2015). http://benadcock.org/wp-content/uploads/2015/11/AdcockBeijingTalk1.pdf
4. B. Adcock, A.C. Hansen, C. Poon, B. Roman, *Breaking the Coherence Barrier: Asymptotic Incoherence and Asymptotic Sparsity in Compressed Sensing* (Department of Mathematics, Purdue University, 2013). arXiv preprint arXiv:1302.0561
5. M.V. Afonso, J.M. Bioucas-Dias, M.A. Figueiredo, Fast image recovery using variable splitting and constrained optimization. IEEE Trans. Image Process. **19**(9), 2345–2356 (2010)
6. M. Aharon, M. Elad, A. Bruckstein, K-SVD: An algorithm for designing overcomplete dictionaries for sparse representation. IEEE T. Signal Process. **54**(11), 4311–4322 (2006)
7. S.I. Amari, A. Cichocki, Information geometry of divergence functions. Bull. Polish Acad. Sci.: Tech. Sci. **58**(1), 183–195 (2010)
8. F.J.A. Artacho, M. Jonathan, Recent results on Douglas-Rachford methods. Serdica Math. J. **39**, 313–330 (2013)
9. H. Attouch, *Alternating Minimization and Projection Algorithms. From Convexity to Nonconvexity* (2009). http://events.math.unipd.it/nactde09/sites/default/files/attouch.pdf
10. J.F. Aujol, G. Aubert, L. Blanc-Féraud, A. Chambolle, Image decomposition into a bounded variation component and an oscillating component. J. Math. Imag. Vision **22**(1), 71–88 (2005)
11. J.F. Aujol, G. Gilboa, Implementation and parameter selection for BV-Hilbert space regularization. Technical report, UCLA, 2004. CAM04-66

12. J.F. Aujol, G. Gilboa, T. Chan, S. Osher, Structure-texture image decomposition–modeling, algorithms, and parameter selection. Int. J. Comput. Vision **67**(1), 111–136 (2006)
13. M. Ayazoglu, M. Sznaier, O.I. Camps, Fast algorithms for structured robust principal component analysis, in *Proceedings IEEE Conference on Computer Vision and Pattern Recognition,(CVPR)*, pp. 1704–1711 (2012)
14. W.U. Bajwa, J. Haupt, A.M. Sayeed, R. Nowak, Compressed channel sensing: a new approach to estimating sparse multipath channels. Proc. IEEE **98**(6), 1058–1076 (2010)
15. A. Banerjee, S. Merugu, I.S. Dhillon, J. Ghosh, Clustering with Bregman divergences. J. Mach. Learn. Res. **6**, 1705–1749 (2005)
16. C. Bao, J.F. Cai, H. Ji, Fast sparsity-based orthogonal dictionary learning for image restoration, in *Proceedings International Conference Computer Vision*, 2013
17. T. Baran, D. Wei, A.V. Oppenheim, Linear programming algorithms for sparse filter design. IEEE T. Signal Process. **58**(3), 1605–1617 (2010)
18. R.G. Baraniuk, Compressive sensing. IEEE Signal Process. Mag. **24**(4) (2007)
19. M. Baumann, Real-time robust principal component analysis for video surveillance. Master's thesis, ETH Zurich, 2010
20. A. Beck, M. Teboulle, A fast iterative shrinkage-thresholding algorithm for linear inverse problems. SIAM J. Imag. Sci. **2**(1), 183–202 (2009)
21. S. Becker, J. Bobin, E.J. Candès, Nesta: a fast and accurate first-order method for sparse recovery. SIAM J. Imag. Sci. **4**(1), 1–39 (2011)
22. C.R. Berger, Z. Wang, J. Huang, S. Zhou, Application of compressive sensing to sparse channel estimation. IEEE Commun. Mag. **48**(11), 164–174 (2010)
23. J.M. Bioucas-Dias, M.A. Figueiredo, A new twist: two-step iterative shrinkage/thresholding algorithms for image restoration. IEEE Trans. Image Process. **16**(12), 2992–3004 (2007)
24. J. Bobin, J.L. Starck, J.M. Fadili, Y. Moudden, D.L. Donoho, Morphological component analysis: an adaptive thresholding strategy. IEEE T. Image Process. **16**(11), 2675–2681 (2007)
25. J.D. Boissonnat, F. Nielsen, R. Nock, Bregman Voronoi diagrams. Discrete Computat. Geometry **44**(2), 281–307 (2010)
26. S. Boyd, N. Parikh, E. Chu, B. Peleato, J. Eckstein, Distributed optimization and statistical learning via the alternating direction method of multipliers. Found. Trends Mach. Learn. **3**(1), 1–122 (2010)
27. P. Breen, Algorithms for sparse approximation. Master's thesis, University of Edinburgh, 2009. Year 4 Project, School of Mathematics
28. L.M. Bregman, The relaxation method of finding the common points of convex sets and its application to the solution of problems in convex programming. USSR Comput. Math. Math. Phys. **7**(3), 200–217 (1967)
29. A. Buades, T.M. Le, J.-M. Morel, L.A. Vese, Fast cartoon + texture image filters. IEEE T. Image Process. **19**(8), 1978–1986 (2010)
30. J. Bush, Bregman algorithms. Master's thesis, University of Santa Barbara, CA, USA, 2011. Senior Thesis
31. C.F. Cadieu, B.A. Olshausen, Learning transformational invariants from natural movies, in *Proceedings NIPS*, pp. 209–216 (2008)
32. J.F. Cai, E.J. Candès, Z. Shen, A singular value thresholding algorithm for matrix completion. SIAM J. Optim. **20**(4), 1956–1982 (2010)
33. X. Cai, R. Chan, T. Zeng, A two-stage image segmentation method using a convex variant of the Mumford-Shah model and thresholding. SIAM J. Imag. Sci. **6**(1), 368–390 (2013)
34. E. Candes, J. Romberg, T. Tao, Robust uncertainty principles: exact signal reconstruction from highly incomplete frequency information. IEEE Trans. Inf. Theory **52**(2), 489–509 (2006)
35. E. Candes, T. Tao, Decoding by linear programming. IEEE T. Inf. Theory **51**(12), 4203–4215 (2005)
36. E. Candes, T. Tao, The Dantzig selector: statistical estimation when P is much larger than N. Ann. Stat 2313–2351 (2007)
37. E.J. Candes, The restricted isometry property and its implications for compressed sensing. C. R. Math. **346**(9), 589–592 (2008)

38. E.J. Candès, X. Li, Y. Ma, J. Wright, Robust principal component analysis? J. ACM (JACM) **58**(3), 11 (2011)
39. E.J. Candes, Y. Plan, Matrix completion with noise. Proc. IEEE **98**(6), 925–936 (2010)
40. E.J. Candès, B. Recht, Exact matrix completion via convex optimization. Found. Comput. Math. **9**(6), 717–772 (2009)
41. E.J. Candes, T. Tao, Near-optimal signal recovery from random projections: Universal encoding strategies? IEEE T. Inf. Theory **52**(12), 5406–5425 (2006)
42. E.J. Candes, M.B. Wakin, An introduction to compressive sampling. IEEE Signal Process. Mgz. 21–30 (2008)
43. S. Cao, Q. Wang, Y. Yuan, J. Yu, Anomaly event detection method based on compressive censing and iteration in wireless sensor networks. J. Netw. **9**(3), 711–718 (2014)
44. C. Caramanis, S. Sanghavi, *Large Scale Optimization* (The University of Texas at Austin, Lecture 24 of EE381V Course, 2012). http://users.ece.utexas.edu/~sanghavi/courses/scribed_notes/Lecture_24_Scribe_Notes.pdf
45. V. Caselles, A. Chambolle, M. Novaga, Total variation in imaging, in *Handbook of Mathematical Methods in Imaging*, pp. 1016–1057 (Springer Verlag, 2011)
46. Y. Censor, A. Lent, An iterative row-action method for interval convex programming. J. Optim. Theory Appl. **34**(3), 321–353 (1981)
47. A. Chambolle, P.L. Lions, Image recovery via total variation minimization and related problems. Numer. Math. **76**(2), 167–188 (1997)
48. T. Chan, S. Esedoglu, F. Park, A. Yip, Recent developments in total variation image restoration. Math. Models Comput. Vision **17**, (2005)
49. T.F. Chan, S. Esedoglu, Aspects of total variation regularized L1 function approximation. SIAM J. Appl. Math. **65**(5), 1817–1837 (2005)
50. T.F. Chan, S. Esedoglu, F.E. Park, Image decomposition combining staircase reduction and texture extraction. J. Visual Commun. Image Represent. **18**(6), 464–486 (2007)
51. V. Chandrasekaran, S. Sanghavi, P.A. Parrilo, A.S. Willsky, Rank-sparsity incoherence for matrix decomposition. SIAM J. Optim. **21**(2), 572–596 (2011)
52. P. Chatterjee, I.P.P. Milanfar, *Denoising using the K-SVD Method* (EE 264 Course, University of California at Santa Cruz, 2007). https://users.soe.ucsc.edu/~priyam/ksvd_report.pdf
53. K.M. Cheman, Optimization techniques for solving basis pursuit problems. Master's thesis, North Carolina State Univ., Raleigh, NC, USA, 2006
54. F. Chen, C.C.P. Wei, Y.C. Wang, Low-rank matrix recovery with structural incoherence for robust face recognition, in *Proceedings IEEE Conference on Computer Vision and Pattern Recognition, (CVPR)*, pp. 2618–2625 (2012)
55. S.S. Chen, D.L. Donoho, M.A. Saunders, Atomic decomposition by basis pursuit. SIAM J. Sci. Comput. **20**(1), 33–61 (1998)
56. E.C. Chi, H. Zhou, G.K. Chen, D.O. Del Vecchyo, K. Lange, Genotype imputation via matrix completion. Genome Res. **23**(3), 509–518 (2013)
57. M.G. Christensen, S.H. Jensen, On compressed sensing and its application to speech and audio signals, in *Proceedings 43th IEEE Asilomar Conference on Signals, Systems and Computers*, pp. 356–360 (2009)
58. I. Cimrak, Analysis of the bounded variation and the G regularization for nonlinear inverse problems. Math. Meth. Appl. Sci. **33**(9), 1102–1111 (2010)
59. A. Cohen, W. Dahmen, I. Daubechies, R. DeVore, Harmonic analysis of the space BV. Revista Matematica Iberoamericana **19**(1), 235–262 (2003)
60. R. Coifman, F. Geshwind, Y. Meyer, Noiselets. Appl. Comput. Harmonic Anal. **10**, 27–44 (2001)
61. P.L. Combettes, J.C. Pesquet, Proximal splitting methods in signal processing, in *Fixed-point Algorithms for Inverse Problems in Science and Engineering*, pp. 185–212 (Springer, 2011)
62. S.B. Damelin, W. Jr, Miller, *The Mathematics of Signal Processing* (Cambridge University Press, 2012)
63. I. Daubechies, M. Fornasier, I. Loris, Accelerated projected gradient method for linear inverse problems with sparsity constraints. J. Fourier Anal. Appl. **14**(5–6), 764–792 (2008)

64. M.A. Davenport, M.F. Duarte, Y.C. Eldar, G. Kutyniok, Introduction to compressed sensing, eds. by Y.C. Eldar, G. Kutyniok. *Compressed Sensing* (Cambridge University Press, 2013)
65. F. De la Torre, M.J. Black, Robust principal component analysis for computer vision, in *Proceedings Eighth IEEE International Conference on Computer Vision, (ICCV)*, vol. 1, pp. 362–369 (2001)
66. G. Dogan, P. Morin, R.H. Nochetto, A variational shape optimization approach for image segmentation with a Mumford-Shah functional. SIAM J. Sci. Comput. **30**(6), 3028–3049 (2008)
67. D. Donoho, J. Tanner, Observed universality of phase transitions in high-dimensional geometry, with implications for modern data analysis and signal processing. Philos. Trans. Roy. Soc. A: Math. Phys. Eng. Sci. **367**(1906), 4273–4293 (2009)
68. D.L. Donoho, Compressed sensing. IEEE T. Inf. Theory **52**(4), 1289–1306 (2006)
69. D.L. Donoho, M. Gavish, A. Montanari, The phase transition of matrix recovery from gaussian measurements matches the minimax MSE of matrix denoising. Proc. Natl. Acad. Sci. **110**(21), 8405–8410 (2013)
70. D.L. Donoho, J. Tanner, Precise undersampling theorems. Proc. IEEE **98**(6), 913–924 (2010)
71. D.L. Donoho, Y. Tsaig, Fast solution of L1-norm minimization problems when the solution may be sparse. IEEE T. Inf. Theory **54**(11), 4789–4812 (2006)
72. I. Drori, D.L. Donoho, Solution of L1 minimization problems by LARS/homotopy methods, in *Proceedings IEEE International Conference Acoustics, Speech and Signal Processing, ICASSP*, vol. 3 (2006)
73. M.F. Duarte, Y.C. Eldar, Structured compressed sensing: from theory to applications. IEEE Trans. Signal Process. **59**(9), 4053–4085 (2011)
74. J. Eckstein, Splitting methods for monotone operators with applications to parallel optimization. Ph.D. thesis, MIT, 1989
75. B. Efron, T. Hastie, I. Johnstone, R. Tibshirani, Least angle regression. Ann. Stat. **32**(2), 407–451 (2004)
76. M. Elad, *Sparse and Redundant Representations* (Springer Verlag, 2010)
77. M. Elad, Sparse and redundant representation modeling–what next? IEEE Signal Process. Lett. **19**(12), 922–928 (2012)
78. M. Elad, M. Aharon, Image denoising via sparse and redundant representations over learned dictionaries. IEEE T. Image Process. **15**(12), 3736–3745 (2006)
79. M. Elad, M.A. Figueiredo, Y. Ma, On the role of sparse and redundant representations in image processing. Proc. IEEE **98**(6), 972–982 (2010)
80. J. Ender, A brief review of compressive sensing applied to radar, in *Proceedings 14thInternational Radar Symposium (IRS)*, pp. 3–16 (2013)
81. M.A. Figueiredo, R.D. Nowak, S.J. Wright, Gradient projection for sparse reconstruction: application to compressed sensing and other inverse problems. IEEE J. Selected Topics in. Signal Process. **1**(4), 586–597 (2007)
82. S. Foucart, H. Rauhut, *A Mathematical Introduction to Compressive Sensing* (Birkhäuser, 2010)
83. K. Fountoulakis, J. Gondzio, P. Zhlobich, Matrix-free interior point method for compressed sensing problems. Math. Programm. Comput. 1–31 (2012)
84. J. Friedman, T. Hastie, R. Tibshirani, *A note on the group Lasso and a sparse group Lasso* (Dept. Statistics, Stanford University, 2010) arXiv preprint arXiv:1001.0736
85. L. Gan, Block compressed sensing of natural images, in *Proceedings IEEE International Conference Digital Signal Processing*, pp. 403–406 (2007)
86. S. Gandy, I. Yamada, Convex optimization techniques for the efficient recovery of a sparsely corrupted low-rank matrix. J. Math-for-Indus. **2**(5), 147–156 (2010)
87. V. Ganesan, A study of compressive sensing for application to structural helath monitoring. Master's thesis, University of Central Florida, 2014
88. J. Gao, Q. Shi, T.S. Caetano, Dimensionality reduction via compressive sensing. Pattern Recogn. Lett. **33**(9), 1163–1170 (2012)
89. P. Getreuer, Chan-Vese segmentation. Image Processing On Line (2012)

90. P. Getreuer, Rudin-osher-fatemi total variation denoising using split Bregman. Image Processing On Line **10** (2012)
91. J.R. Gilbert, C. Moler, R. Schreiber, Sparse matrices in MATLAB: design and implementation. SIAM J. Matrix Anal. Appl. **13**(1), 333–356 (1992)
92. J. Gilles, Noisy image decomposition: a new structure, texture and noise model based on local adaptivity. J. Math. Imag. Vision **28**(3), 285–295 (2007)
93. J. Gilles, Image decomposition: theory, numerical schemes, and performance evaluation. Adv. Imag. Electron Phys. **158**, 89–137 (2009)
94. T. Goldstein, B. O'Donoghue, S. Setzer, R. Baraniuk, Fast alternating direction optimization methods. SIAM J. Imag. Sci. **7**(3), 1588–1623 (2014)
95. T. Goldstein, S. Osher, The split Bregman method for L1 regularized problems. SIAM J. Imag. Sci. **2**(2), 323–343 (2009)
96. Y. Gousseau, J.M. Morel, Are natural images of bounded variation? SIAM J. Math. Anal. **33**(3), 634–648 (2001)
97. X. Guo, S. Li, X. Cao, Motion matters: a novel framework for compressing surveillance videos, in *Proceedings 21st ACM International Conference on Multimedia*, pp. 549–552 (2013)
98. M. Hardt, R. Meka, P. Raghavendra, B. Weitz, *Computational Limits for Matrix Completion* (IBM Research Almaden, 2014). arXiv preprint arXiv:1402.2331
99. J. Haupt, W.U. Bajwa, M. Rabbat, R. Nowak, Compressed sensing for networked data. IEEE Signal Process. Mag. **25**(2), 92–101 (2008)
100. K. Hayashi, M. Nagahara, T. Tanaka, A user's guide to compressed sensing for communications systems. IEICE Trans. Commun. **96**(3), 685–712 (2013)
101. A.E. Hoerl, R.W. Kennard, Ridge regression: biased estimation for nonorthogonal problems. Technometrics **12**(1), 55–67 (1970)
102. H. Huang, S. Misra, W. Tang, H. Barani, H. Al-Azzawi, *Applications of Compressed Sensing in Communications Networks* (New Mexico State University, USA, 2013). arXiv preprint arXiv:1305.3002
103. Y. Huang, J.L. Beck, S. Wu, H. Li, Robust Bayesian compressive sensing for signals in structural health monitoring. Comput.-Aid. Civil Infrastruct. Eng. **29**(3), 160–179 (2014)
104. H. Jung, K. Sung, K.S. Nayak, E.Y. Kim, J.C. Ye, K-t FOCUSS: a general compressed sensing framework for high resolution dynamic MRI. Magn. Reson. Med. **61**(1), 103–116 (2009)
105. O. Kardani, A.V. Lyamin, K. Krabbenhoft, A comparative study of preconditioning techniques for large sparse systems arising in finite element analysis. IAENG Intl. J. Appl. Math. **43**(4), 1–9 (2013)
106. S.J. Kim, K. Koh, M. Lustig, S. Boyd, D. Gorinevsky, An interior-point method for large-scale L1-regularized least squares. IEEE J. Sel. Top. Sign. Process. **1**(4), 606–617 (2007)
107. M. Kowalski, B. Torrésani, Structured sparsity: from mixed norms to structured shrinkage, in *Proceedings SPARS'09-Signal Processing with Adaptive Sparse Structured Representations* (2009)
108. G. Kutyniok, Theory and applications of compressed sensing. GAMM-Mitteilungen **36**(1), 79–101 (2013)
109. V. Le Guen, *Cartoon+ Texture Image Decomposition by the TV-L1 Model*. IPOL, Image Processing On Line (2014). http://www.ipol.im/pub/algo/gjmr_line_segment_detector/
110. F. Lenzen, F. Becker, J. Lellmann, Adaptive second-order total variation: an approach aware of slope discontinuities, in *Scale Space and Variational Methods in Computer Vision*, pp. 61–73 (Springer, 2013)
111. Z. Li, Y. Zhu, H. Zhu, M. Li, Compressive sensing approach to urban traffic sensing, in *Proceedings IEEE International Conference Distributed Computing Systems, (ICDCS)*, pp. 889–898 (2011)
112. H. Liebgott, A. Basarab, D. Kouame, O. Bernard, D. Friboulet, Compressive sensing in medical ultrasound, in *Proceedings IEEE International Ultrasonics Symposium, (IUS)*, pp. 1–6 (2012)
113. F. Lin, Z. Hu, S. Hou, J. Yu, C. Zhang, N. Guo, K. Currie, Cognitive radio network as wireless sensor network (ii): Security consideration, in *Proceedings IEEE National Aerospace and Electronics Conference,(NAECON)*, pp. 324–328 (2011)

114. Z. Lin. *Some Software Packages for Partial SVD Computation* (School of EECS, Peking University, 2011). arXiv preprint arXiv:1108.1548
115. Z. Lin, M. Chen, Y. Ma, *The Augmented Lagrange Multiplier Method for Exact Recovery of Corrupted Low-rank Matrices* (Microsoft Research Asia, 2010). arXiv preprint arXiv:1009.5055
116. Z. Lin, A. Ganesh, J. Wright, L. Wu, M. Chen, Y. Ma. *Fast Convex Optimization Algorithms for Exact Recovery of a Corrupted Low-rank Matrix* (Microsoft Research Asia, 2009). http://yima.csl.illinois.edu/psfile/rpca_algorithms.pdf
117. D. Liu, T. Zhou, H. Qian, C. Xu, Z. Zhang, A nearly unbiased matrix completion approach, in *Machine Learning and Knowledge Discovery in Databases*, pp. 210–225 (Springer Verlag, 2013)
118. G. Liu, W. Kang, IDMA-Based compressed sensing for ocean monitoring information acquisition with sensor networks. Math. Probl. Eng. **2014**, 1–13 (2014)
119. C. Luo, F. Wu, J. Sun, C.W. Chen, Compressive data gathering for large-scale wireless sensor networks, in *Proceedings 15th ACM Annual International Conference on Mobile Computing and Networking*, pp. 145–156 (2009)
120. M. Lustig, D. Donoho, J.M. Pauly, Sparse MRI: the application of compressed sensing for rapid MR imaging. Magn. Reson. Med. **58**(6), 1182–1195 (2007)
121. M. Lysaker, X.C. Tai, Iterative image restoration combining total variation minimization and a second-order functional. Int. J. Comput. Vision **66**(1), 5–18 (2006)
122. J. Mairal, M. Elad, G. Sapiro, Sparse representation for color image restoration. IEEE T. Image Process. **17**(1), 53–69 (2008)
123. J. Mairal, B. Yu, *Complexity Analysis of the Lasso Regularization Path* (Department of Statistics, University of California at Berkeley, 2012) arXiv preprint arXiv:1205.0079
124. D. Mascarenas, A. Cattaneo, J. Theiler, C. Farrar, Compressed sensing techniques for detecting damage in structures. Struct. Health Monit. (2013)
125. P. Maurel, J.F. Aujol, G. Peyré, Locally parallel texture modeling. SIAM J. Imag. Sci. **4**(1), 413–447 (2011)
126. D. McMorrow, Compressive sensing for DoD sensor systems. Technical report (MITRE Corp, 2012)
127. J. Meng, H. Li, Z. Han, Sparse event detection in wireless sensor networks using compressive sensing, in *Proceedings IEEE 43rd Annual Conference Information Sciences and Systems,(CISS)*, pp. 181–185 (2009)
128. Y. Meyer, Oscillating patterns in image processing and nonlinear evolution equations: the fifteenth Dean Jacqueline B. Lewis memorial lectures. AMS Bookstore **22** (2001)
129. M. Michenkova, *Numerical Algorithms for Low-rank Matrix Completion Problems* (Swiss Federal Institute of Technology, Zurich, 2011). http://sma.epfl.ch/~anchpcommon/students/michenkova.pdf
130. D. Mumford, J. Shah, Optimal approximations by piecewise smooth functions and associated variational problems. Comm. Pure Appl. Math. **42**(5), 577–685 (1989)
131. S. Mun, J.E. Fowler, Block compressed sensing of images using directional transforms, in *Proceedings IEEE International Conference Image Processing, (ICIP)*, pp. 3021–3024 (2009)
132. M. Nagahara, T. Matsuda, K. Hayashi, Compressive sampling for remote control systems. IEICE Trans. Fundam. Electron. Commun. Comput. Sci. **95**(4), 713–722 (2012)
133. D. Needell, J.A. Tropp, CoSaMP: iterative signal recovery from incomplete and inaccurate samples. Appl. Comput. Harmon. Anal. **26**(3), 301–321 (2009)
134. Y. Nesterov, Smooth minimization of non-smooth functions. Math. Programm. **103**(1), 127–152 (2005)
135. M. Nikolova, A variational approach to remove outliers and impulse noise. J. Math. Imag. Vision **20**(1–2), 99–120 (2004)
136. R. Nowak, M. Figueiredo, Fast wavelet-based image deconvolution using the EM algorithm, in *Proceedings 35th Asilomar Conference Signals, Systems and Computers* (2001)
137. S. Oh, *Matrix Completion: Fundamental Limits and Efficient Algorithms*. Ph.D. thesis, Stanford University, 2010

138. B.A. Olshausen, D.J. Field, Emergence of simple-cell receptive field properties by learning a sparse code for natural images. Nature **381**(6583), 607–609 (1996)
139. B.A. Olshausen, D.J. Field, Sparse coding with an overcomplete basis set: a strategy employed by V1? Vision Res. **37**(23), 3311–3325 (1997)
140. B.A. Olshausen, D.J. Field, Sparse coding of sensory inputs. Curr. Opin. Neurobiol. **14**(4), 481–487 (2004)
141. M.R. Osborne, B. Presnell, B.A. Turlach, A new approach to variable selection in least squares problems. IMA J. Numer. Anal. **20**(3), 389–403 (2000)
142. S. Osher, M. Burger, D. Goldfarb, J. Xu, W. Yin, An iterative regularization method for total variation-based image restoration. Multiscale Model. Simul. **4**(2), 460–489 (2005)
143. S. Osher, J.A. Sethian, Fronts propagating with curvature-dependent speed: algorithms based on Hamilton-Jacobi formulations. J. Comput. Phys. **79**(1), 12–49 (1988)
144. I. Papusha, *Fast Automatic Background Extraction Via Robust PCA* (Caltech, 2011). http://www.cds.caltech.edu/~ipapusha/pdf/robust_pca_apps.pdf
145. N. Parikh, S. Boyd, Proximal algorithms. Found. Trends Optim. **1**(3), 123–231 (2013)
146. J.C. Park, B. Song, J.S. Kim, S.H. Park, H.K. Kim, Z. Liu, W.Y. Song, Fast compressed sensing-based CBCT reconstruction using Barzilai-Borwein formulation for application to on-line IGRT. Med. Phys. **39**(3), 1207–1217 (2012)
147. S. Patterson, Y. C. Eldar, I. Keidar, *Distributed Compressed Sensing for Static and Time-varying Networks* (Rensselaer Polytechnic Institute, 2013). arXiv preprint arXiv:1308.6086
148. Y. Peng, A. Ganesh, J. Wright, W. Xu, Y. Ma, RASL: robust alignment by sparse and low-rank decomposition for linearly correlated images. IEEE Trans. Pattern Anal. Mach. Intell. **34**(11), 2233–2246 (2012)
149. P. Perona, J. Malik, Scale space and edge detection using anisotropic diffusion, in *Proceedings IEEE Computer Society Workshop on Computer Vision*, pp. 16–22 (1987)
150. P. Perona, J. Malik, Scale space and edge detection using anisotropic diffusion. IEEE T. Pattern Anal. Mach. Intell. **12**, 629–639 (1990)
151. G. Peyre, *Matrix Completion with Nuclear Norm Minimization*. A Numerical Tour of Signal Processing (2014). http://gpeyre.github.io/numerical-tours/matlab/sparsity_3_matrix_completion/
152. T. Pock, D. Cremers, H. Bischof, A. Chambolle, An algorithm for minimizing the Mumford-Shah functional, in *Proceedings IEEE 12th International Conference on Computer Vision*, pp. 1133–1140 (2009)
153. S. Qaisar, R.M. Bilal, W. Iqbal, M. Naureen, S. Lee, Compressive sensing: from theory to applications, a survey. J. Commun. Netw. **15**(5), 443–456 (2013)
154. C. Quinsac, A. Basarab, D. Kouamé, Frequency domain compressive sampling for ultrasound imaging. Adv. Acoust. Vib. **1–16**, 2012 (2012)
155. I. Ram, M. Elad, I. Cohen, Image processing using smooth ordering of its patches. IEEE T. Image Process. **22**(7), 2764–2774 (2013)
156. M.A. Rasmussen, R. Bro, A tutorial on the Lasso approach to sparse modeling. Chemom. Intell. Lab. Syst. **119**, 21–31 (2012)
157. H. Rauhut, Compressive sensing and structured random matrices, ed. by Formasier. *Theoretical Foundations and Numerical Methods for Sparse Recovery* (Walter de Gruyter, Berlin, 2010)
158. J. Richy, Compressive sensing in medical ultrasonography. Master's thesis, Kungliga Tekniska Högskolan, 2012
159. P. Rodríguez, B. Wohlberg, Fast principal component pursuit via alternating minimization, in *Proceedings IEEE International Conference on Image Processing, (ICIP)*, pp. 69–79 (2013)
160. J. Romberg, Imaging via compressive sampling (introduction to compressive sampling and recovery via convex programming). IEEE Signal Process. Magz. **25**(2), 14–20 (2008)
161. R. Rubinstein, A.M. Bruckstein, M. Elad, Dictionaries for sparse representation modeling. Proc. IEEE **98**(6), 1045–1057 (2010)
162. R. Rubinstein, M. Zibulevsky, M. Elad, Efficient implementation of the K-SVD algorithm and the Batch-OMP method. Technical report, Department of Computer Science, Technion, Israel, 2008. Technical CS–08

163. L.I. Rudin, S. Osher, E. Fatemi, Nonlinear total variation based noise removal algorithms. Physica D: Nonlinear Phenomena **60**(1), 259–268 (1992)
164. L. Ryzhik, *Lecture Notes for Math 221* (Stanford University, 2013). http://math.stanford.edu/~ryzhik/STANFORD/STANF221-13/stanf221-13-notes.pdf
165. M. Saunders, *PDCO – Primal-dual Interior Methods* (CME 338 Lecture Notes 7, Stanford University, 2013). https://web.stanford.edu/group/SOL/software/pdco/pdco.pdf
166. M Schmidt, Least squares optimization with L1-norm regularization. Technical report, University of British Columbia, 2005. Project Report
167. I. Selesnick, *Total Variation Denoising (an MM Algorithm)* (New York University, 2012). http://eeweb.poly.edu/iselesni/lecture_notes/TVDmm/TVDmm.pdf
168. I.W. Selesnick, *Sparse Signal Restoration* (New York University, 2010). http://eeweb.poly.edu/iselesni/lecture_notes/sparse_signal_restoration.pdf
169. I.W. Selesnick, I. Bayram, *Total Variation Filtering* (New York University, 2010). http://eeweb.poly.edu/iselesni/lecture_notes/TVDmm/
170. Y. Shen, Z. Wen, Y. Zhang, Augmented Lagrangian alternating direction method for matrix separation based on low-rank factorization. Optim. Meth. Softw. **29**(2), 239–263 (2014)
171. E.P. Simoncelli, B.A. Olshausen, Natural image statistics and neural representation. Ann. Rev. Neurosci. **24**(1), 1193–1216 (2001)
172. J.L. Starck, M. Elad, D. Donoho, Redundant multiscale transforms and their application for morphological component separation. Adv. Imag. Electron Phys. **132**(82), 287–348 (2004)
173. J.L. Starck, Y. Moudden, J. Bobin, M. Elad, D.L. Donoho, Morphological component analysis. Opt. Photon. 1–15 (2005)
174. J.L. Starck, F. Murtagh, J.M. Fadilii, *Sparse Image and Signal Processing* (Cambridge University Press, 2010)
175. D. Sundman, Compressed sensing: algorithms and applications. Master's thesis, KTH Electrical Engineering, Stockholm, 2012. Licenciate thesis
176. R. Tibshirani, Regression shrinkage and selection via the Lasso. J. Royal Stat. Soc. **58**, 267–288 1996). series B
177. J.A. Tropp, S.J. Wright, Computational methods for sparse solution of linear inverse problems. Proc. IEEE **98**(6), 948–958 (2010)
178. R. Valentine, *Image Segmentation with the Mumford Shah Functional* (2007). http://coldstonelabs.org/files/science/math/Intro-MS-Valentine.pdf
179. S.A. Van De Geer, P. Bühlmann, On the conditions used to prove oracle results for the Lasso. Electron. J. Stat. **3**, 1360–1392 (2009)
180. L.A. Vese, S.J. Osher, Modeling textures with total variation minimization and oscillating patterns in image processing. J. Sci. Comput. **19**(1–3), 553–572 (2003)
181. S. Wang, Z. Zhang, Colorization by matrix completion, in *Proceedings 26th AAAI Conference on Artificial Intelligence*, pp. 1169–1175 (2012)
182. Z. Wang, M.J. Lai, Z. Lu, J. Ye, *Orthogonal Rank-one Matrix Pursuit for Low Rank Matrix Completion* (The Biodesign Institue, Arizona State University, 2014) arXiv preprint arXiv:1404.1377
183. S.J. Wei, X.L. Zhang, J. Shi, G. Xiang, Sparse reconstruction for SAR imaging based on compressed sensing. Prog. Electromagn. Res. **109**, 63–81 (2010)
184. J. Weickert, *Anisotropic Diffusion in Image Processing*, vol. 1 (Teubner, Stuttgart, 1998)
185. S. Weisberg, *Applied Linear Regression* (Wiley, 1980)
186. S.J. Wright, R.D. Nowak, A.T. Figueiredo, Sparse reconstruction by separable approximation. IEEE Trans. Signal Process. **57**(7), 2479–2493 (2009)
187. H. Xu, C. Caramanis, S. Mannor, Robust regression and Lasso. IEEE T. Inf. Theory **56**(7), 3561–3574 (2010)
188. X. Xu, Online robust principal component analysis for background subtraction: a system evaluation on Toyota car data. Master's thesis, University of Illinois at Urbana-Champaign, 2014
189. Y. Xu, W. Yin, A fast patch-dictionary method for whole-image recovery. Technical report, UCLA, 2013. CAM13-38

190. M. Yan, Y. Yang, S. Osher, Exact low-rank matrix completion from sparsely corrupted entries via adaptive outlier pursuit. J. Sci. Comput. **56**(3), 433–449 (2013)
191. S. Yan, C. Wu, W. Dai, M. Ghanem, Y. Guo, Environmental monitoring via compressive sensing, in *Proceedings ACM International Workshop on Knowledge Discovery from Sensor Data*, pp. 61–68 (2012)
192. J. Yang, Y. Zhang, Alternating direction algorithms for L1 problems in compressive sensing. SIAM J. Sci. Comput. **33**(1), 250–278 (2011)
193. W. Yin, D. Goldfarb, S. Osher, Total variation based image cartoon-texture decomposition. Technical report, Columbia Univ., Dep. Industrial Eng. & Operation Res., 2005. Rep. CU-CORC-TR-01
194. W. Yin, D. Goldfarb, S. Osher, A comparison of three total variation based texture extraction models. J. Vis. Commun. Image Represent. **18**(3), 240–252 (2007)
195. M. Yuan, Y. Lin, Model selection and estimation in regression with grouped variables. J. Royal Stat Soc **68**(1), 49–67 (2007). series B
196. X. Yuan, J. Yang, *Sparse and low-rank matrix decomposition via alternating direction methods* (Nanjing University, China, 2009). http://math.nju.edu.cn/~jfyang/files/LRSD-09.pdf
197. B. Zhang, X. Cheng, N. Zhang, Y. Cui, Y. Li, Q. Liang, Sparse target counting and localization in sensor networks based on compressive sensing, in *Proceedings IEEE INFOCOM*, pp. 2255–2263 (2011)
198. J. Zhang, T. Tan, Brief review of invariant texture analysis methods. Pattern Recogn. **35**(3), 735–747 (2002)
199. Z. Zhang, A. Ganesh, X. Liang, Y. Ma, TILT: Transform invariant low-rank textures. Int. J. Comput. Vis. **99**(1), 1–24 (2012)
200. C. Zhao, X. Wu, L. Huang, Y. Yao, Y.C. Chang, Compressed sensing based fingerprint identification for wireless transmitters. Sci. World J. **1–9**, 2014 (2014)
201. P. Zhao, B. Yu, On model selection consistency of Lasso. J. Mach. Learn. Res. **7**, 2541–2563 (2006)
202. S. Zhou, L. Kong, N. Xiu, New bounds for RIC in compressed sensing. J. Oper. Res. Soc. China **1**(2), 227–237 (2013)
203. X. Zhou, W. Yu, Low-rank modeling and its applications in medical image analysis. SPIE Defense Security Sens. 87500V (2013)
204. Z. Zhou, X. Li, J. Wright, E. Candes, Y. Ma, Stable principal component pursuit, in *Proceedings IEEE International Symposium on Information Theory, (ISIT)*, pp. 1518–1522 (2010)
205. J. Zhu, D. Baron. Performance regions in compressed sensing from noisy measurements, in *Proceedings IEEE 47th Annual Conference on Information Sciences and Systems, (CISS)*, pp. 1–6 (2013)
206. H. Zou, The adaptive Lasso and its oracle properties. J. Am. Stat. Assoc. **101**(476), 1418–1429 (2006)
207. H. Zou, T. Hastie, Regularization and variable selection via the elastic net. J. Royal Stat. Soc. **67**(2), 301–320 (2005). series B

Appendix A
Selected Topics of Mathematical Optimization

A.1 Introduction

This appendix is mostly intended for supporting the chapter on sparse representations. Therefore, some pertinent optimization topics have been selected for this purpose.

Presently, optimization is a broad discipline covering several types of problems. The usual way to introduce the subject in education activities, is by finding the points of an objective function where the gradient is zero, and then using second order derivatives (the Hessian matrix) to classify these points: if the Hessian is positive definite, then it is a local minimum; if the Hessian is negative definite, it is a local maximum; if indefinite, it is a kind of saddle point. Of course, this analysis corresponds to twice-differentiable functions. The points where the gradient is zero are called stationary points.

In real life, there are constraints to be considered. For instance, one could want to go non-stop with the car a very long distance, but the fuel tank has a certain size that poses a limit.

Constrained optimization problems could be converted to unconstrained ones by using Lagrange multipliers.

When both the objective function and the constraints are linear, it is possible to apply Linear Programming, which is a very important topic in the world of economy and industrial production.

As it will be treated in this appendix, there are more general scenarios that can be considered, based on Quadratic Programming, Convex Programming, Conic Programming, Semi-definite Programming, or Non-smooth Optimization.

In some cases, there are non-continuous variables. For instance, in integer programming, there are variables that only can take integer values; like for example how many shoes in a container.

There are several computational techniques that can be used for optimization purposes; like for instance the famous simplex algorithm (see next subsection). A number of iterative methods exist, like the Newton's method. Some of these iterative methods are based on evaluating gradients and Hessians, others just evaluate

J.M. Giron-Sierra, *Digital Signal Processing with Matlab Examples, Volume 3*,
Signals and Communication Technology, DOI 10.1007/978-981-10-2540-2

gradients, and in the case of heuristic methods they simply use function values at certain points.

In this appendix only a subset of optimization topics have been selected. There are other topics that have been not mentioned, like, for instance, a classical branch of optimization theory which is based on variational approaches. Instead of looking for stationary points, the question is to find functions (extremals) that optimize a certain criterion. Important names in this context are Euler, Weierstrass, Jacobi, Legendre, Mayer, Bolza, Pontryagin, etc. An important algorithm belonging to this area is Dynamic Programming, by Bellman. Typical applications of this algorithm are shortest path or critical-path scheduling problems.

The next section, devoted to Linear Programming, serves also for the introduction of some important concepts and terms.

The matters of this appendix are treated at introductory level and trying to include illustrative examples. Some of the many books that could be recommended for background are [24, 87].

A.2 Linear Programming and the Simplex Algorithm

The term '*linear programming*' is due to G.B. Dantzig, who published the simplex algorithm in 1947. This algorithm is considered as one of the ten most important algorithms of the previous century. Much of the theory on which it is based was introduced by Kantorovich in 1939. It is also worth mentioning the work of J. von Neumann on the theory of duality (also in 1947).

Consider the following optimization problem:

Maximize:

$$z = c_1 x_1 + c_2 x_2 + \ldots + c_n x_n \tag{A.1}$$

Subject to (constraints):

$$a_{11} x_1 + a_{12} x_2 + \ldots + a_{1n} x_n \leq b_1 \tag{A.2}$$

$$a_{21} x_1 + a_{22} x_2 + \ldots + a_{2n} x_n \leq b_2 \tag{A.3}$$

- - - - - - - - - - - - - - - - - - - -

$$a_{m1} x_1 + a_{m2} x_2 + \ldots + a_{mn} x_n \leq b_m \tag{A.4}$$

(the variables take continuous real values; inequalities can also be $\geq$)

This is a '*linear programming*' problem.

A.2.1 Analysis Based on Geometry

Perhaps the most important thing to do now, in order to solve the problem, is to adopt the point of view of geometry. Let us start by saying that the following expression:

$$a_1 x_1 + a_2 x_2 + \ldots + a_n x_n = b \tag{A.5}$$

corresponds to a hyperplane in n-dimensions. In the case of $n = 2$, this is a line. An equivalent expression of the hyperplane is the following:

$$\mathbf{a}^T \mathbf{x} = b \tag{A.6}$$

Two hyperplanes with the following equations:

$$\mathbf{a}^T \mathbf{x} = b_1 \tag{A.7}$$

$$\mathbf{a}^T \mathbf{x} = b_2 \tag{A.8}$$

are parallel.

A hyperplane separates the n-dimensional space into two half spaces.

In the case of an inequality:

$$\mathbf{a}^T \mathbf{x} \leq b \tag{A.9}$$

this also separates the n-dimensional space into two half spaces; one of which satisfies the inequality.

The constraints of the linear programming problem define a set of points **x** inside an n-dimensional region formed by the intersection of hyperplanes and half spaces.

For example, suppose the following constraints:

$$3x_1 + 5x_2 \geq 15$$

$$4\,x_1 + 9\,x_2 \leq 36$$

$$x_1\,,\; x_2 \geq 0$$

Let us plot in Fig. A.1 the region defined by these constraints. This region is called the feasible solution set.

Usually the feasible solution region is bounded, in which case it is a polygon (2D), or a polyhedron (3D), or a polytope (nD).

Now, suppose that the function to maximize is:

$$z = 2\,x_1 - 4\,x_2 \tag{A.10}$$

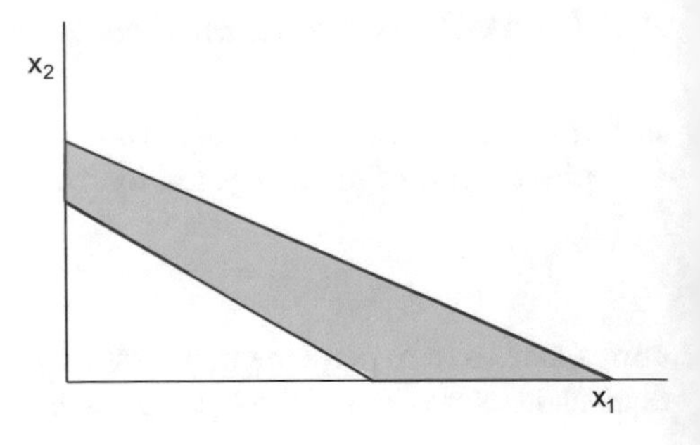

Fig. A.1 The feasible solution set

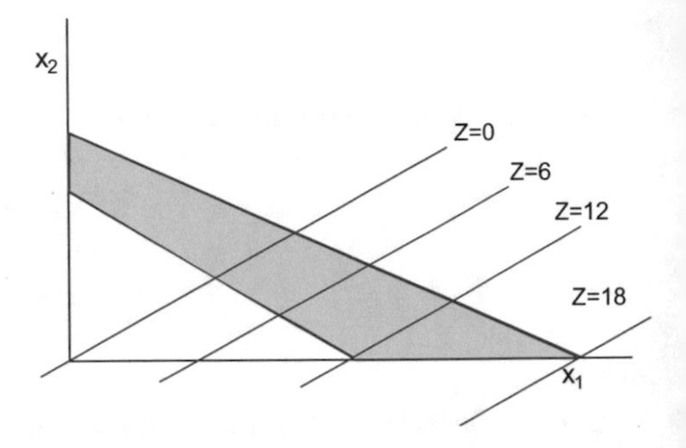

Fig. A.2 Some values of the objective function

subject to the constraints above. Figure A.2 shows some parallel lines obtained for different values of z.

The key observation is that the maximum of z is attained at a vertex of the feasible solution region.

Let us say in advance that the simplex algorithm tests adjacent vertices in sequence, so that at each new vertex the objective function increases, or at least remains unchanged. Therefore, it is convenient to study in more detail vertices and segments.

A segment between two points $\mathbf{x}_1$ and $\mathbf{x}_2$ is the set of points $\mathbf{x}$ given by:

$$\mathbf{x} = \lambda x_2 + (1 - \lambda)\, x_1 \;, \;\; 0 \leq \lambda \leq 1 \tag{A.11}$$

And now a very important concept. A set C is ***convex*** if for all $\mathbf{x}_1, \mathbf{x}_2 \in C$ the segment between $\mathbf{x}_1$ and $\mathbf{x}_2$ is composed entirely of points of C.

The hyperplanes and the half-spaces are convex. The intersection of a finite number of convex sets is convex.

A point $\mathbf{x} \in, C$ is an *extreme point* of C if there is no segment between two points of C containing $\mathbf{x}$.

A point $\mathbf{x} \in C$ is an *interior point* of C if around it there is a ball made entirely of points of C.

By adding artificial variables, all inequalities can be converted to equalities. This way, the linear programming problem can be compactly expressed as follows:

Maximize:

$$z = \mathbf{c}^T \cdot \mathbf{x} \tag{A.12}$$

Subject to:

$$A\,\mathbf{x} = \mathbf{b}\;,\;\; \mathbf{x} \geq \mathbf{0} \tag{A.13}$$

Already one has the following facts. The constraints define a set F of feasible solutions. It is a convex set. The maximum of the objective function takes place on an extreme point of F.

Now, it can be shown that if there is a set of linearly independent columns of A, $\mathbf{a}_1, \mathbf{a}_2, \mathbf{a}_k,\;\; (k \leq m)$, such that:

$$x_1\,\mathbf{a}_1 + x_2\,\mathbf{a}_2 + \ldots + x_k\,\mathbf{a}_k = \mathbf{b}\,,\;\; \forall x_j \geq 0 \tag{A.14}$$

then the point $\mathbf{x} = (x_1,\, x_2,\, \ldots,\, x_k,\, 0,\, \ldots,\, 0)$ is an extreme point of F.

There are at most $\begin{pmatrix} n \\ m \end{pmatrix}$ sets of m linearly independent columns of A. One could check all of them, to determine the optimum value of the objective function, but this is ordinarily an excessive effort. There are optimization problems involving hundreds or even thousands of variables. Fortunately, the simplex algorithm provides a more economic way.

A.2.2 *The Simplex Algorithm*

The basis of a matrix is a maximal linearly independent set of columns of that matrix. Suppose that a basis of the matrix A has been found, and form the matrix B with the columns $\mathbf{a}_1, \mathbf{a}_2, \mathbf{a}_m$ of this basis. Any other column $\mathbf{a}_j$ of A can be written in function of the basis:

$$\mathbf{a}_j = y_{1j}\,\mathbf{a}_1 + y_{2j}\,\mathbf{a}_2 + \ldots + y_{m\,j}\,\mathbf{a}_m \tag{A.15}$$

Or, equivalently:

$$\mathbf{a}_j = B\;\mathbf{y}_j \tag{A.16}$$

As seen before, there is an extreme point associated to B, and in this point the objective function has the value z_B. Now, we want to substitute one of the columns of B by another column of A, so a new basis B' was obtained and such that $z'_B \geq z_B$ (z'_B is the value of the objective function at the extreme point associated to B'). There are two questions, what column to extract from B, and what column to take from A.

Consider a partition of the constraint equations:

$$A\,\mathbf{x} = [B, N]\,\mathbf{x} = [B,\; N] \cdot \begin{bmatrix} \mathbf{x}_B \\ \mathbf{x}_N \end{bmatrix} = \mathbf{b} \tag{A.17}$$

Basic feasible solutions of the constraint equations are feasible solutions with no more than m positive entries. One of these solutions will be such that:

$$\mathbf{x}_N = \mathbf{0} \tag{A.18}$$

$$B\,\mathbf{x}_B = \mathbf{b} \tag{A.19}$$

This second equation can be also written as follows:

$$x_{B1}\,\mathbf{a}_1 + x_{B2}\,\mathbf{a}_2 + \ldots + x_{Bm}\,\mathbf{a}_m = \mathbf{b} \tag{A.20}$$

After an algebraic study, the simplex algorithm concludes that the column $\mathbf{a}_r$ to extract from B should be such that:

$$\frac{x_{Br}}{y_{rj}} = \min_i \left\{ \frac{x_{Bi}}{y_{ij}},\; y_{ij} > 0 \right\} \tag{A.21}$$

This was the answer to the first question. For the second question, denote:

$$z_j = \sum_{i=1}^{m} c_i\, y_{ij} \tag{A.22}$$

The column $\mathbf{a}_j$ to take from A should be such that it maximizes the following expression:

$$\frac{x_{Br}}{y_{rj}}\,(c_j - z_j) \tag{A.23}$$

In practice, linear programming problems are handled with a specific tableau format, so the algorithm can be applied in a clear systematic way. Initial solutions are easily forced via artificial variables.

Let us present an example of using the tableau format [105]. The problem to solve is:

Maximize:

$$z = 7\,x_1 + 5\,x_2 \tag{A.24}$$

Subject to:

$$2\,x_1 + x_2 \leq 100 \tag{A.25}$$

$$4\,x_1 + 3\,x_2 \leq 240 \tag{A.26}$$

Slack variables are added:

$$2\,x_1 + x_2 + s_1 = 100 \tag{A.27}$$

$$4\,x_1 + 3\,x_2 + s_2 = 240 \tag{A.28}$$

The first tableau is easily built as follows:

$c_j \rightarrow$		7	5	0	0	
$\downarrow$	Basis	x_1	x_2	s_1	s_2	b
0	s_1	2	1	1	0	100
0	s_2	4	3	0	1	240
	z_j	0	0	0	0	0
	$c_j - z_j$	7	5	0	0	

Notice that the initial tableau is filled with just the problem statement: the objective function and the constraints. If all numbers in the last row were zero or negative, the optimum was reached; but this is not the case, the algorithm should continue.

Now, the effort concentrates in selecting *a pivot*. The pivot column is where $c_j - z_j$ is maximum; so it is the x_1 column. Now, compute 100/2 and 240/4 (the numbers 100 and 240 belong to the b column; the numbers 2 and 4 belong to the pivot column). The smallest non-negative result tell us the pivot row, which in this case is the s_1 row. Therefore, s_1 abandons the basis, and is substituted by x_1. The value of the pivot was 2. The next tableau is the following:

$c_j \rightarrow$		7	5	0	0	
$\downarrow$	Basis	x_1	x_2	s_1	s_2	b
7	x_1	1	1/2	1/2	0	50
0	s_2	0	1	-2	1	40
	z_j	7	7/2	7/2	0	350
	$c_j - z_j$	0	3/2	-7/2	0	

The x_1 row has been obtained from the s_1 row in the previous tableau, dividing it by the pivot.

Take the *kth* entry of the new s_2 row. It has been obtained from the *kth* entry of the old s_2 row minus a quantity u_k; where u_k is the product of the number of the old s_2 row which is below the pivot (4), and the kth entry of the new x_1 row. That is:

$$s_2(1) = 4 - 4 \cdot 1; \;\; s_2(2) = 3 - 4 \cdot 1/2; \;\; s_2(3) = 0 - 4 \cdot 1/2;$$
$$s_2(4) = 1 - 4 \cdot 0; \;\; s_2(5) = 240 - 4 \cdot 50$$

The values of the z_j row are obtained from the column of c_j and the $x_1, x_2, s_1, s_2,$ columns as follows:

$$z_j(1) = 7 \cdot 1 + 0 \cdot 0; \;\; z_j(2) = 7 \cdot 1/2 + 0 \cdot 1; \;\; z_j(3) = 7 \cdot 1/2 + 0 \cdot -2;$$
$$z_j(4) = 7 \cdot 0 + 0 \cdot 1; \;\; z_j(5) = 7 \cdot 50 + 0 \cdot 40$$

The next pivot would be the 1 at the intersection of the x_2 column and the s_2 row. A third tableau should be computed, which results to be the final one with the optimum.

Since this is a brief summary, no space has been conceded to some special situations that may arise, concerning some steps of the algorithm. Also, there are revised versions of the basic algorithm that would deserve a more detailed study. Fortunately, there is a vast literature on linear programming that the interested reader may easily reach [29, 40, 112]

A word of caution: the term 'simplex' is also used to designate a set of $n + 1$ points in a n-dimensional space. There are iterative optimization methods based on the use of simplexes. These methods should not be confused with the simplex algorithm just described.

A.2.3 Application to FIR Filter Design

One of the first papers proposing the use of linear programming for the design of FIR filters was [90] in 1972.

The main idea for writing the design problem in linear programming format was to discretize the desired frequency response along a set of frequencies $\Omega = \{\omega_1, \omega_2, \omega_m\}$. A dense frequency sampling (not necessarily regularly spaced) is recommended.

The frequency response of a FIR filter can be written as follows:

$$H(\omega) = h_0 + 2 \sum_{k=1}^{N} h_k \cos(\omega\, k) \tag{A.29}$$

Suppose that the problem is to minimize a weighted error:

$$E(\omega) = W(\omega)\,[H(\omega) - D(\omega)] \tag{A.30}$$

The minimization target could be specified as follows: find the FIR filter coefficients h_k such that δ is minimized, for $0 \leq \omega \leq \pi, \ |E(\omega)| < \delta$.

According with [102] the problem can now be specified as:

Minimize:

$$\delta \tag{A.31}$$

Subject to:

$$H(\omega_k) - \frac{\delta}{W(\omega_k)} \leq D(\omega_k), \ 1 \leq k \leq m \tag{A.32}$$

$$(-H(\omega_k)) - \frac{\delta}{W(\omega_k)} \leq -D(\omega_k), \ 1 \leq k \leq m \tag{A.33}$$

More or less connected with this approach, a number of proposals have been made for using linear programming or other optimization methods for the design of FIR and IIR filters, [28, 91, 113]. In [8] linear programming is applied for the design of sparse filters.

A.2.4 Special Circumstances

In practical optimization application, some special circumstances may appear. Frequently, they are related to the feasible solution region F.

For instance, a bad specification of constraints may lead to an empty feasible solution set; usually this would be due to unconsistent (contradictory) constraints. In this case, there is no solution.

Other cases are better illustrated with figures. For example, Fig. A.3 shows an unbounded F and an objective function that can increase without limit inside F. Hence, the solution is unbounded.

Figure A.4 shows another case with the same unbounded F, but with an unique solution at a vertex.

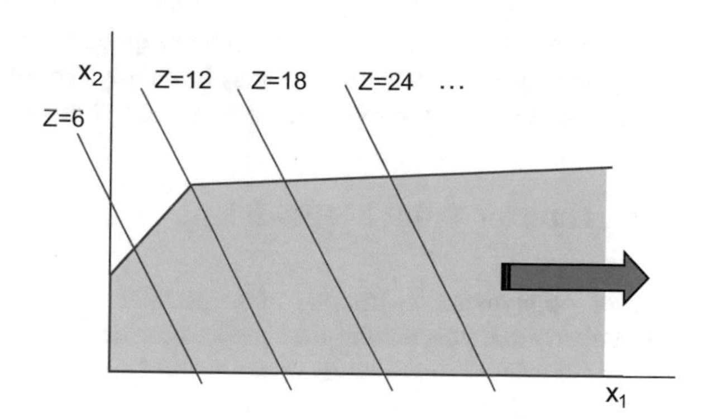

Fig. A.3 A case with unbounded solution

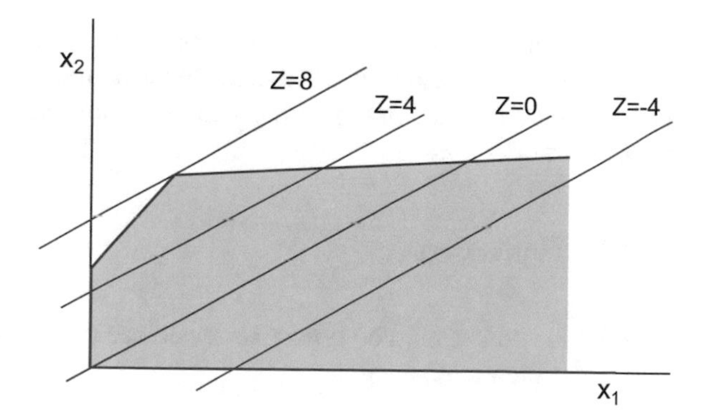

Fig. A.4 A case with unique solution

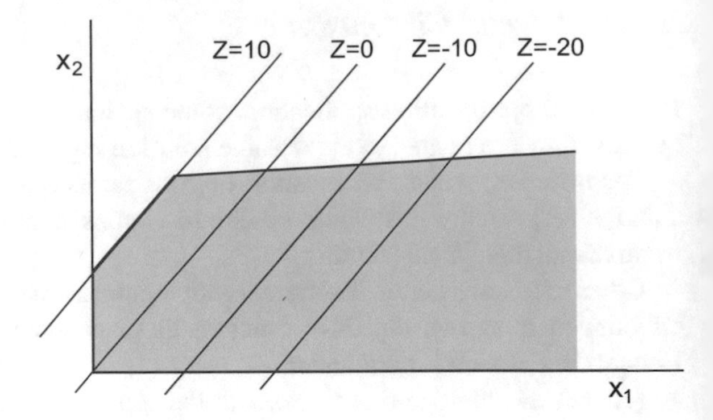

Fig. A.5 A case with infinite solutions

Finally, Fig. A.5 shows a case with infinite (good) solutions, since the optimal objective function is coincident with a line segment joining two vertices.

Just from these first examples it may be clearly sensed that the deep mathematical analysis of optimization topics requires a strong basis of topology.

A.3 Interior Point Methodology

Interior point methods became quite popular after 1984, when Karmarkar [60] announced a fast polynomial-time method for linear programming. In fact, the comment of specialists was that an *interior-point revolution* emerged from then. It also came to headlines of newspapers, like The New York Times or The Wall Street Journal.

The basic method is iterative; it starts from an interior point of F and then it uses a modified Newton's method to reach the solution. Along the algorithm steps, a primal-dual optimization problem is treated. More details of these aspects are given now, as preliminaries.

A.3.1 Preliminaries

Most of the concepts to be briefly reviewed are classic, so more details are easy to find in the literature.

A.3.1.1 The Newton's Method

The iterative Newton's method is used to solve:

$$f(x) = 0 \tag{A.34}$$

where $g(x)$ is a smooth function.

The method is based on a Taylor series approximation:

$$f(x + \Delta x) \approx f(x) + f'(x)\,\Delta x \tag{A.35}$$

The Newton step is given by:

$$\Delta x = -\frac{f(x)}{f'(x)} \tag{A.36}$$

And the iteration is:

$$x^{(k+1)} = x^{(k)} + \Delta x^{(k)} = x^{(k)} - \frac{f(x^{(k)})}{f'(x^{(k)})} \tag{A.37}$$

When the method converges, it does so quadratically. Of course, the method can be effortlessly generalized for n dimensions, using gradients.

Suppose you want to find the minimum of a function $g(x)$ using the Newton's method. Then, you take $f(x) = g'(x)$. In case of n variables, with $g(x_1, x_2, \ldots, x_n)$, the Newton step would be:

$$\Delta \mathbf{x} = -\frac{\nabla g(\mathbf{x})}{\nabla^2 g(\mathbf{x})} \tag{A.38}$$

so there is a gradient in the numerator and a Hessian in the denominator.

A.3.1.2 Lagrange Multipliers

Suppose you have the following optimization problem:

Minimize:

$$f(\mathbf{x}) \tag{A.39}$$

Subject to:

$$g_i(\mathbf{x}) = 0, \; i = 1, 2, \ldots, m \tag{A.40}$$

The idea of Lagrange for solving this problem is to use a set of multipliers λ_i to build the following function:

$$L(\mathbf{x}, \boldsymbol{\lambda}) = f(\mathbf{x}) - \sum_{i=1}^{m} \lambda_i\, g_i(\mathbf{x}) \tag{A.41}$$

And then minimize this function, which is called the Lagrangian function, by solving the following equation system:

$$\frac{\partial L}{\partial x_j} = \frac{\partial f}{\partial x_j} - \sum_{i=1}^{m} \lambda_i \cdot \frac{\partial g_i}{\partial x_j} = 0, \;\; j = 1, 2, \ldots, n \tag{A.42}$$

$$\frac{\partial L}{\partial \lambda_i} = -\, g_i(\mathbf{x}) = 0, \quad i = 1, 2, \ldots, m \tag{A.43}$$

These equations can be solved by Newton's method.

A.3.1.3 Inequalities. Duality

Let us take into account inequalities in the constraints, so the problem becomes:

Minimize:

$$f(\mathbf{x}) \tag{A.44}$$

Subject to:

$$h_i(\mathbf{x}) \le 0, \; i = 1, 2, \ldots, m \tag{A.45}$$

$$l_i(\mathbf{x}) = 0, \; i = 1, 2, \ldots, r \tag{A.46}$$

This problem will be denoted as the '*primal problem*'.

The following Lagrangian is now formed:

$$L(\mathbf{x}, \mathbf{u}, \mathbf{v}) = f(\mathbf{x}) + \sum_{i=1}^{m} u_i \, h_i(\mathbf{x}) + \sum_{i=1}^{r} v_i \, l_i(\mathbf{x}) \tag{A.47}$$

A Lagrangian *dual function* is considered:

$$q(\mathbf{u}, \mathbf{v}) = \min_{\mathbf{x}} L(\mathbf{x}, \mathbf{u}, \mathbf{v}) \tag{A.48}$$

(there are specialists that prefer to use:

$$q(\mathbf{u}, \mathbf{v}) = \inf_{\mathbf{x}} L(\mathbf{x}, \mathbf{u}, \mathbf{v}) \tag{A.49}$$

The infimum is used instead of the minimum, because the Lagrangian might not have a minimum on F).

An important fact is the following: denote as f^* the solution of the primal problem; then $q(\mathbf{u}, \mathbf{v})$ is always a lower bound of f^*. This property is called '*weak duality*'.

Then, it would make sense to look for a set of values $(\mathbf{u}^*, \mathbf{v}^*)$ such that the Lagrangian dual function was maximized. This leads to the Lagrange '*dual problem*':

Maximize:

$$q(\mathbf{u}, \mathbf{v}) \tag{A.50}$$

Subject to:

$$\mathbf{u} \ge \mathbf{0} \tag{A.51}$$

The dual problem is always convex. The primal and dual optimal values, f^* and q^*, always satisfy weak duality: $f^* \geq q^*$.

The difference:

$$\eta(\mathbf{x}, \mathbf{u}, \mathbf{v}) = f(\mathbf{x}) - q(\mathbf{u}, \mathbf{v}) \tag{A.52}$$

is called the *'duality gap'* between **x** and **u**, **v**.

Under certain conditions, it is possible to have $f^* = q^*$; in this case we have *'strong duality'*, and the duality gap is zero. For instance, the Slater's condition for convex primal problems establishes that if there is an **x** such that:

$$h_1(\mathbf{x}) < 0,\ h_2(\mathbf{x}) < 0,\ \ldots,\ h_m(\mathbf{x}) < 0 \ \ ,$$

and

$$l_1(\mathbf{x}) = 0,\ l_2(\mathbf{x}) = 0,\ \ldots,\ l_r(\mathbf{x}) = 0$$

then we have strong duality.

A.3.1.4 KKT Conditions

In case of strong duality, one can write:

$$\begin{aligned} f^*(\mathbf{x}) &= q(\mathbf{u}^*, \mathbf{v}^*) = \inf_{\mathbf{x}}(f(\mathbf{x}) + \sum_{i=1}^{m} u_i^* h_i(\mathbf{x}) + \sum_{i=1}^{r} v_i^* l_i(\mathbf{x})) \\ &\leq f(\mathbf{x}^*) + \sum_{i=1}^{m} u_i^* h_i(\mathbf{x}^*) \leq f(\mathbf{x}^*) \end{aligned} \tag{A.53}$$

Then: $\sum_{i=1}^{m} u_i^* h_i(\mathbf{x}^*) = 0$. This means that:

$$u_i^* h_i(\mathbf{x}^*) = 0, \ \ i = 1, 2, \ldots, m \tag{A.54}$$

which is called the *'complementary slackness'* condition.

Joining together the facts associated with strong duality, one obtains the *Karush-Kuhn-Tucker (KKT)* optimality conditions:

$$h_i(\mathbf{x}^*) \leq 0\,,\ i = 1, 2,\ \ldots, m \tag{A.55}$$

$$l_i(\mathbf{x}^*) = 0\,,\ i = 1, 2,\ \ldots, r \tag{A.56}$$

$$u_i^* \geq 0\,,\ i = 1, 2,\ \ldots, m \tag{A.57}$$

$$u_i^* h_i(\mathbf{x}^*) = 0\,,\ i = 1, 2,\ \ldots, m \tag{A.58}$$

$$\nabla f(\mathbf{x}^*) + \sum_{i=1}^{m} u_i^* \nabla h_i(\mathbf{x}^*) + \sum_{i=1}^{r} v_i^* \nabla l_i(\mathbf{x}^*)) = 0 \tag{A.59}$$

Conversely, if the problem is convex and $\mathbf{x}^*$, $\mathbf{u}^*$, $\mathbf{v}^*$ satisfy the KKT conditions, then these values are (primal, dual) optimal.

Time ago the above conditions were named KT conditions since they appeared in a publication of H.W. Kuhn and A.W. Tucker, in 1951. Later on, it was discovered that they were stated in the unpublished Master's Thesis of W. Karush in 1939.

A.3.1.5 Examples

Next two examples were taken from [13]. The first example is a knapsack problem:

Minimize: $z = 3x_1 + 4x_2 + 5x_3 + 5x_4$
Subject to: $7x_1 + 5x_2 + 4x_3 + 3x_4 \geq 17; \quad \forall x_i \in \{0, 1\}$

The Lagrangian would be:

$$L(\mathbf{x}, \lambda) = (3 - 7\lambda)\,x_1 + (4 - 5\lambda)\,x_2 + (5 - 4\lambda)\,x_3 + (5 - 3\lambda)\,x_4 + 17\lambda$$

Clearly, in order to minimize L the solution is:

$x_1 = 1,\ if\ (3 - 7\lambda) < 0$; or $x_1 = 0$, otherwise
$x_2 = 1,\ if\ (4 - 5\lambda) < 0$; or $x_2 = 0$, otherwise
$x_3 = 1,\ if\ (5 - 4\lambda) < 0$; or $x_3 = 0$, otherwise
$x_4 = 1,\ if\ (5 - 3\lambda) < 0$; or $x_4 = 0$, otherwise

If we now consider the values of λ, the optimal solutions would be:

$\lambda < 3/7 \Rightarrow x = (0, 0, 0, 0);\ L = 17\lambda$
$3/7 \leq \lambda < 4/5 \Rightarrow x = (1, 0, 0, 0);\ L = 3 + 10\lambda$
$4/5 \leq \lambda < 5/4 \Rightarrow x = (1, 1, 0, 0);\ L = 7 + 5\lambda$
$5/4 \leq \lambda < 5/3 \Rightarrow x = (1, 1, 1, 0);\ L = 12 + \lambda$
$5/3 \leq \lambda \Rightarrow x = (1, 1, 1, 1);\ L = 17 - 2\lambda$

The largest value of L would be: $L(5/3) = 12 + (5/3)$. This is the best (highest) lower bound of the optimization problem. Actually, the solution of this problem is $z = 12 + 5/3$ at the point $x = (1, 1, 1, 1/3)$. Therefore, in this case the best lower bound was equal to the optimum value.

The second example considers a nonlinear programming problem:

Minimize: $z = x^2 + 2y^2$
Subject to:

$$x + y \geq 3$$

$$y - x^2 \geq 1$$

The Lagrangian would be:

$$L(\mathbf{x}, \boldsymbol{\lambda}) = x^2 + 2y^2 - \lambda_1(x + y - 3) - \lambda_2(y - x^2 - 1)$$

Then, the KKT conditions would be:

$$x + y \geq 3$$

$$y - x^2 \geq 1$$

$$\lambda_1(x + y - 3) = 0$$

$$\lambda_2(y - x^2 - 1) = 0$$

$$\frac{\partial L}{\partial x} = 2x - \lambda_1 + 2\lambda_2 x = 0$$

$$\frac{\partial L}{\partial y} = 4y - \lambda_1 - \lambda_2 = 0$$

$$\lambda_1, \lambda_2 \geq 0$$

If one analyzes the cases of ($\lambda_1 = 0,\ \lambda_2 = 0$), ($\lambda_1 > 0,\ \lambda_2 = 0$), ($\lambda_1 = 0,\ \lambda_2 > 0$), none of them can satisfy all conditions.

In the case of ($\lambda_1 > 0,\ \lambda_2 > 0$), the study of:

$$(x + y - 3) = 0$$

$$(y - x^2 - 1) = 0$$

gives two possible solutions of (x, y): $(-2, 5)$ or $(1, 2)$, of which only $(1, 2)$ can satisfy the other equations. Therefore, the optimum is found at the point $(1, 2)$, being $z = 9$.

A.3.1.6 Barrier Functions. Central Path

Usually some of the constraints are of the type $x_i \geq 0$ (non-negativity of variables). The idea of the barrier function is to keep solutions at the interior of F building a barrier that prevents variables from reaching the boundary $x_i = 0$. For instance, in the case of:

minimize $f(\mathbf{x})$ subject to $\mathbf{x} \geq 0$

an equivalent unconstrained problem is written as:

minimize $B(\mathbf{x}|\mu) = f(\mathbf{x}) - \mu \sum_{i=1}^{m} \log(x_i)$

A simple and interesting example is given in [76]:

minimize $(x_1 + 1)^2 + (x_2 + 1)^2$ subject to $x_1 \geq 0,\ x_2 \geq 0$

The unconstrained minimum is at $(-1, -1)$, but the constrained minimum is at $(0, 0)$.

The barrier function is introduced as follows:

$$B = (x_1 + 1)^2 + (x_2 + 1)^2 - \mu \log(x_1) - \mu \log(x_2) \tag{A.60}$$

The equations to find the minimum would be:

$$\frac{\partial B}{\partial x_1} = 2(x_1 + 1) - \frac{\mu}{x_1} = 0 \tag{A.61}$$

$$\frac{\partial B}{\partial x_2} = 2(x_2 + 1) - \frac{\mu}{x_2} = 0 \tag{A.62}$$

The solution would be:

$$x_1(\mu) = x_2(\mu) = -\frac{1}{2} + \frac{1}{2}\sqrt{1 + 2\mu} \tag{A.63}$$

Notice that the solution tends to $(0, 0)$ as $\mu \to 0$. This is a general result of the barrier function method [37]. In iterative schemes, there would be a sequence of solutions $\mathbf{x}^*(\mu_1),\ \mathbf{x}^*(\mu_2),\ \ldots,\ \mathbf{x}^*(\mu_n)$ as $\mu \to 0$. This sequence is called the *central trajectory* or the *central path*.

In particular, in the case of linear programming, the optimal solution will be located at the boundary of F. The central path would be a series of interior points, tending to this optimal solution at the boundary.

In the general case of a $B(\mathbf{x}|\mu)$ to be minimized, the set of equations $\partial B / \partial x_i = 0,\ i = 1, \ldots, m$ would be written, and then one could use the Newton's method to solve them.

The barrier function technique for linear programming was introduced in 1967, in the [37] book with subtitle: *Sequential Unconstrained Minimization Techniques* (also known as SUMT).

A.3.2 Interior Point Methods

Let us apply the mechanisms already introduced to the linear programming problem.

The primal problem was:

Maximize $\mathbf{c}^T \cdot \bar{x}$ subject to $A\,\mathbf{x} = \mathbf{b}\ ,\ \mathbf{x} \geq \mathbf{0}$.

The dual problem would be:

Maximize $\mathbf{b}^T \cdot \boldsymbol{\lambda}$ subject to $A^T \boldsymbol{\lambda} + \mathbf{s} = \mathbf{c}\ ,\ \mathbf{s} \geq \mathbf{0}$.

(the vector $\mathbf{s}$ is an artificial variable called the *dual slack*).

The duality gap would be: $\eta = \mathbf{c}^T\mathbf{x} - \mathbf{b}^T\boldsymbol{\lambda}$.

Now, barrier functions are introduced and two Lagrangians are built, corresponding to the primal and the dual problem:

$$L_p(\mathbf{x}, \boldsymbol{\lambda}) = \mathbf{c}^T\mathbf{x} - \mu \sum_{i=1}^{m} \log(x_i) - \boldsymbol{\lambda}^T(A\mathbf{x} - \mathbf{b}) \tag{A.64}$$

$$L_d(\mathbf{x}, \boldsymbol{\lambda}, \mathbf{s}) = \mathbf{b}^T\boldsymbol{\lambda} + \mu \sum_{i=1}^{n} \log(s_i) - \mathbf{x}^T(A^T\boldsymbol{\lambda} + \mathbf{s} - \mathbf{c}) \tag{A.65}$$

After setting the derivatives of the Lagrangians to zero, one obtains just three equations:

$$A\,\mathbf{x} = \mathbf{b} \tag{A.66}$$

$$A^T\,\boldsymbol{\lambda} + \mathbf{s} = \mathbf{c} \tag{A.67}$$

$$x_i\, s_i = \mu, \;\; i = 1, \ldots, n \tag{A.68}$$

(with $\mathbf{x} \geq \mathbf{0},\;\; \mathbf{s} \geq \mathbf{0}$)

In order to simplify the notation, the following diagonal matrices will be used:

$$X = diag\{x_1,\, x_2, \ldots, x_n\} \tag{A.69}$$

$$S = diag\{s_1,\, s_2, \ldots, s_n\} \tag{A.70}$$

Then, the complementary slackness condition can be written as:

$$X\,S\,\mathbf{e} = \mu\,\mathbf{e} \tag{A.71}$$

where: $\mathbf{e}^T = [1,\, 1,\, \ldots,\, 1]$

Hence, one could apply the Newton's method to solve the following equations:

$$A\,\mathbf{x} - \mathbf{b} = \mathbf{0} \tag{A.72}$$

$$A^T\,\boldsymbol{\lambda} + \mathbf{s} - \mathbf{c} = \mathbf{0} \tag{A.73}$$

$$X\,S\,\mathbf{e} - \mu\,\mathbf{e} = \mathbf{0}, \tag{A.74}$$

In the case of $\mathbf{f}(\mathbf{x}) = \mathbf{0}$ (a vector function $\mathbf{f}$ of several variables), the Newton step can be computed from:

$$J(\mathbf{x}^{(k)})\,\Delta\mathbf{x} = -\mathbf{f}(\mathbf{x}^{(k)}) \tag{A.75}$$

where $J\,(.)$ is the Jacobian.

After obtaining the Jacobian of the three equations given before, one has:

$$\begin{pmatrix} A & 0 & 0 \\ 0 & A^T & I \\ S & 0 & X \end{pmatrix} \begin{bmatrix} \Delta \mathbf{x} \\ \Delta \boldsymbol{\lambda} \\ \Delta \mathbf{s} \end{bmatrix} = \begin{bmatrix} \mathbf{0} \\ \mathbf{0} \\ \mu \mathbf{e} - XZ\mathbf{e} \end{bmatrix} \tag{A.76}$$

This equation gives a Newton step. However, the actual next iterate to be applied would be:

$$(\mathbf{x}, \boldsymbol{\lambda}, \mathbf{s}) + \alpha(\Delta \mathbf{x}, \Delta \boldsymbol{\lambda}, \Delta \mathbf{s}) \tag{A.77}$$

where $\alpha \in (0, 1]$ is chosen so the next solution is feasible.

A question under active research is what strategy should be adopted for choosing values of α. Actually, several types of interior point algorithms can be recognized, like the *short-step* algorithms that set α close to 1 so the solutions keep near the central path, or the *long-step* algorithms that choose α close to 0 so the solutions move more resolutely towards the optimal point. There are also *predictor-corrector* algorithms that take alternating 0 and 1 values. A step with $\alpha = 1$ is said to be a centering step; and a step with $\alpha = 0$ is called an affine-scaling step.

Once a sketch of the methodology has been done, it is convenient to add some comments on its rationale.

An important reason for using a parameter μ is the following: if you set $\mu = 0$ then some of the components of **x** and **s** become zero (on the boundary of the problem) and then it cannot be guaranteed that the Jacobian is non-singular. In such a case, the algorithm gets stuck. The consequence is that it is important to always use interior points.

The direct application of KKT conditions leads to a formulation equivalent to having $\mu = 0$, but the central path approach prefers to use a perturbed version of KKT by using the parameter μ.

There is another group of interior point algorithms, called potential-reduction algorithms, which use a potential function that is reduced in each iteration. A popular potential function, proposed by Tanabe [106], Todd and Ye [107], is the following:

$$\Phi(\mathbf{x}, \mathbf{s}) = \rho \log \mathbf{x}^T \mathbf{s} - \sum_{i=1}^{n} \log x_i s_i \tag{A.78}$$

with $\rho > n$.

Karmarkar's algorithm is a potential-reduction algorithm, based on the following potential function:

$$\Phi(\mathbf{x}) = \rho \log(\mathbf{c}^T \mathbf{x} - Z) - \sum_{i=1}^{n} \log x_i \tag{A.79}$$

with $\rho = n + 1$ and Z a lower bound of the optimal objective value.

It has been noticed in many applications that interior-point algorithms only need a few tens of iterations to obtain good approximations of the optimum, being sufficient to terminate the iteration process.

An extensive academic exposition of the interior point methodology is provided by [44]. Other publications of interest would be [49, 63, 64], the book [93] and the tutorial [20].

A.4 Quadratic Programming

A linearly constrained optimization problem with a quadratic objective function is called a quadratic program (QP). The general case can be written as:

Minimize:

$$z = \frac{1}{2}\mathbf{x}^T Q\,\mathbf{x} + \mathbf{c}^T \cdot \mathbf{x} \tag{A.80}$$

Subject to:

$$A\,\mathbf{x} \le \mathbf{b}\ ,\ \mathbf{x} \ge \mathbf{0} \tag{A.81}$$

where Q is a symmetric matrix.

An example of application could be the management of certain devices such that the energy spent is minimized. Energy is typically a quadratic quantity (never negative).

If there were no constraints, the solution could be found as follows:

$$\Delta z^T = Q\,\mathbf{x} + \mathbf{c} \tag{A.82}$$

$$\mathbf{x}^* = -Q^{-1}\,\mathbf{c} \tag{A.83}$$

The computation of the inverse of Q could be done using a Cholesky factorization or any other appropriate scheme. Likewise gradient descent methods, or the Newton's method (to solve $Q\,\mathbf{x} + \mathbf{c} = \mathbf{0}$) could be used.

Figure A.6 shows a typical scenario when there are constraints. The point M marks the minimum of the objective function. The solution of the constrained problem takes place at the point P, which is where the distance between point M and the feasible region F is measured.

The scenario depicted in Fig. A.6 corresponds to Q being positive definite. In the case of Q being indefinite, things are not so simple and several local minima may appear. If Q is positive definite or positive semidefinite, the optimization problem is convex. If Q is negative definite or negative semidefinite the problem is concave.

The KKT optimally conditions are necessary conditions, and in the case of convex optimization problems are also sufficient. These conditions provide ways for solving different constrained QP problems.

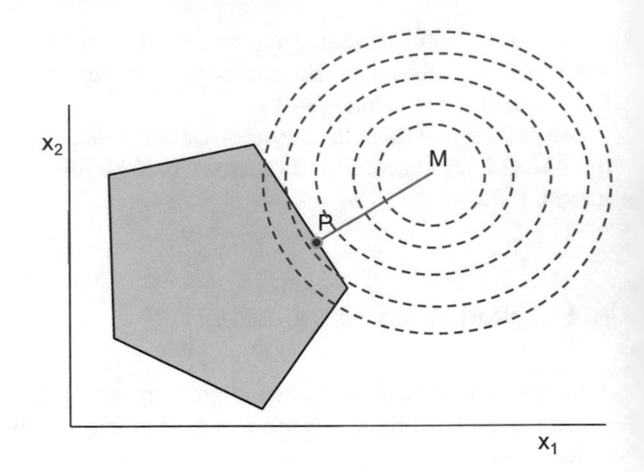

Fig. A.6 A typical quadratic programming scenario

A.4.1 The Case of Equality Constraints

Consider that the constraints are given by:

$$A\,\mathbf{x} = 0 \tag{A.84}$$

To solve the optimization problem, the following simple method could be applied. First, a change of variables:

$$Z\,\mathbf{y} = \mathbf{x} \tag{A.85}$$

where Z is chosen so that $AZ = 0$ (Z is a basis for the null space of A).

When the change of variables is done, the constraints are automatically satisfied, since $A\,Z\,\mathbf{y}$ is zero, and one has an unconstrained optimization problem:

Minimize: $\frac{1}{2}\mathbf{y}^T\,Z^T\,Q\,Z\,\mathbf{y} + \mathbf{c}^T Z\,\mathbf{y}$

Taking derivatives equal to zero, the solution is found to be:

$$Z^T Q\,Z\,\mathbf{y} = -Z^T\,\mathbf{c}$$

If the matrix $Z^T Q\,Z$ (which is called *'the reduced Hessian'*) is positive definite this is the unique solution.

Apart from the null-space approach, one could choose the Lagrange scheme. The Lagrangian for the problem with constraints $A\,\mathbf{x} = \mathbf{b}$ would be:

$$L(\mathbf{x},\,\boldsymbol{\lambda}) = \frac{1}{2}\mathbf{x}^T Q\,\mathbf{x} + \mathbf{c}^T\mathbf{x} - \boldsymbol{\lambda}^T(A\mathbf{x} - \mathbf{b}) \tag{A.86}$$

Taking partial derivatives, one obtains the following equations:

$$Q\,\mathbf{x} + \mathbf{c} + A^T\,\boldsymbol{\lambda} = 0 \tag{A.87}$$

$$A\,\mathbf{x} = \mathbf{b} \tag{A.88}$$

Which can be written as:

$$\begin{pmatrix} Q & A^T \\ A & 0 \end{pmatrix} \begin{bmatrix} \mathbf{x} \\ \boldsymbol{\lambda} \end{bmatrix} = \begin{bmatrix} -\mathbf{c} \\ b \end{bmatrix} \tag{A.89}$$

The matrix at the left hand side of (A.89) is called *the KKT matrix*. Assume that A is full row rank $m < n$, and that the reduced Hessian is positive definite; then the KKT matrix is nonsingular, and so the equations (A.89) have a unique solution, which is the optimum.

A.4.2 *The Case of Inequality Constraints*

There is a number of approaches for the case in which the constraints were:

$$A\,\mathbf{x} \leq \mathbf{b}\;,\;\; \mathbf{x} \geq \mathbf{0} \tag{A.90}$$

A.4.2.1 The Wolfe's Method (Extended Simplex)

The linear equations provided by the derivatives of the Lagrangian can be used to establish a kind of linear programming problem, so the simplex algorithm can be applied in a special way. These equations would be:

$$Q\,\mathbf{x} + \mathbf{c} + A^T\,\boldsymbol{\lambda} \geq 0 \tag{A.91}$$

$$A\,\mathbf{x} \leq \mathbf{b} \tag{A.92}$$

The idea is to use two sets of artificial variables for creating a linear problem. The first set is added to convert equations into equalities. The second set is added to obtain a linear objective function.

For instance, consider the following problem [56]:

Minimize: $z = -8\,x_1 - 16\,x_2 + x_1^2 + 4\,x_2^2$

Subject to: $x_1 + x_2 \leq 5$ and $x_1 \leq 3$

with all variables ≥ 0

After building the Lagrangian and taking derivatives, one obtains:

$$2\,x_1 + \lambda_1 + \lambda_2 - e_1 = 8$$

$$8\,x_2 + \lambda_1 - e_2 = 16$$

$$x_1 + x_2 + \nu_1 = 5$$

$$x_1 + \nu_2 = 3$$

Notice that a first set of artificial variables $e_1,\ e_2,\ \nu_1,\ \nu_2$ has been added to obtain the equalities.

Now, more variables are added and the following LP problem is formulated:

Minimize: $z = a_1 + a_2 + a_3 + a_4$

Subject to:

$$2x_1 + \lambda_1 + \lambda_2 - e_1 + a_1 = 8$$

$$8x_2 + \lambda_1 - e_2 + a_2 = 16$$

$$x_1 + x_2 + \nu_1 + a_3 = 5$$

$$x_1 + \nu_2 + a_4 = 3$$

And then the simplex algorithm is applied taking into account complementary slackness:

x_j and e_j are complementary
λ_j and ν_j are complementary

Recall that during the simplex steps, when exchanging columns, a maximization criterion is applied for choosing the entering variable (column). In the present case (Wolfe's method), the entering variable should not have its complementary variable in the basis (or would leave the basis on the same iteration). Therefore, in each step, some variables must be excluded from the maximization criterion.

The history of the simplex method application to this example can be summarized as follows:

Iteration	Basis	Solution	Objective value	Entering variable	Leaving variable
1	(a_1, a_2, a_3, a_4)	(8, 16, 5, 3)	32	x_2	a_2
2	(a_1, x_2, a_3, a_4)	(8, 2, 3, 3)	14	x_1	a_3
3	(a_1, x_2, x_1, a_4)	(2, 2, 3, 0)	2	λ_1	a_4
4	$(a_1, x_2, x_1, \lambda_1)$	(2, 2, 3, 0)	2	λ_2	a_1
5	$(\lambda_2, x_2, x_1, \lambda_1)$	(2, 2, 3, 0)	0		

A.4.2.2 Active Set Method

Along iterative methods a series of solutions $\mathbf{x}^{(j)}$ is obtained, until reaching the optimum $\mathbf{x}^*$. Usually, these solutions $\mathbf{x}^{(j)}$ are on the boundary of F, and so equality holds for some of the equations in $A\mathbf{x} \leq \mathbf{b}$; the set of these equations (constraints) is called the '*active set*', which will be denoted as $\tilde{\text{A}}$.

The active set method iteratively changes the set of constraints that are active, which are contained in *a working set* W_K. In each step a constraint is added or removed from W_K, to decrease the objective function.

For each step k, the constraints outside W_K are omitted and then an equality QP is solved. The solution $\mathbf{x}_W$ give us a direction to follow:

$$\mathbf{p}_K = \mathbf{x}_W - \mathbf{x}^{(K-1)}$$

If $\mathbf{p}_K = \mathbf{0}$, the optimum with current W_K cannot be decreased. If all Lagrange multipliers were non-negative that means the final solution was obtained. Otherwise, a constraint with negative λ is dropped from W_K and a new QP is solved, to obtain a new $\mathbf{p}_{K+1}$.

If $\mathbf{p}_K \neq \mathbf{0}$, a step length must be specified so that $\mathbf{x}^{(K-1)} + \alpha\,\mathbf{p}_K$ was feasible (with $0 \leq \alpha \leq 1$). The parameter α is increased until a constraint not in W_K was reached. This blocking constraint is included in W_K. If it was not possible to reach such a constraint, then α is set to one.

Let us consider an example [69]:

Minimize: $z = (x_1 - 1)^2 + (x_2 - 2.5)^2$

Subject to:

$x_1 - 2x_2 + 2 \geq 0$ (A)
$-x_1 - 2x_2 + 6 \geq 0$ (B)
$-x_1 + 2x_2 + 2 \geq 0$ (C)
with $x_1 \geq 0$ (D), and $x_2 \geq 0$ (E).

Figure A.7 shows the geometry of the problem.

Let us start with $\mathbf{x}^{(0)} = \begin{bmatrix} 2 \\ 0 \end{bmatrix}$

The working set would be: $W_0 = \{C, E\}$

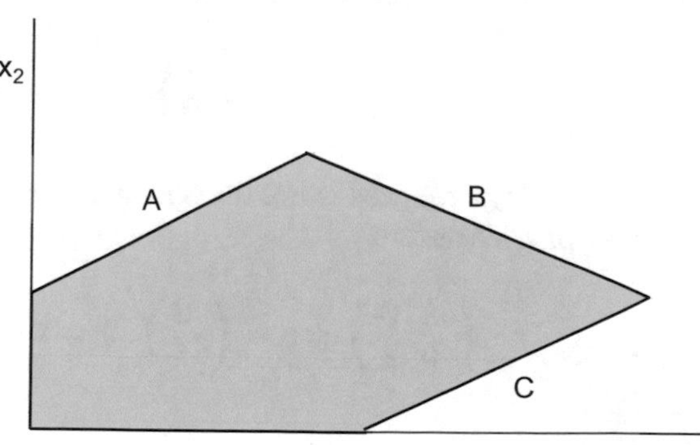

Fig. A.7 The feasible region of the QP example

and the equality QP would be:

$$\text{Minimize: } z = \mathbf{x}^T \begin{pmatrix} 2 & 0 \\ 0 & 2 \end{pmatrix} \mathbf{x} + \begin{pmatrix} -2 \\ -5 \end{pmatrix}^T \mathbf{x} + \frac{29}{4}$$

Subject to:

$$\begin{pmatrix} 1 & 2 \\ 0 & 1 \end{pmatrix} \mathbf{x} = \begin{pmatrix} -2 \\ 0 \end{pmatrix}$$

The KKT equations to solve the QP would be:

$$\begin{pmatrix} 2 & 0 & -1 & 0 \\ 0 & 2 & 2 & 1 \\ -1 & 2 & 2 & 1 \\ 0 & 1 & 0 & 0 \end{pmatrix} \begin{pmatrix} \mathbf{x} \\ -\boldsymbol{\lambda} \end{pmatrix} = \begin{pmatrix} 2 \\ 5 \\ -2 \\ 0 \end{pmatrix}$$

The solution is:

$$\mathbf{x}_W = \begin{pmatrix} 2 \\ 0 \end{pmatrix} \ ; \ \boldsymbol{\lambda}_W = \begin{pmatrix} -2 \\ -1 \end{pmatrix}$$

Therefore, $\mathbf{p}_1 = \mathbf{0}$. Since all Lagrange multipliers are negative, this is not the optimum. One of the constraints, (C), is removed, so $W_1 = \{E\}$. A new equality QP should be solved, whose KKT equations are:

$$\begin{pmatrix} 2 & 0 & 0 \\ 0 & 2 & 1 \\ 0 & 1 & 0 \end{pmatrix} \begin{pmatrix} \mathbf{x} \\ -\boldsymbol{\lambda} \end{pmatrix} = \begin{pmatrix} 2 \\ 5 \\ -2 \end{pmatrix}$$

The solution is:

$$\mathbf{x}_W = \begin{pmatrix} 1 \\ 0 \end{pmatrix} \ ; \ \boldsymbol{\lambda}_W = -5$$

Then, $\mathbf{p}_2 = \begin{pmatrix} 1 \\ 0 \end{pmatrix} - \begin{pmatrix} 2 \\ 0 \end{pmatrix} = \begin{pmatrix} -1 \\ 0 \end{pmatrix}$ is a search direction. The parameter α is set to one, and so the new solution would be:

$$\mathbf{x}^{(2)} = \begin{pmatrix} 1 \\ 0 \end{pmatrix} \ ; \ \boldsymbol{\lambda}^{(2)} = -5$$

Since $\lambda^{(2)}$ is negative, this is not the optimum. Another constraint is removed, so $W_2 = empty$. Therefore:

$$\begin{pmatrix} 2 & 0 \\ 0 & 2 \end{pmatrix} (\mathbf{x}) = \begin{pmatrix} 2 \\ 5 \end{pmatrix}, \ \Rightarrow \mathbf{x}_W = \begin{pmatrix} 1 \\ 2.5 \end{pmatrix}$$

The new search direction would be: $\mathbf{p}_3 = \begin{pmatrix} 1 \\ 2.5 \end{pmatrix} - \begin{pmatrix} 1 \\ 0 \end{pmatrix} = \begin{pmatrix} 0 \\ 2.5 \end{pmatrix}$

It is not possible to set α to one, since constraint (A) would not be satisfied. Actually, from:

$$x_1 - 2\,x_2 + 2 \geq 0(\mathrm{A})$$

one concludes that $\alpha \leq 0.6$.

Therefore, one moves to: $\mathbf{x}^{(3)} = \begin{pmatrix} 1 \\ 1.5 \end{pmatrix}$. And adds (A) to the working set: $W_3 = \{A\}$. The KKT conditions of the new QP problem would be:

$$\begin{pmatrix} 2 & 0 & 1 \\ 0 & 2 & -2 \\ 1 & -2 & 0 \end{pmatrix} \begin{pmatrix} \mathbf{x} \\ -\boldsymbol{\lambda} \end{pmatrix} = \begin{pmatrix} 2 \\ 5 \\ -2 \end{pmatrix}$$

The solution is:

$$\mathbf{x}_W = \begin{pmatrix} 1.4 \\ 1.7 \end{pmatrix} \; ; \; \boldsymbol{\lambda}_W = 0.8$$

Since λ_W is non-negative, this is the optimal solution.

The same example has been considered in [66]. The evolution of the algorithm was concisely summarized using Fig. A.8.

There is a quadratic programming web site (see the Resources section) with a lot of information and links to software. A suitable tutorial and review of quadratic programming is [42].

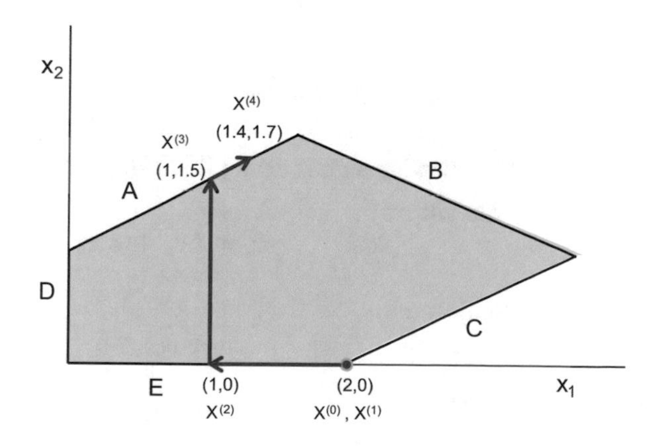

Fig. A.8 The evolution of the active set algorithm for the QP example

A.5 Integer Programming

This section has been introduced because some of the techniques related to integer programming have connections with methods employed in sparse representation problems.

- The canonical form of integer linear programming (ILP) is the following:

  Maximize:
$$z = \mathbf{c}^T \cdot \mathbf{x} \tag{A.93}$$

  Subject to:
$$A\,\mathbf{x} \leq \mathbf{b}\ ,\ \mathbf{x} \geq \mathbf{0} \tag{A.94}$$

  where all variables included in $\mathbf{x}$ are integer.

- The standard form of ILP is:

  Maximize:
$$z = \mathbf{c}^T \cdot \mathbf{x} \tag{A.95}$$

  Subject to:
$$A\,\mathbf{x} = \mathbf{b}\ ,\ \mathbf{x} \geq \mathbf{0} \tag{A.96}$$

  where all the entries of A, $\mathbf{c}$, $\mathbf{b}$, and $\mathbf{x}$, are integer.

A naïve approach for solving this optimization problem is rounding. Conventional linear programming is a first step, and then, in a second step, the result is rounded. Some care is recommended, especially about possible violation of constraints.

A.5.1 Cutting Plane Techniques

Consider the following simple example [119]:

Maximize:
$$z = 5\,x_1 + 8\,x_2$$

Subject to:
$$x_1 + x_2 \leq 6$$
$$5\,x_1 + 9\,x_2 \leq 45$$

with x_1, x_2 being non-negative integers.

First, one could try the simplex algorithm, treating the case as a linear programming problem. The feasible region is represented in Fig. A.9.

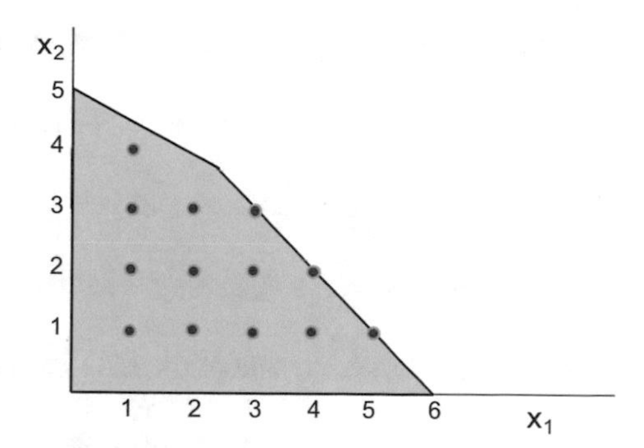

Fig. A.9 An integer programming example

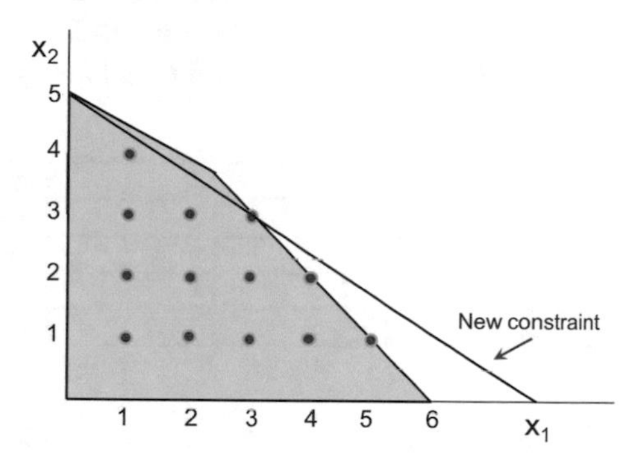

Fig. A.10 The new constraint (*the cut*)

If the solution was an integer, then that's all. But no, the solution takes place at the point $\mathbf{x}^* = (2.25,\ 3.75)$, so it is not integer.

Let us generate a *cut*, which is a constraint that is satisfied by all feasible integers, but not by $\mathbf{x}^*$. For example:

$$2x_1 + 3x_2 \leq 15 \tag{A.97}$$

Figure A.10 shows this new constraint.

The new constraint is added to the linear programming problem. Now, the solution is $\mathbf{x}^* = (3,\ 3)$ which is the desired optimal integer solution.

Notice that the cut only removed the non-integer solution; cuts never remove feasible integer solutions.

Gomory presented in 1958 a method for introducing cuts in the simplex algorithm [47]. Choose for instance one of the constraints:

$$a_1 x_1 + a_2 x_2 + \ldots + a_n x_n = b \tag{A.98}$$

where b and the coefficients a_j are not necessarily integers.

Denote $floor(\eta)$ as $\lfloor \eta \rfloor$. Then, the chosen constraint can be written as:

$$[\lfloor a_1 \rfloor + (a_1 - \lfloor a_1 \rfloor)]\, x_1 + \ldots + [\lfloor a_n \rfloor + (a_n - \lfloor a_n \rfloor)]\, x_n = [\lfloor b \rfloor + (b - \lfloor b \rfloor)] \tag{A.99}$$

Denoting as f_i the fractional terms, an equivalent expression would be:

$$[[a_1| + f_1)]x_1 + \ldots + [|a_n| + f_n]x_n = [|b| + f] \quad \text{(A.100)}$$

Separating fractional and integer terms:

$$f_1 x_1 + \ldots + f_n x_n - f = |b| - |a_1|x_1 - \ldots - |a_n|x_n \quad \text{(A.101)}$$

Therefore, the left-hand side of (A.101) must be an integer; and, since $0 \leq f < 1$, it must be non-negative.

Based on these observations, adequate constraints could be added to the simplex algorithm in order to introduce a cut. For example, suppose you arrived to a non integer optimal in a certain problem, having the following simplified tableau:

Basis	x_1	x_2	s_1	s_2	b
x_1	1	0	0.6	−0.4	3.2
x_2	0	1	−0.4	0.6	3.2
z_j	1.4	1	0	0	
$c_j - z_j$	0	0	0.2	0.2	

Select for instance the x_1 row. The corresponding constraint would be:

$$x_1 + 0.6\, s_1 - 0.4\, s_2 = 3.2 \quad \text{(A.102)}$$

In other terms:

$$x_1 - 3 = 0.2 + 0.6\, s_1 - 0.4\, s_2 \quad \text{(A.103)}$$

Therefore an adequate constraint can be:

$$0.2 + 0.6\, s_1 - 0.4\, s_2 \leq 0 \quad \text{(A.104)}$$

This constraint is satisfied by every feasible integer solution, but is not feasible for the current optimal solution (which includes $s_1 = 0;\ s_2 = 0$). By adding this constraint to the problem, a different optimal solution would be reached, which might be integer.

A.5.2 *Branch and Bound*

It is typical of integer programming scenarios to have decision variables, of the type yes or not (1 or 0). It would be convenient, then, to see what happens if yes, or what happens if not.

Branch and bound methods divide optimization problems into sub-problems. A simple first example is the following [34]:

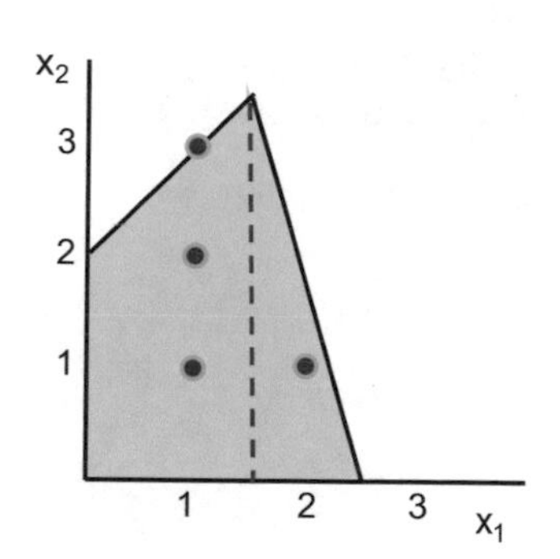

Fig. A.11 An example for branch and bound

Maximize:

$$z = x_1 + x_2$$

Subject to:

$$x_2 - x_1 \le 2$$

$$8x_1 + 2x_2 \le 19$$

with x_1, x_2 being non-negative integers.

Figure A.11 depicts the situation. If real valued variables were accepted, the solution would take place at *(1.5, 3.5)*; but only integer variables are allowed, and so it is not the desired solution.

Let us decompose the problem into two sub-problems, one for $x_1 \le 1$ and the other with $x_1 \ge 2$. Both sub-problems are treated as linear programming (LP) cases, with the conventional simplex algorithm. The solutions are:

- sub-problem 1: *(1, 3)*, $z = 4$
- sub-problem 2: *(2, 1.5)*, $z = 3.5$

When an integer solution is found in a sub-problem, no further branching of this sub-problem will yield better results. Therefore, our optimal solution has been found, in the sub-problem 1.

A more complicated example is the following [31]:

Maximize:

$$z = 3x_1 + 2x_2$$

Subject to:

$$4x_1 + 2x_2 \le 15$$

$$x_1 + 2x_2 \le 8$$

$$x_1 + x_2 \le 5$$

with x_1, x_2 being non-negative integers.

By using the simplex algorithm, one finds the optimum at *(2.5, 2.5)* with $z = 12$. Let us branch on x_1. Two sub-problems are obtained:

- The first, subject to:

$$4x_1 + 2x_2 \leq 15$$

$$x_1 + 2x_2 \leq 8$$

$$x_1 + x_2 \leq 5$$

$$x_1 \geq 3$$

- The second, subject to:

$$4x_1 + 2x_2 \leq 15$$

$$x_1 + 2x_2 \leq 8$$

$$x_1 + x_2 \leq 5$$

$$x_1 \leq 2$$

Suppose one chooses the first sub-problem. Using the simplex algorithm for this LP, the solution is (3, 1.5), which is not valid. Then, let us branch on x_2. Two more sub-problems are obtained:

- The third, subject to:

$$4x_1 + 2x_2 \leq 15$$

$$x_1 + 2x_2 \leq 8$$

$$x_1 + x_2 \leq 5$$

$$x_1 \geq 3$$

$$x_2 \geq 2$$

- The fourth, subject to:

$$4x_1 + 2x_2 \leq 15$$

$$x_1 + 2x_2 \leq 8$$

$$x_1 + x_2 \leq 5$$

$$x_1 \leq 2$$

$$x_2 \leq 1$$

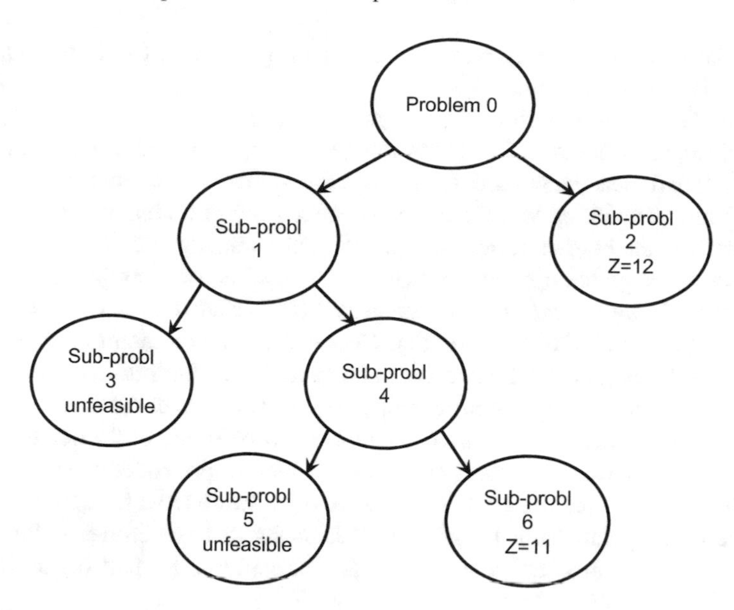

Fig. A.12 Branch and bound exploration of an example

The third sub-problem is unfeasible. The solution for the fourth sub-problem is (3.25, 1), not valid. If one takes the fourth sub-problem and branches on x_1, two more sub-problems are generated, the 5th with $x_1 \geq 4$, and the 6th with $x_1 \leq 3$. The 5th sub-problem is unfeasible, while the 6th gives the solution (3, 1) with $z = 11$. This is a lower bound.

Let us come back, and see the second sub-problem. The optimal solution found with the simplex algorithm is (2, 3) with $z = 12$. This is the optimum.

Figure A.12 summarizes the process that has been followed in this example.

In general there are two exploration alternatives. In the example, *the depth-first* alternative has been chosen. The other alternative was *breadth-first* search. Had we chosen this second way, the searching had been much shorter (just the sub-problems 1 and 2).

There are several web sites on Integer Programming (see the Resources section). One of the good tutorials available from Internet is [71]. Most cited papers on this topic are [43, 38]. The article [67] offers a review of fifty years of solution approaches.

A.6 A Wider Range of Problems

In addition to LP, QP and IP problems there are much more types of optimization problems waiting for a solution. This is well-known in the industrial and economics worlds. In the last decades, the range of problems that now can be tackled has substantially increased. Large part of this advancement is due to the three fields contemplated in this section, namely Convex Programming, Conic Programming and Semidefinite Programming (SDP).

One of the aspects of optimization in general that causes more worries is that there are applications with several (perhaps many) local optima. One could imagine the objective function as many mountains (like around the Everest), so if you are lucky you may find yourself on top of the highest, after much climbing, or perhaps not on the highest. In the case of convex optimization problems, you don't have to worry, there is only one global optimum, and there are no other local optima.

The term *'convex analysis'* was suggested by Tucker. But it was the book of Rockafellar [95] in 1970, with the title *Convex Analysis* the point of departure of new theory developments. Indeed the characteristics of the problem successfully considered by Linear Programming invited to study in more detail convexity.

The advent of interior-point methods implied a broadening of the horizon. These methods are not only applicable for LP, they can also be applied for convex problems [85]. This fact was potentiated by a new wave of software tools for optimization.

However, as pointed out by Bryson, it happens that sometimes is difficult to determine that my problem is a convex problem. Therefore, part of the focus of Convex Optimization was put on convexity itself.

After some initial years of applying new software tools based on interior-points, it was recognized that problems of convergence and numerical error sensitivity may appear. A deeper analysis of the convergence of the Newton's method was done, and the theory of *self-concordant* functions was developed along several papers, together with the proposal of *conic programming*. The book of [86] in 1994 helped definitively to establish this new methodology.

An important specific branch of convex programming is SDP. The topic was introduced in 1963, by Bellman and Fan [12]. The generalization of interior-point methods from LP to SDP was introduced in [4, 57, 58].

A self-concordant (convex) function $f(t)$ obeys:

$$|f'''(t)| \leq 2\, f''(t)^{3/2} \tag{A.105}$$

What it is important in this expression (A.105) is that the third derivative is bounded at every point by the second derivative.

A main idea of [85] was to replace the inequality constraints by self-concordant barrier terms in the objective function. By using self-concordant functions one can define closeness to the central path and the policy to update μ in order that the Newton steps stay close to the central path, with not much fluctuation (this is implied by the derivative bounds). The result is a better convergence of the interior-point method.

All linear and quadratic functions are self-concordant. The sum of two self-concordant functions is also self-concordant. An important example of self-concordant function is $f(t) = -\log(t)$, for which the expression (A.105) holds with equality.

An interesting property of self-concordant functions is that they inform on how close to an optimal point is the current algorithm step, and you can obtain a upper bound on the number of iterations required to approach the optimum within a certain ε.

During last years is being realized that better schemes can be obtained, based on barrier functions other than the logarithmic one [7].

A.6.1 Convex Programming

Consider a generic optimization problem:

Minimize:

$$f_0(\mathbf{x}) \tag{A.106}$$

Subject to:

$$f_i(\mathbf{x}) \leq 0, \quad i = 1, 2, \ldots, m \tag{A.107}$$

$$h_j(\mathbf{x}) = 0, \quad j = 1, 2, \ldots, r \tag{A.108}$$

$$\mathbf{x} \in S \tag{A.109}$$

The optimization problem is said to be *convex* if all functions $f_k(\mathbf{x})$, $k = 0, 1, 2, \ldots, m$ are convex and the functions $h_j(\mathbf{x})$, $j = 1, 2, \ldots, r$ are affine (of the form $A\mathbf{x} + \mathbf{b}$), and S is a convex set.

A set S is convex if for any two points $\mathbf{x}$, $\mathbf{y} \in S$, the line segment joining $\mathbf{x}$ and $\mathbf{y}$ lies in S.

In general a convex set must be solid, with no holes, and with curves outward. Figure A.13 shows examples of convex and not-convex sets. The intersection of convex sets is a convex set.

The separating hyperplane theorem states that if two convex sets S and T are disjoint ($S \cap T = 0$), then there exists a hyperplane which separates them. Figure A.14 illustrates this theorem.

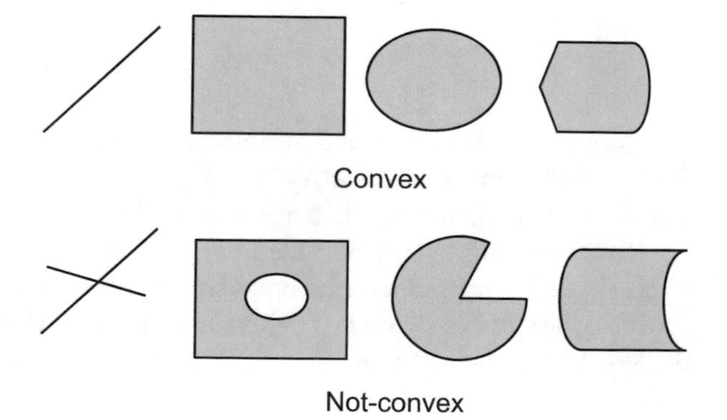

Fig. A.13 Examples of convex and not-convex sets

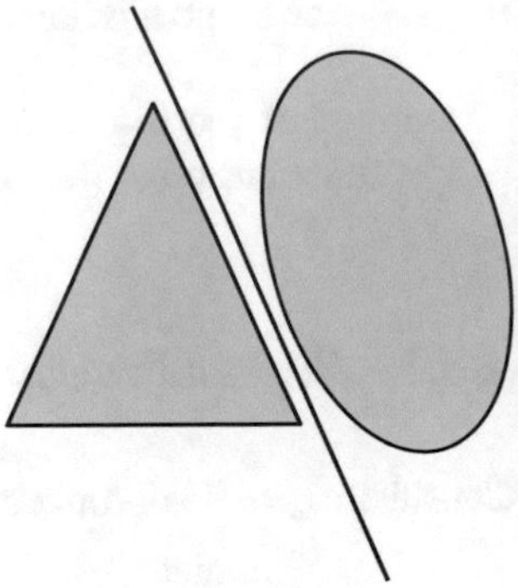

Fig. A.14 A separating hyperplane

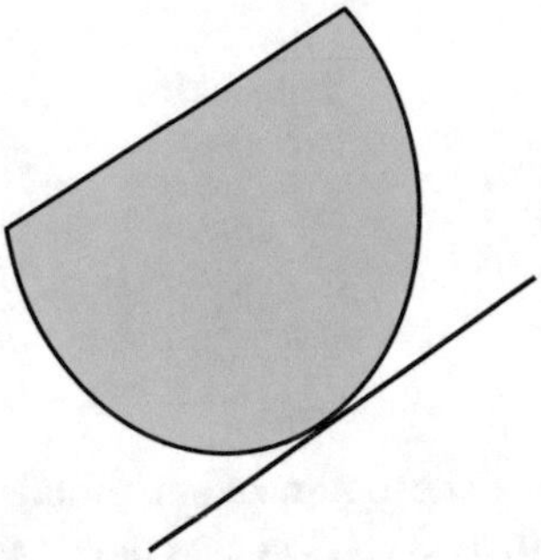

Fig. A.15 A supporting hyperplane

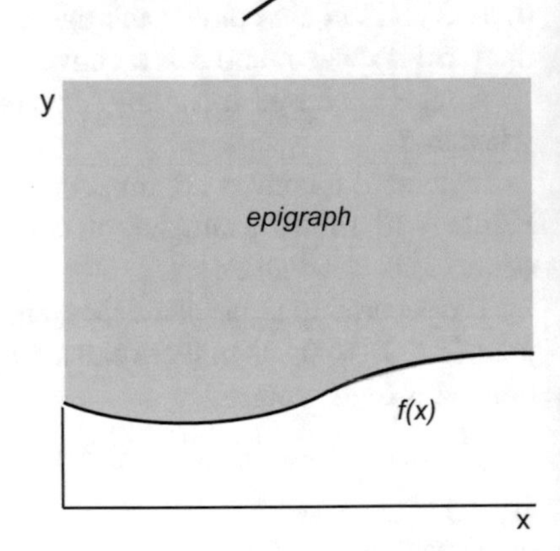

Fig. A.16 Epigraph of a function f(x)

Given a hyperplane $\mathbf{a}^T\mathbf{x}$ and a convex set S, it is said that the hyperplane *supports* S at $\mathbf{x}_0$ if all points $\mathbf{x}$ of S satisfy: $\mathbf{x} \in S \Rightarrow \mathbf{a}^T\mathbf{x} \leq \mathbf{a}^T\mathbf{x}_0$. The hyperplane is called *supporting* hyperplane, and is tangent to S. Figure A.15 depicts an example.

The supporting hyperplane theorem establishes that there exists a supporting hyperplane at every point of the boundary of a convex set.

The *epigraph* of a function $f(x)$ is the set (x, y) such that $y \geq f(x)$. Figure A.16 depicts the epigraph of a function $f(x)$; the epigraph being the area above the curve $f(x)$.

Similarly, the *hypograph* of the function $f(x)$ is the area below the curve.

The function $f(x)$ is a convex function if its epigraph is a convex set.

The convexity of a differentiable function can also be characterized by its gradient and Hessian. Based on the first order Taylor approximation a first order condition can be enounced as follows, $f(x)$ is convex if:

$$f(x) \geq f(x_0) + f'(x_0)(x - x_0) \tag{A.110}$$

Likewise, a second order condition can be enounced, so $f(x)$ is convex if its Hessian is positive semidefinite (≥ 0).

Examples of convex functions are univariate functions like $|x|$, e^x, x^2, or multivariate functions of the form $a^T x + b$, with a and b being a constant, vector or matrix. Also norms are convex. And, x^α(and $|x|^\alpha$ with $\alpha \geq 1$ is convex), $x \log x$ with $x > 0$ is convex, etc.

Affine functions are convex and concave (the Hessian is zero).

Quadratic functions $\mathbf{x}^T P\mathbf{x}$ are convex if P is positive semidefinite.

There are several operations that preserve convexity, like for instance:

- Non-negative scaling: f convex, $\alpha \geq 0 \Rightarrow \alpha f$ convex
- Affine composition: f convex, $\Rightarrow f(A\mathbf{x} + \mathbf{b})$ convex
- Perspective transformation: f convex, $\Rightarrow tf(x/t)$ convex for t>0

The sum of convex functions is also convex.

It has been recently shown that deciding the convexity of quartic polynomials is NP-hard [2]. So, looking at more general contexts, it seems that it could be really difficult, in certain situations, to determine that the problem at hand is convex. In consequence, it is recommendable to specify the optimization problems, whenever possible, so they were convex by construction.

An important book on convex optimization is [18]. More details on this topic can be found in [53, 14, 74].

A.6.2 Conic Programming

An important generalization of the type of constraints included in optimization problems is based on cones.

A n-dimensional set K is a *cone* if:

$$\mathbf{x} \in K \Rightarrow \theta\mathbf{x} \in K, \ \forall\theta \geq 0 \tag{A.111}$$

Figure A.17 shows some examples of 2D cones [118]. Obviously, one of them is not convex.

The intersection of two cones is a cone.

A set K is a *convex cone* if K is a cone and K is convex.

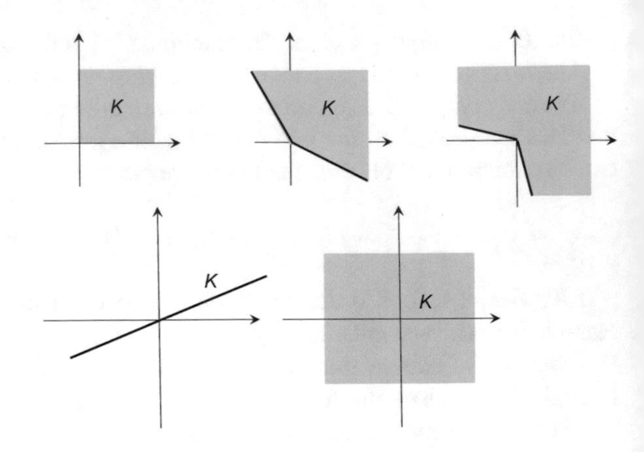

Fig. A.17 Examples of 2D cones

A cone K is a *proper cone* if it is convex, closed, and it is pointed (if $\mathbf{x} \in K$ and $-\mathbf{x} \in K$ then $\mathbf{x} = \mathbf{0}$).

Important examples of proper cones are the following:

- The non-negative orthant:

$$K = \{\mathbf{x} \mid x_i \geq 0,\ i = 1 = 1, 2 \ldots, n\} \tag{A.112}$$

- The cone of positive semidefinite (PSD) matrices:

$$K = \{A \in S^n \mid \mathbf{u}^T A\, \mathbf{u} \geq 0,\ \forall \mathbf{u} \in \Re^n\} \tag{A.113}$$

with S^n being the set of $n \times n$ symmetric matrices

- Norm cones:

$$K = \{(\mathbf{x}, t) \mid \ \|\mathbf{x}\| \leq t\} \tag{A.114}$$

An example of norm cones is the *second order cone (SOC)*, which corresponds to $\|\mathbf{x}\| = \|\mathbf{x}\|_2$ (that is: $\sqrt{x^2 + y^2} \leq z$). Figure A.18 depicts a second order cone. This cone is also called the Lorentz cone or, more colloquially, the ice-cream cone.

Evidently, a particular case of convex programming is the following problem:

Minimize:

$$f_0(\mathbf{x}) \tag{A.115}$$

Subject to:

$$f_i(\mathbf{x}) \leq 0,\ \ i = 1, 2, \ldots, m \tag{A.116}$$

$$h_j(\mathbf{x}) = 0,\ \ j = 1, 2, \ldots, r \tag{A.117}$$

$$\mathbf{x} \in K \tag{A.118}$$

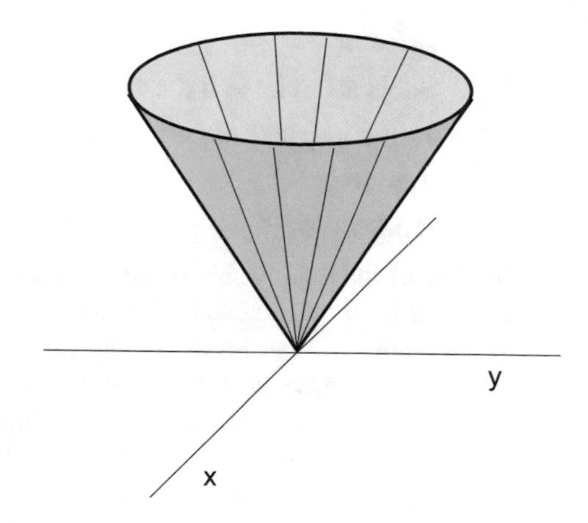

Fig. A.18 Second order cone

where K is a convex cone.

This problem may adopt more specific forms, depending on the type of cone and on the other aspects of the problem. An important example corresponds to K being a second order cone; the problem is then called *second order cone programming* (SOCP).

The theory also considers dual cones. Given a cone K, its *dual cone* K^* is the following:

$$K^* = \{\mathbf{y} \mid \mathbf{x}^T\mathbf{y} \geq 0, \ \forall \mathbf{x} \in K\} \tag{A.119}$$

(recall the equation of the perpendicular to a line).

Figure A.19 shows an example of 2D cone and dual cone.

The dual cone is always convex, even when K is not. If K is a proper cone, then $K = K^{**}$.

A cone is self-dual if $K^* = K$. The positive orthant, the SOC, and the cone of PSD matrices are self-dual cones.

Like in the case of LP, it is possible to work with primal-dual problems based on the Lagrange approach. Concretely, in the following conic programming case [44]:

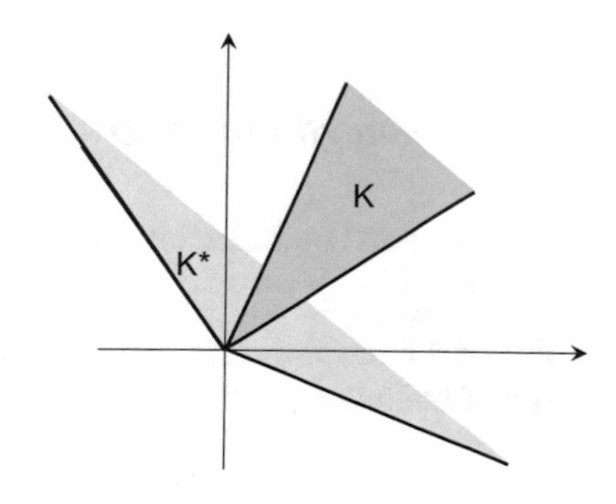

Fig. A.19 Example of cone K and dual cone K^*

Inf: $\mathbf{c}^T\mathbf{x}$

Subject to: $A\mathbf{x} = \mathbf{b}\,,\ \mathbf{x} \in K$

If K is a solid, pointed, closed convex cone, the dual problem would be:

Sup: $\mathbf{b}^T\mathbf{y}$

Subject to: $A^T\mathbf{y} + \mathbf{s} = \mathbf{c}\,,\ \mathbf{s} \in K^*$

The dual of the dual problem is equivalent to the primal problem.

Clearly, it is advantageous to deal with scenarios based on self-dual cones.

Once the conic programming fundamentals has been introduced, it seems opportune to complete the view and add some observations.

It can be shown that any convex programme can be written as a conic programming [36]. So conic programming is important, and it provides also a convenient framework from the mathematical point of view.

In geometry courses, an important lesson is the one devoted to the different curves generated by the intersection of a plane and a cone: the conic sections. In the case of conic programming this is extended to n-dimensions. The feasible region can be the intersection of polyhedrals, ellipsoids, paraboloids, and hyperboloids. Quadratic programming can be easily treated by adding a new variable and considering the quadratic objective function as included in the set of constraints.

There are some peculiarities to have in mind. For instance, the problem:

$$\min\{x_3 \mid x_1 = 1,\ x_2 = 1,\ x_3 \geq \sqrt{x_1^2 + x_2^2}\} \tag{A.120}$$

has an irrational solution: $x_3 = \sqrt{2}$

Another case:

$$\inf\{x_3 - x_2 \mid x_1 = 1,\ x_3 \geq \sqrt{x_1^2 + x_2^2}\} \tag{A.121}$$

Although $x_3 - x_2$ can be made arbitrarily close to 0, it is not possible to actually reach 0. Therefore, in some cases the infimum may not be attained.

Some relevant publications on cone programming and its applications are [5, 72]. A review of current conic optimization software is offered by [78].

It may be of interest for the reader to explore copositive programming [23].

A.6.3 Semidefinite Programming

The semidefinite optimization problem is specified in terms of positive semidefinite matrices.

- A $n \times n$ real symmetric matrix A is said to be *positive semidefinite* if it satisfies one of the following properties:

- For all $\mathbf{x} \in \Re^n$, $\mathbf{x}^T A\,\mathbf{x} \geq 0$
- All eigenvalues of A are positive.

The set of positive semidefinite matrices is denoted as S^n_+.

- A matrix P is *positive definite* if for all $\mathbf{x} \in \Re^n$, $\mathbf{x}^T P\,\mathbf{x} > 0$, $\mathbf{x} \neq 0$.
The set of positive definite matrices is denoted as S^n_{++}.

All eigenvalues of a symmetric matrix are real. The corresponding eigenvectors can be chosen so that they are orthogonal. The determinant of the matrix is the product of the eigenvalues; and the sum of all entries is equal to the sum of the eigenvalues (the trace of the matrix).

A.6.3.1 The SDP Problem

The SDP problem is defined in [111] as follows:

Minimize:

$$\mathbf{c}^T \mathbf{x} \tag{A.122}$$

Subject to:

$$F(\mathbf{x}) \geq 0 \tag{A.123}$$

where:

$$F(\mathbf{x}) = F_0 + \sum_{i=1}^{m} x_i\, F_i \tag{A.124}$$

and the $m + 1$ matrices $F_0, F_1, \ldots, F_m \in \Re^{n \times n}$ are symmetric.

The expression $F(\mathbf{x}) \geq 0$ means that F is positive semidefinite. This expression is also called a *linear matrix inequality* (LMI). Some topics of modern control theory are directly linked with the use of LMIs.

The SDP problem is a convex optimization problem.

Figure A.20 shows a simple example of the feasible region of a SDP problem. It consists of two line segments and two smooth curved segments. The optimal solution will be at the boundary of the feasible set F. In this sense, the problem has evident similarities with the LP problem.

A.6.3.2 An Abstract Example

Consider the following simple example of a nonlinear convex optimization problem [111], which can be handled as SDP but not as LP:

Minimize: $\frac{(\mathbf{c}^T\mathbf{x})^2}{\mathbf{d}^T\mathbf{x}}$ (with $\mathbf{d}^T\mathbf{x} > 0$)

Subject to: $A\mathbf{x} + \mathbf{b} \geq 0$

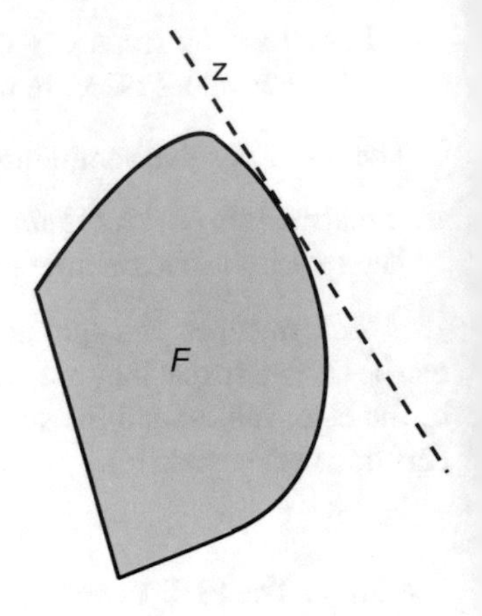

Fig. A.20 Example of SDP feasibility region

Using a typical trick, it can be written as:

Minimize: t

Subject to:

$$A\mathbf{x} + \mathbf{b} \geq 0$$
$$\frac{(\mathbf{c}^T\mathbf{x})^2}{\mathbf{d}^T\mathbf{x}} \leq t$$

In LMI form:

Minimize: t

Subject to:

$$\begin{pmatrix} diag(A\mathbf{x} + \mathbf{b}) & 0 & 0 \\ 0 & t & \mathbf{c}^T\mathbf{x} \\ 0 & \mathbf{c}^T\mathbf{x} & \mathbf{d}^T\mathbf{x} \end{pmatrix} \geq 0$$

In this way, a SDP formulation has been obtained. Notice that the matrix has a block-diagonal form. The matrix inequality:

$$\begin{pmatrix} t & \mathbf{c}^T\mathbf{x} \\ \mathbf{c}^T\mathbf{x} & \mathbf{d}^T\mathbf{x} \end{pmatrix} \geq 0 \tag{A.125}$$

is equivalent to $\mathbf{d}^T\mathbf{x} \geq 0$ and $t - \frac{(\mathbf{c}^T\mathbf{x})^2}{\mathbf{d}^T\mathbf{x}} \geq 0$.

The term, $t - \frac{(\mathbf{c}^T\mathbf{x})^2}{\mathbf{d}^T\mathbf{x}} \geq 0$, is called the *Schur complement* of $\mathbf{d}^T\mathbf{x}$ in the matrix inequality (A.125).

A.6.3.3 The Dual SDP Problem

The dual SDP problem would be:

Maximize:

$$- tr(F_0 \, Z) \tag{A.126}$$

Subject to:

$$tr(F_i \, Z) = c_i \; , \; i = 1, 2, \; \ldots, m \tag{A.127}$$

$$Z \geq 0 \tag{A.128}$$

where $tr(.)$ is the trace of the matrix, and $Z = Z^T$ is the variable added for the Lagrangian.

The duality gap would be:

$$\mathbf{c}^T \mathbf{x} + tr(F_0 Z) = tr(F(\mathbf{x}) Z) \tag{A.129}$$

A.6.3.4 Another Statement of the Primal SDP Problem

Let C and X be two symmetric matrices. Denote:

$$\langle C, X \rangle = \sum_{i=1}^{n} \sum_{j=1}^{n} c_{i\,j} \, x_{i\,j} \tag{A.130}$$

Another statement of the SDP problem is the following:

Minimize:

$$\langle C, X \rangle \tag{A.131}$$

Subject to:

$$\langle A_i, X \rangle = b_i \; , \; i = 1, \, 2, \; \ldots, m \tag{A.132}$$

$$X \geq 0 \tag{A.133}$$

The SDP problem is a conic programming case, since the set of positive semidefinite matrices X is a cone.

A.6.3.5 Example of SDP Problem

Here is an example of SDP problem [35]:

$$A_1 = \begin{pmatrix} 1 & 0 & 1 \\ 0 & 3 & 7 \\ 1 & 7 & 5 \end{pmatrix} \quad , \quad A_2 = \begin{pmatrix} 0 & 2 & 8 \\ 2 & 6 & 0 \\ 8 & 0 & 4 \end{pmatrix} \quad ; \quad b = \begin{pmatrix} 11 \\ 9 \end{pmatrix}; \quad C = \begin{pmatrix} 1 & 2 & 3 \\ 2 & 9 & 0 \\ 3 & 0 & 7 \end{pmatrix}$$

The variable X:

$$X = \begin{pmatrix} x_{11} & x_{12} & x_{13} \\ x_{21} & x_{22} & x_{23} \\ x_{31} & x_{32} & x_{33} \end{pmatrix}$$

Therefore:

$$\langle C, X \rangle = x_{11} + 2x_{12} + 3x_{13} + 2x_{21} + 9x_{22} + 0x_{23} + 3x_{31} + 0x_{32} + 7x_{33}$$

and the constraints $\langle A_i, X \rangle = b_i \ , \ i = 1, 2, \ldots, m$:

$$x_{11} + 0x_{12} + 1x_{13} + 0x_{21} + 3x_{22} + 7x_{23} + 1x_{31} + 7x_{32} + 5x_{33} = 11$$

$$0x_{11} + 2x_{12} + 8x_{13} + 2x_{21} + 6x_{22} + 0x_{23} + 8x_{31} + 0x_{32} + 4x_{33} = 9$$

All these polynomials can be further simplified taking into account that X is symmetric.

The dual problem of the above primal SDP problem would be:

Maximize:

$$\sum_{i=1}^{m} y_i b_i \tag{A.134}$$

Subject to:

$$\sum_{i=1}^{m} y_i A_i + S = C \tag{A.135}$$

$$S \geq 0 \tag{A.136}$$

So, in the case of the previous example:

Maximize:

$$11y_1 + 9y_2 \tag{A.137}$$

Subject to:

$$y_1 \begin{pmatrix} 1 & 0 & 1 \\ 0 & 3 & 7 \\ 1 & 7 & 5 \end{pmatrix} + y_2 \begin{pmatrix} 0 & 2 & 8 \\ 2 & 6 & 0 \\ 8 & 0 & 4 \end{pmatrix} + S = \begin{pmatrix} 1 & 2 & 3 \\ 2 & 9 & 0 \\ 3 & 0 & 7 \end{pmatrix} \tag{A.138}$$

$$S \geq 0 \tag{A.139}$$

These constraints can be also written as:

$$\begin{pmatrix} 1-1y_1-0y_2 & 2-0y_1-2y_2 & 3-1y_1-8y_2 \\ 2-0y_1-2y_2 & 9-3y_1-6y_2 & 0-7y_1-0y_2 \\ 3-1y_1-8y_2 & 0-7y_1-0y_2 & 7-5y_1-4y_2 \end{pmatrix} \geq 0 \tag{A.140}$$

A.6.3.6 Another Statement of the Duality Gap

Given a feasible solution X of the primal and a feasible solution (y, S) of the dual, the duality gap is:

$$\langle C, X\rangle - \sum_{i=1}^{m} y_i b_i = \langle S, X\rangle \geq 0 \tag{A.141}$$

If the duality gap was zero, then the solutions X and (y, S) would be optimal solutions of the primal and the dual problem respectively.

A.6.3.7 The Dual Problem is Also SDP

An equivalent expression of the dual SDP problem is:

Maximize:

$$\mathbf{b}^T\mathbf{y} \tag{A.142}$$

Subject to:

$$C - \sum_{i=1}^{m} y_i A_i \geq 0 \tag{A.143}$$

and it happens that this is also a positive semidefinite problem, similar to:

Minimize:

$$\mathbf{c}^T\mathbf{x} \tag{A.144}$$

Subject to:

$$F(\mathbf{x}) \geq 0 \tag{A.145}$$

with:

$$F(\mathbf{x}) = F_0 + \sum_{i=1}^{m} x_i F_i \tag{A.146}$$

A.6.3.8 Primal-Dual Optimization

In most SDP problems the optimum of the primal problem is coincident with the optimum of the dual problem, so the duality gap is zero. A combined view of the optimization problem could be to consider:

Minimize:

$$\eta = \mathbf{c}^T \mathbf{x} + tr(F_0 Z) \tag{A.147}$$

Subject to:

$$F(\mathbf{x}) \geq 0, \ \ Z \geq 0, \tag{A.148}$$

$$tr(F_i \, Z) = c_i, \ \ i = 1, 2, \ldots, m \tag{A.149}$$

This form, which intends to minimize the duality gap η, is called the primal-dual optimization approach.

A.6.3.9 A Barrier Function

The following barrier function is frequently used for SDP problems:

$$\phi(\mathbf{x}) = \begin{cases} \log \det \ F(\mathbf{x})^{-1}, \ \ when \, F(\mathbf{x}) > 0 \\ \infty, \ \ otherwise \end{cases} \tag{A.150}$$

(other barrier functions can be used)

Figure A.21 illustrates how is the barrier inside F, mostly flat in the interior with significant increase near the boundary.

The gradient of the barrier function would be:

$$(\nabla \phi(\mathbf{x}))_i = -tr(F(\mathbf{x})^{-1} F_i) \tag{A.151}$$

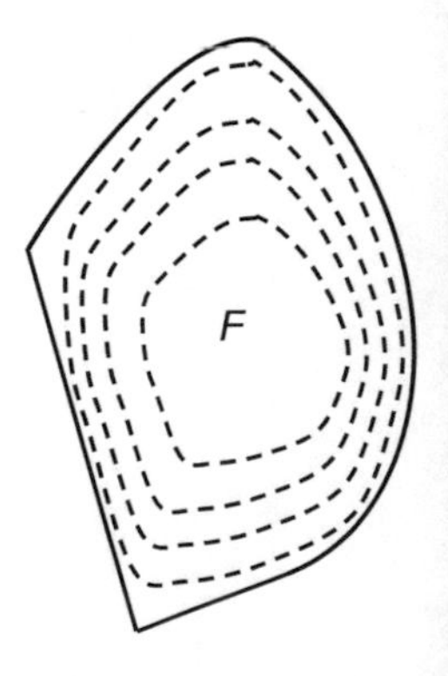

Fig. A.21 Example of barrier function

and the Hessian:

$$(\nabla^2 \phi(\mathbf{x}))_{ij} = -tr(F(\mathbf{x})^{-1} F_i \, F(\mathbf{x})^{-1} F_j) \tag{A.152}$$

In the case of linear inequalities $\mathbf{a}_i^T \mathbf{x} + b_i \geq 0, \; i = 1, \ldots, n$, the barrier function would be:

$$\phi(\mathbf{x}) = \begin{cases} -\sum\limits_{i=1}^{n} \log(\mathbf{a}_i^T \mathbf{x} + b_i), & if \; \mathbf{a}_i^T \mathbf{x} + b_i > 0, \; i = 1, \ldots, n \\ \infty, & otherwise \end{cases} \tag{A.153}$$

According to [111], the function $\phi()$ can be regarded as a potential function corresponding to a repelling force from each constraint hyperplane. Note that:

$$\nabla \phi(\mathbf{x}) = \sum_{i=1}^{n} \frac{1}{\mathbf{a}_i^T \mathbf{x} + b_i} \mathbf{a}_i = \sum_{i=1}^{n} \frac{-1}{r_i \, \|\mathbf{a}_i\|} \mathbf{a}_i \tag{A.154}$$

where r_i is the distance from $\mathbf{x}$ to the *i-th* constraint hyperplane.

A.6.3.10 Analytic Center

The '*analytic center*' of the LMIs $F(\mathbf{x}) \geq 0$, is the following:

$$\mathbf{x}_A = \arg\min \, \phi(\mathbf{x}) \tag{A.155}$$

It can be deduced from the gradient that:

$$tr(F(\mathbf{x}_A)^{-1} F_i) = 0, \; i = 1, \, 2, \, \ldots, m \tag{A.156}$$

In the case of LMIs:

$$\mathbf{x}_A = \arg\max \prod_{i=1}^{n} (\mathbf{a}_i^T \mathbf{x} + b_i) \tag{A.157}$$

So the analytic center is the feasible point that maximizes the product of distances to the constraint hyperplanes. It is also the equilibrium point of the "repulsive forces".

The Newton's method can be used to compute the analytic center of a given set of constraints, departing from an initial strictly feasible point (a point such $F(\mathbf{x}) > 0$). The Newton direction would be [111]:

$$\delta \, \mathbf{x}^{(n)} = \arg\min_{v} \left\| -I + \sum_{i=1}^{m} v_i \, F^{-1/2} \, F_i \, F^{-1/2} \right\|_F \tag{A.158}$$

where the Frobenius norm has been used:

$$\|A\|_F = (tr(A^T A))^{1/2} = (\sum_{ij} A_{1j}^2)^{1/2}$$

The iterative procedure will be:

(a) compute the Newton direction $\delta\,\mathbf{x}$,
(b) find $\mathbf{p} = \arg\min\ \phi(\mathbf{x} + \mathbf{p}\,\delta\,\mathbf{x})$,
(c) update $\mathbf{x} \leftarrow \mathbf{x} + \mathbf{p}\,\delta\,\mathbf{x}$,
back to (a).

A.6.3.11 Central Path

Recall that the SDP problem was:

Minimize: $\mathbf{c}^T\mathbf{x}$; subject to: $F(\mathbf{x}) \geq 0$

Consider now the following equations:

$$F(\mathbf{x}) > 0 \tag{A.159}$$

$$\mathbf{c}^T\mathbf{x} = \gamma \tag{A.160}$$

Figure A.22 depicts the situation. The line $\mathbf{c}^T\mathbf{x} = \bar{p}$ crosses the point O, and the line $\mathbf{c}^T\mathbf{x} = p^*$ crosses the point P (the optimal).

The analytic center corresponding to (A.159) and (A.160), would be:

$$\mathbf{x}_A(\gamma) = \arg\min\,(\log\det\ F(\mathbf{x})^{-1}) \tag{A.161}$$

subject to: $F(\mathbf{x}) > 0$ and $\mathbf{c}^T\mathbf{x} = \gamma$.

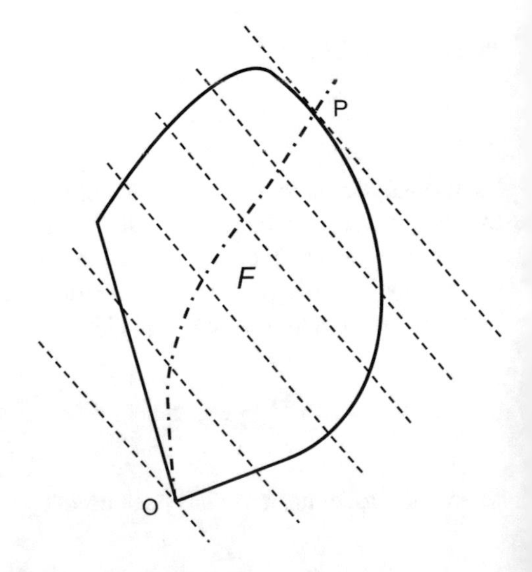

Fig. A.22 Central path and $\mathbf{c}^T\mathbf{x} = \gamma$ lines

The curve described by $\mathbf{x}_A(\gamma)$, for $\gamma : \bar{p} \rightarrow p^*$, is the central path, which is the curve (dash-point-dash) that in the figure comes from O to P.

The solution of (A.161) satisfies:

$$tr(F(\mathbf{x}_A(\gamma))^{-1} F_i) = \lambda c_i \ , \ i = 1, 2, \ldots, m \tag{A.162}$$

where λ is a Lagrange multiplier.

It can be shown that $\mathbf{u}(\gamma) = F(\mathbf{x}_A(\gamma))^{-1} / \lambda$ minimizes the barrier function $\log \det Z^{-1}$ [111]. Hence, there is a correspondence between points $\mathbf{x}_A(\gamma)$ of the primal central path, and points $\mathbf{u}\,(\gamma)$ of the dual central path. Actually, the duality gap associated with the primal-dual feasible pair $\mathbf{x}_A(\gamma)$ and $\mathbf{u}\,(\gamma)$, is $\eta = n/\lambda$.

Most interior-point methods follow the central path, returning periodically to it or measuring the deviation from it.

See [111] for more details on potential reduction methods, which combine the duality gap and the deviation from the central path.

For background literature, it would be recommended the tutorial [52] about LMI optimization, and the presentation [32] that concisely introduces linear programming (LP), quadratic programming (QP), second-order cone programming (SOCP), semi-definite programming (SDP), and others. According with [32], the generality of these methods are related as follows:

$$LP \subset QP \subset SOCP \subset SDP$$

It would also be recommended to visit the web site of El Ghaoui (see the resources section).

A.6.4 Duality

It is usual, when using interior-point methods, to deal with dual problems. This is one of the reasons that have motivated much interest on duality issues. In this context, it is illuminating to consider the Fenchel's duality theorem, which we introduce now by means of two simple figures.

The scenario consists of a convex and a concave curve. The problem is to find the minimum vertical distance between the two curves. Figure A.23 depicts the case.

A second problem is to find two separated parallel tangents, so their distance is maximal. Figure A.24 depicts this other case.

The Fenchel's duality theorem establishes that he points having the minimal vertical separation are also the tangency points for the maximally separated parallel tangents.

Let us devote some efforts to formalize this result. Of course, what makes it interesting for us is because of primal-dual solutions and duality gap issues.

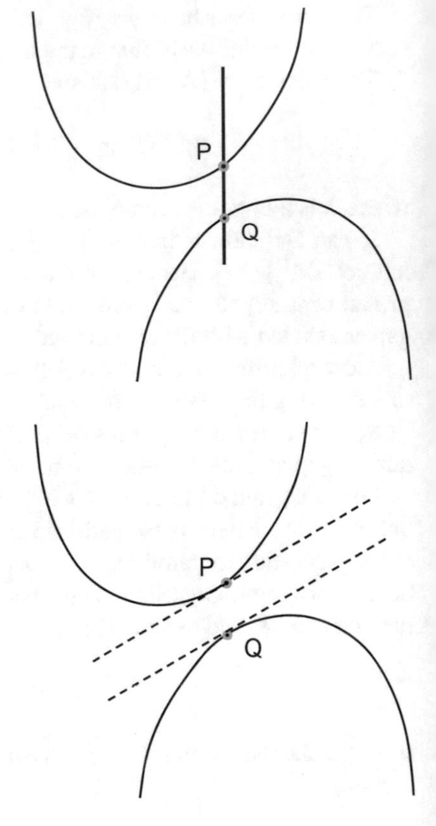

Fig. A.23 Minimal vertical distance between two curves

Fig. A.24 Maximally separated parallel tangents

A.6.4.1 The Legendre-Fenchel Transform

A first important ingredient is the Legendre-Fenchel (LF) transform, which is a generalization of the Legendre transform. The Legendre transform applies for convex, differentiable functions, while the LF transform can be applied to a continuous but not necessarily differentiable functions [108].

The concept of the Legendre transform can be intuitively introduced by considering the exercise represented in Fig. A.25. A certain slope a is chosen. Now, one moves a line with slope a until touching the curve $f(x)$. This tangent intersects the vertical axis at b.

The value of the Legendre transform $f^*(a)$ will be $-b$ (notice the change of sign).

Given a convex, differentiable function $f(x)$, its Legendre transform is a continuous set of values $-b$ obtained for different values of a. The definition of the Legendre transform (LF) is:

$$f^*(a) = \max_x (a\,x - f(x)) \tag{A.163}$$

The LF transform of a continuous function $f(x)$ is defined as follows:

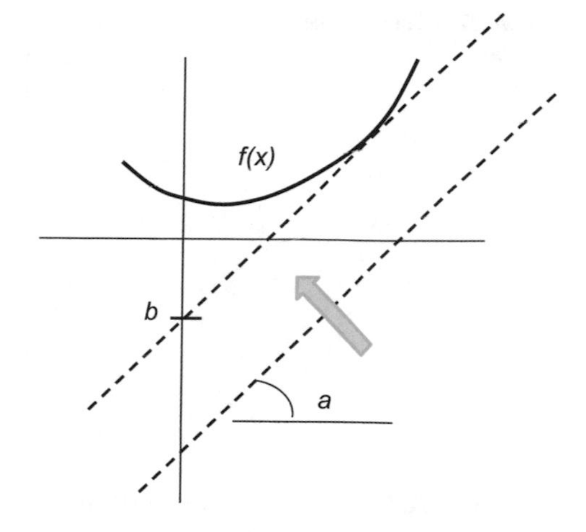

Fig. A.25 Obtaining a value of $f^*(a)$ (the Legendre transform)

$$f^*(a) = \sup_x (a\,x - f(x)) \tag{A.164}$$

The LF transform of $f^*(a)$ is:

$$f^{**}(x) = \sup_a (ax - f^*(a)) \tag{A.165}$$

The reader may suspect that mathematical difficulties may arise when functions are not convex and/or non-differentiable. Of course, it is not our mission here to try a complete study of this question; but it is convenient to include some hints.

For instance, Fig. A.26 shows how the intersection of two differentiable functions may give a non-differentiable (non-smooth) function.

Given a function $f(x)$ it is said that it admits a '*supporting line*' at x, if there exist a parameter p such that:

$$f(y) \geq f(x) + p(y - x), \quad \forall y \in \Re \tag{A.166}$$

A supporting line is strictly supporting if:

$$f(y) > f(x) + p(y - x), \quad \forall y \neq x \tag{A.167}$$

For differentiable functions, a supporting line is also a tangent line.

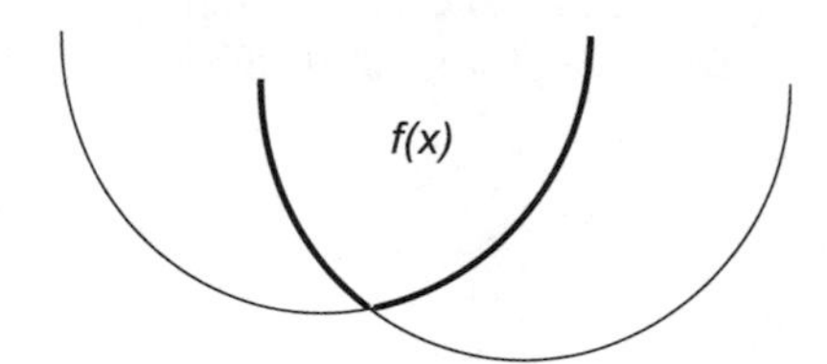

Fig. A.26 Example of non-smooth function

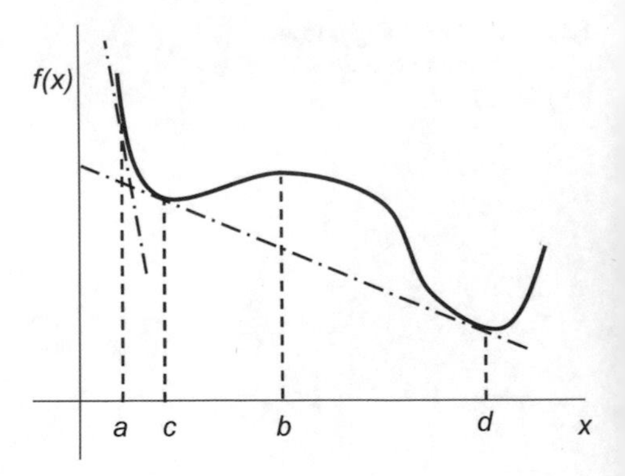

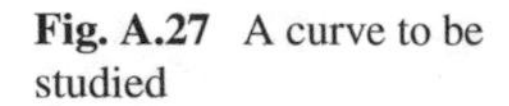
Fig. A.27 A curve to be studied

Figure A.27 depicts a curve with a number of interesting points. Two supporting lines have been plotted (dash-point-dash lines).

Let us do the following remarks:

- The point (a) admits a strictly supporting line.
- The point (b) does not admit any supporting line.
- The point (c) admits a non-strictly supporting line, as it touches another point (d) of $f(x)$.

For the general cases, it can be shown that if $f(x)$ admits a supporting line at x with slope k, then $f^*(k)$ admits a supporting line at k with slope x. Moreover, if $f(x)$ admits a strictly supporting line at x with slope k, then $f^*(k)$ admits a tangent line at k with slope $f^{*\prime}(k) = x$ (therefore, $f^*(k)$ is differentiable).

It also can be shown that $f(x) = f^{**}(x)$ iff $f(x)$ admits a supporting line at x. Also, if $f^*(k)$ is differentiable at k, then $f(x) = f^{**}(x)$ at $x = f^{*\prime}(k)$. Indeed, if $f^*(k)$ is everywhere differentiable, then $f(x) = f^{**}(x)$ for all x.

An important property of the LF transform is that $f^*(k)$ and $f^{**}(x)$ are convex functions (they are U-shaped).

Two interesting properties:

- A convex function can always be written as the LF transform of another function.
- $f^{**}(x)$ is the largest convex function that satisfies: $f^{**}(x) \leq f(x)$.

And finally, concerning non-differentiable points, suppose that $f(x)$ has one of these points (like the one represented in Fig. A.26) and denote it as x_C. This point admits infinitely many supporting lines, with slopes in the range $k_1 \ldots k_2$. The corresponding $f^*(k)$ would be a line of constant slope x_C in the interval $k_1 \ldots k_2$. Figure A.28 depicts an example.

See [108] for more details on the LF transform.

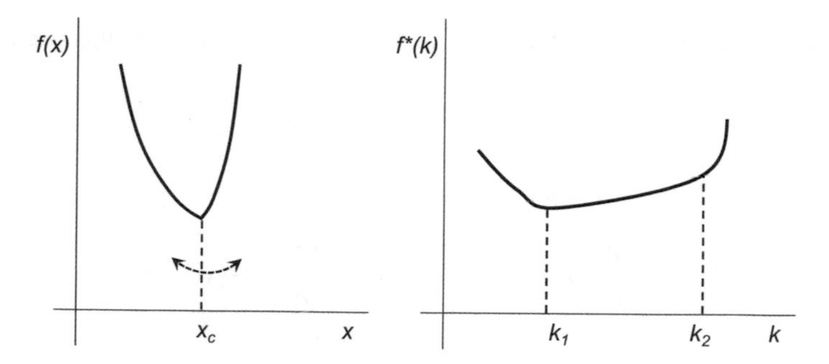

Fig. A.28 LF transform of a non-differentiable point

A.6.4.2 The Fenchel's Duality Theorem

Let us express the Fenchel's duality theorem using the formalism already introduced. Suppose that $f(\mathbf{x})$ is a proper convex function, $g(\mathbf{x})$ is a proper concave function, and regularity conditions are satisfied. Then:

$$\min_{\mathbf{x}}(f(\mathbf{x}) - g(\mathbf{x})) = \max_{\mathbf{a}}(g_*(\mathbf{a}) - f^*(\mathbf{a})) \tag{A.168}$$

where:

$$f^*(\mathbf{a}) = \sup_{\mathbf{x}}(\mathbf{a}^T\mathbf{x} - f(\mathbf{x})) \tag{A.169}$$

$$g_*(\mathbf{a}) = \inf_{\mathbf{x}}(\mathbf{a}^T\mathbf{x} - g(\mathbf{x})) \tag{A.170}$$

The function $f^*(\mathbf{x})$ is called the *convex conjugate* of $f(\mathbf{x})$; and the function $g_*(\mathbf{x})$ is called the *concave conjugate* of $g(\mathbf{x})$.

A.6.4.3 Other Relationships

From the very definition of LF transform, it is possible to write the following:

$$f(\mathbf{x}) + f^*(\mathbf{a}) \geq \mathbf{a}^T\mathbf{x} \tag{A.171}$$

This is called the *Fenchel's inequality*, and it is a generalization of the Young's inequality.

Under certain regularity conditions, if $f(\mathbf{x}) \geq g(\mathbf{x})$, $\forall x \in \Re^n$, then there exists a *separation plane* between $f(\mathbf{x})$ and $g(\mathbf{x})$ given by:

$$f(\mathbf{x}) \geq \mathbf{a}^T\mathbf{x} + b \geq g(\mathbf{x}) \tag{A.172}$$

(actually, this result is established by the hyperplane separating theorem [15]).

If $f(\mathbf{x})$ is convex and differentiable, one can consider the Legendre transform,

$$f^*(a) = \max_x(a\,x - f(x)) \tag{A.173}$$

and compute the maximization by taking the derivative:

$$\frac{d}{dx}(a\,x - f(x)) = a - \frac{df(x)}{dx} = 0 \tag{A.174}$$

The usual procedure is to write:

$$a = \frac{df(x)}{dx} \tag{A.175}$$

and to obtain from this equation x as a function of a. Then:

$$f^*(a) = a\,x(a) - f(x(a)) \tag{A.176}$$

Another definition of Legendre transform says that $f(x)$ and $f^*(a)$ are Legendre transforms of each other if $D(f(x)) = (D(f^*(a))^{-1}$ (where D means derivative). Actually, from the definition it can be derived that:

$$a = \frac{df(x)}{dx}; \quad x = \frac{df^*(a)}{da} \tag{A.177}$$

These expressions make you recall Hamiltonian mechanics and other main topics of Physics.

Notice that based on (A.175), in the n-dimensional case, one can write:

$$f^*(\nabla f(\mathbf{x})) = \nabla f(\mathbf{x})^T\mathbf{x} - f(\mathbf{x}) \tag{A.178}$$

An economics' interpretation of the LF transform is that in:

$$f^*(a) = \sup_{\mathbf{x}}(a\,x - f(x))$$

$f^*(a)$ can represent the gain, a is the price, x the level of production, and $f(x)$ the cost of this production. It may happen that the price depends on x, so our formula should be modified. This is in accordance with the initiative of Moreau, who extended the Fenchel's conjugation using coupling functions:

$$f^*(a) = \sup_x\,(c\,(a, x) - f(x)) \tag{A.179}$$

(*Moreau's conjugate*)

A.6.4.4 Saddle-Value

An important reference concerning duality is the book of Rokafellar [96]. It introduces rigorous mathematical concepts and terminology. The central problem started with a certain function $K(x, y)$, and the following pair of functions:

$$f(x) = \sup_y K(x, y) \tag{A.180}$$

$$g(y) = \inf_x K(x, y) \tag{A.181}$$

The variables x and y belong to certain sets X and Y. Two optimization problems were considered:

$$\text{minimize } f(x) \text{ over all } x \in X$$
$$\text{maximize } g(x) \text{ over all } y \in Y$$

Clearly:

$$f(x) \geq K(x, y) \geq g(y) \tag{A.182}$$

And then:

$$\inf_x f(x) \geq \sup_y g(y) \tag{A.183}$$

If equality holds, the common value is called the *saddle-value* of K, which exists if there is a *saddle-point* of K. A point (x_s, y_s) is a saddle-point of K if:

$$K(x, y_s) \geq K(x_s, y_s) \geq K(x_s, y), \quad \forall x \in X, \forall y \in Y \tag{A.184}$$

The saddle-point solves the two optimization problems.

According with [96] many different functions $K(x, y)$ can be proposed, and also given an optimization problem many duals could be constructed: the question is how to obtain desired properties.

The principal theorems about the existence of the saddle-value of K require that $K(x, y)$ be convex in x and concave in y, or almost so [96].

According with simple searching on Internet, a lot of scientific contributions have been made (several thousands, many of them related to image processing) on primal-dual optimization and the saddle-value of an appropriate $K(x, y)$. As it is done in [96], one could choose a Lagrangian for this function.

A.6.4.5 Examples of Conjugate Functions

A rapid overview of conjugate function examples may help for an evaluation of the duality approach usefulness.

Let us start with the indicator function and norms.

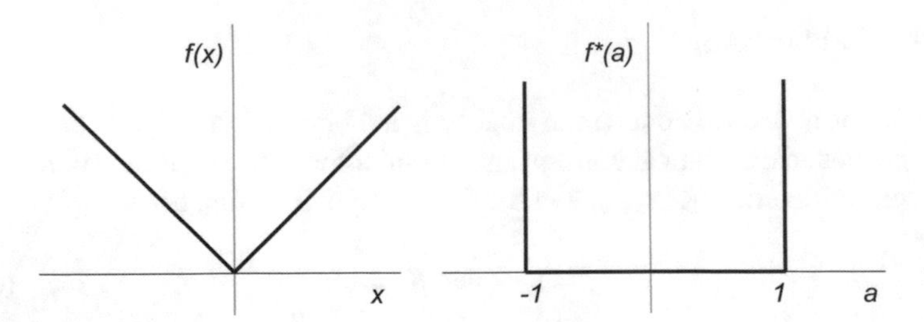

Fig. A.29 The norm $\|x\|$ and its conjugate

Given a set C, its indicator function $i_C(\mathbf{x})$ is 0 if $\mathbf{x} \in C$, and ∞ otherwise. The conjugate of $i_C(\mathbf{x})$ is:

$$\sup_{x \in C} \mathbf{a}^T \mathbf{x}$$

which is the support function of C and is sometimes denoted as $\|a\|_*$.

Figure A.29 shows the norm $\|x\|$ and its conjugate, which is 0 for $\|a\| \leq 1$, and ∞ otherwise. Note that $\|x\|$ is not differentiable at the origin.

In the case of:

$$f(x) = \frac{1}{p} \|x\|^p \ , \ 1 < p < \infty \tag{A.185}$$

the conjugate is:

$$f^*(a) = \frac{1}{q} \|a\|^q \tag{A.186}$$

with: $\frac{1}{p} + \frac{1}{q} = 1$

Now, let us devote some space for curves.

Suppose that $f(x)$ is a parabola: $f(x) = x^2$. Its conjugate is: $f^*(a) = \frac{1}{4}a^2$. Figure A.30 visualizes this case.

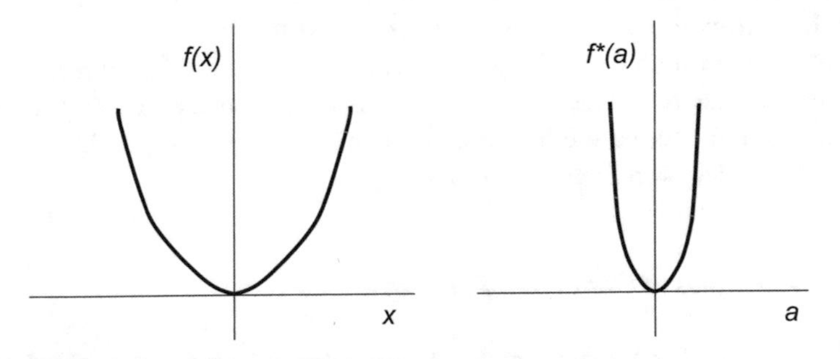

Fig. A.30 The conjugate of a parabola

If $f(x)$ is a general quadratic curve:

$$f(\mathbf{x}) = \frac{1}{2}\mathbf{x}^T A\,\mathbf{x} \tag{A.187}$$

the conjugate is:

$$f^*(\mathbf{a}) = \frac{1}{2}\mathbf{a}^T A^{-1}\mathbf{a} \tag{A.188}$$

A more general case could be:

$$f(\mathbf{x}) = \frac{1}{2}\mathbf{x}^T A\,\mathbf{x} + \mathbf{c}^T\mathbf{x} + b \tag{A.189}$$

the conjugate is:

$$f^*(\mathbf{a}) = \frac{1}{2}(\mathbf{a}-\mathbf{c})^T A^{-1}(\mathbf{a}-\mathbf{c}) + \mathbf{c}^T A^{-1}(\mathbf{a}-\mathbf{c}) + b \tag{A.190}$$

The conjugate of the exponential function $f(x) = e^x$ is:

$$f^*(a) = a\,(\ln\, a - 1) \tag{A.191}$$

The conjugate of $f(x) = -\log\, x$ is:

$$f^*(a) = -1 - \log(-a) \tag{A.192}$$

(only valid if $a < 0$)

In case of matrices, the Frobenius norm is self-conjugate, and the conjugate of the spectral norm (the largest singular value) is the trace norm (sum of singular values).

Coming to optimization matters, let us consider:

$$\min_x f(x) + g(x) \tag{A.193}$$

which is equivalent to:

$$\min_{x,y} f(x) + g(y) \text{ subject to } x = y$$

Then, the dual problem would be:

$$\max_u\, -f^*(a) - g^*(-a) \tag{A.194}$$

A.6.4.6 The Fenchel-Rockafellar Duality

Rockafellar extended the Fenchel's duality theorem considering as primal problem the following:

$$\min_{\mathbf{x}} f(\mathbf{x}) + g(A\mathbf{x}) \tag{A.195}$$

and writing the corresponding dual problem as:

$$\min_{\mathbf{y}} f^*(-A^*\mathbf{y}) + g^*(\mathbf{y}) \tag{A.196}$$

where A^* is the adjoint.

Denote:

$$\mu = \inf_{\mathbf{x}} f(\mathbf{x}) + g(A\mathbf{x}) \tag{A.197}$$

$$\mu^* = \inf_{\mathbf{y}} f^*(-A^*\mathbf{y}) + g^*(\mathbf{y}) \tag{A.198}$$

Then: $\mu \geq -\mu^*$

Moreover, if a (primal) *constraint qualification* is satisfied, such as $0 \in int\ (dom\ g - A \cdot dom f)$, then: $\mu = -\mu^*$. The unique solution $\mathbf{x}_M$ of the primal problem can be derived from a (non-necessarily unique) solution $\mathbf{y}_M$ of the dual problem as:

$$\mathbf{x}_M = \nabla f^*(-A^*\,\mathbf{y}_M) \tag{A.199}$$

In [77] this duality was further extended to the context of saddle functions and dual mini-max problems.

A.6.4.7 Mini-Max Results

As the reader surely had noticed, some of the previous expressions involve a chaining of min (or inf) and max (or sup). Actually this is connected with mini-max results, which in turn are representative of game theory [115].

Let us select some theorems.

The first (weak duality) says that:

$$\min_x \max_y\ g(x, y) \geq \max_y\ \min_x g(x, y) \tag{A.200}$$

(then, in some games it matters who plays first)

A second theorem (strong duality) due to J. von Neumann in 1928, establishes that given a continuous $g(x, y)$, convex on $x \in X$ and concave on $y \in Y$, with X and Y convex and compact, then:

$$\min_x \max_y g(x, y) = \max_y \min_x g(x, y) \tag{A.201}$$

The conditions can be relaxed a bit, so a third theorem (strong duality) due to Sion in 1958, establishes that given a $g(x, y)$, lower semi-continuous quasi-convex on $x \in X$ and upper semi-continuous quasi-concave on $y \in Y$, with X and Y convex and one of them compact, then:

$$\min_x \max_y g(x, y) = \max_y \min_x g(x, y) \tag{A.202}$$

There are several theorems and mathematical results in connection with the mini-max context. For instance, [39] includes up to 14 mini-max theorems, showing that they form an equivalent chain.

In a very enjoyable article [26], the authors describe an historic encounter of von Neumann and Dantzig in 1947, at Princeton. Those were the times when it was conjectured that close relationships existed between game theory, duality, and linear programming. Now it is well known, see for instance [1], that any zero-sum game can be reduced to a linear programming problem (and it can be used to prove the mini-max theorem based on strong duality), and vice-versa: a linear programming problem can be reduced to a zero-sum game.

In order to give some more details on this aspect, we could introduce a couple of rapid examples.

A standard problem is the prisoner's dilemma, which can be represented with the following tables (Figs. A.31 and A.32).

In each cell, the payoff of each prisoner is given in terms of year's reduction.

The sum of payoffs in each cell of a zero-sum game is zero. That means, in a two-person game, that if one player wins 100, then the other player loses 100. This type of games can be represented by just one payoff matrix (the other having term

Fig. A.31 Prisoner's dilemma

dilemma		Prisoner B (column)	
		Confess	Don't
Prisoner A (row)	Confess	-10, -10	0, -20
	Don't	-20, 0	-1, -1

Fig. A.32 Negotiation

contract		Employer (column)	
		A	B
Player (row)	1	35000	40000
	2	30000	20000

by term the opposite sign). Take for example the case of a negotiation between a football player and his employer:

Usually, zero-sum tables put on the left the agent that wants to maximize the outcome (choosing between strategies 1 or 2), and on top the agent that wants to minimize it (selecting strategy A or B). If the player chooses 1, the employer will choose A. If the player chooses 2, the employer will choose B. The best for the athlete would be 1-A. In mathematical terms:

$$\max_i \min_j P = 35000 \tag{A.203}$$

where P is a matrix with the four profit entries of the table.

From the point of view of the employer, if he chooses A the player will choose 1. Instead, if he chooses B, the player will choose also 1. The best for the employer would be A-1. In mathematical terms:

$$\min_j \max_i P = 35000 \tag{A.204}$$

And so, a mini-max result has been found. The entry 1-A, with the value 35,000, is a **saddle-point** of the game. Because of the saddle-point, neither player can take advantage of the rival's strategy; it is said that one has a *stable solution* (or *equilibrium solution*).

Some games do not possess a saddle-point. In such a case, players should avoid a predictable strategy, which may suppose an advantage for the opponent. Hence, each player should choose at random among their alternatives, according with some probability distributions $\mathbf{x} = (x_1, x_2, \ldots, x_n)$, of player 1, and $\mathbf{y} = (y_1, y_2, \ldots, y_m)$, of player 2. This is called mixed-strategies games. The equivalence between games and linear programming was originally established in this context [26]. In particular, given the LP problem:

$$\min \mathbf{c}^T\mathbf{x} \text{ subject to } A\mathbf{x} \geq \mathbf{b},\ \mathbf{x} \geq 0$$

and the dual:

$$\max \mathbf{b}^T\mathbf{y} \text{ subject to } A^T\mathbf{y} \leq \mathbf{c},\ \mathbf{y} \geq 0$$

Dantzig suggested to reduce this pair of LP problems to a symmetric zero-sum game by using the following payoff matrix:

$$P = \begin{pmatrix} 0 & A & -\mathbf{b} \\ -A^T & 0 & \mathbf{c} \\ \mathbf{b}^T & -\mathbf{c}^T & 0 \end{pmatrix} \tag{A.205}$$

(when P is skew-symmetric, i.e. $P^T = -P$, the game is called symmetric game).

This reduction may have some difficulties, as discussed in [1].

A.6.4.8 The Farkas' Lemma

The Farkas'lemma can be expressed in several ways [55, 99]. In geometrical terms, it says that a vector is either in a given convex cone, or there is a hyperplane separating the vector from the cone. The lemma can be regarded as a particular case of the separating hyperplane theorem.

Another formulation is the following: let A be a $m \times n$ matrix and $\mathbf{b}$ a column vector of size n. Only one of the next alternatives holds:

- There exists $\mathbf{x} \in \Re^n$ such that $A\,\mathbf{x} = \mathbf{b}$ and $\mathbf{x} \geq 0$
- There exists $\mathbf{y} \in \Re^m$ such that $A^T\,\mathbf{y} \geq 0$ and $\mathbf{b}^T\mathbf{y} < 0$

Note that this second alternative is equivalent to:

- There exists $\mathbf{y} \in \Re^m$ such that $A^T\,\mathbf{y} \leq 0$ and $\mathbf{b}^T\mathbf{y} > 0$

An important application of the lemma is for the analysis of Linear Programming problems, concerning in particular feasibility aspects.

For example, in the case of the LP problem:

$$\min\ \mathbf{c}^T\mathbf{x} \text{ subject to } A\mathbf{x} \geq \mathbf{b},\ \ \mathbf{x} \geq 0$$

and the dual:

$$\max\ \mathbf{b}^T\mathbf{y} \text{ subject to } A^T\mathbf{y} \leq \mathbf{c},\ \ \mathbf{y} \geq 0$$

The Farkas' lemma implies that exactly one of the following cases occurs:

- Both the primal and the dual have optimal solutions $\mathbf{x}^*$ and $\mathbf{y}^*$ with equal values $\mathbf{c}^T\mathbf{x}^* = \mathbf{b}^T\mathbf{y}^*$
- The dual is unfeasible, and the primal is unbounded ($\mathbf{c}^T\mathbf{x} \to \infty$)
- The primal is unfeasible, and the dual is unbounded ($\mathbf{b}^T\mathbf{y} \to \infty$)
- Both primal and dual are unfeasible

The lemma can be used as a certificate of infeasibility [6]. In particular, the primal problem:

$$\min\ \mathbf{c}^T\mathbf{x} \text{ subject to } A\mathbf{x} \geq \mathbf{b},\ \ \mathbf{x} \geq 0$$

is unfeasible if and only if

- There exists $\mathbf{y} \in \Re^m$ such that $A^T\,\mathbf{y} \leq 0$ and $\mathbf{b}^T\mathbf{y} > 0$

A.7 Gradients. Trust Regions

This section is devoted to complete our collection of selected optimization topics. Some of them are directly related to optimization methods and techniques that have been incorporated in the MATLAB Optimization Toolbox. Most of the materials in this section are covered in [61].

A.7.1 Gradient Descent

The use of gradients is important in signal processing, in particular for iterative adaptation and/or optimization schemes. This is evident from the widespread presence of gradients in the chapters of this book.

The iterative gradient descent method can be summarized with a simple equation:

$$\mathbf{x}_{k+1} = \mathbf{x}_k - \alpha_k \nabla f(\mathbf{x}_k) \tag{A.206}$$

where α_k is the step size.

One takes the direction of minus the gradient for steepest descent. Of course, one could take the direction of the gradient for steepest ascent.

The steepest descent method was proposed in 1847 by Cauchy. The step size α_k^* was computed so as:

$$f(\mathbf{x}_k - \alpha_k^* \mathbf{g}_k) = \min_{\alpha > 0} f(\mathbf{x}_k - \alpha\, \mathbf{g}_k) \tag{A.207}$$

where $\mathbf{g}_k = \nabla f(\mathbf{x}_k)$.

As asserted by [11, 48], gradient-based methods are attracting a renewed interest, since they can be competitive for large scale optimization problems [25]. A main effort nowadays is devoted to obtain first-order methods able to capture curvature information. First order methods only use derivatives; while second order methods use the Hessian (which effectively obtains curvature information, at the price of higher complexity).

This subsection focuses on first order methods, and the next subsection extends the view to Newton related methods.

A natural analogy can be established between steepest ascent and hill climbing, as already mentioned in this book. The basic idea is to choose in each step the direction of largest ascent. However this may drive you to a stationary point which is not the highest peak, but a local maximum instead. It is pertinent to determine—as far as possible—if a function to be optimized offers good opportunities for gradients or not. Some ideas for it come next.

A.7.1.1 Types of Functions. Bounds

Given an objective function to be minimized, it is important to establish upper and lower bounds.

Let us start with an upper bound, which is based on Lipschitz functions. A function $f(\mathbf{x})$ is said to satisfy a Lipschitz condition if a constant $L > 0$ exists with:

$$\| f(\mathbf{x}) - f(\mathbf{y}) \|_2 \leq L \, \| \mathbf{x} - \mathbf{y} \|_2 \,, \; \forall \mathbf{x}, \, \mathbf{y} \in \Re^n \tag{A.208}$$

This expression says that the slope of the line connecting the points $f(\mathbf{x})$ and $f(\mathbf{y})$ is no greater than L.

Functions satisfying this condition are called *Lipschitz functions*; also, colloquially, one could say that a function $f(\mathbf{x})$ is Lipschitz.

The Lipschitz condition is stronger than continuity. Indeed, Lipschitz functions are continuous functions.

The function $|\mathbf{x}|$ is Lipschitz, but not differentiable. The function $\sqrt{x}$ is a continuous function but it is not a Lipschitz function because it becomes infinitely steep as $x \to 0$. Likewise, the function x^2 is a continuous function but is not Lipschitz, since it becomes arbitrarily steep as $x \to \infty$. See [101] for more examples.

In the case of differentiable functions, gradients are defined and so they can be object of study. The gradient of a function is Lipschitz if a constant $L > 0$ exists with:

$$\|\nabla f(\mathbf{x}) - \nabla f(\mathbf{y})\|_2 \leq L \, \|\mathbf{x} - \mathbf{y}\|_2 \, , \; \forall \mathbf{x}, \, \mathbf{y} \in \Re^n \tag{A.209}$$

On the basis of the Cauchy-Schwarz inequality, it can be shown (lemma) that if the gradient is Lipschitz then:

$$|f(\mathbf{y}) - f(\mathbf{x}) - \nabla f(\mathbf{x})^T (\mathbf{y} - \mathbf{x})| \leq \frac{L}{2} \, \|\mathbf{y} - \mathbf{x}\|_2^2 \tag{A.210}$$

Put in other way:

$$f(\mathbf{y}) \leq f(\mathbf{x}) + \nabla f(\mathbf{x})^T (\mathbf{y} - \mathbf{x})| + \frac{L}{2} \, \|\mathbf{y} - \mathbf{x}\|_2^2 \tag{A.211}$$

A consequence of this lemma is that:

$$\frac{L}{2} \, \|\mathbf{x}\|_2^2 - f(\mathbf{x}) \tag{A.212}$$

is convex.

For twice differentiable $f(\mathbf{x})$, if the gradient is Lipschitz: $\nabla^2 f(\mathbf{x}) \leq L \, I$

Figure A.33 depicts the upper bound in case the gradient of $f(\mathbf{x})$ is Lipschitz. Now, for lower bounds one considers some more definitions related to convexity.

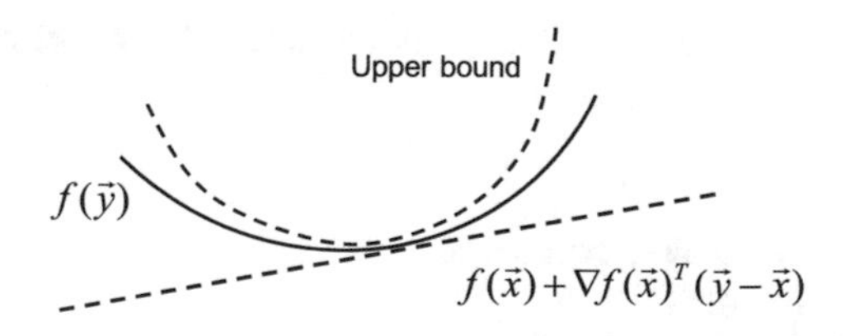

Fig. A.33 Upper bound corresponding to the gradient being Lipschitz

- A function $f(\mathbf{x})$ is *convex* iff:

$$f(\alpha\,\mathbf{x} + (1-\alpha)\,\bar{y}) \le \alpha\, f(\mathbf{x}) + (1-\alpha)\, f(\mathbf{y}), \quad \forall \mathbf{x}, \mathbf{y};\ 0 \le \alpha \le 1 \tag{A.213}$$

- The function $f(\mathbf{x})$ is *strictly convex* if the equality holds only when $\mathbf{x} = \mathbf{y}$:

$$f(\alpha\,\mathbf{x} + (1-\alpha)\,\bar{y}) < \alpha\, f(\mathbf{x}) + (1-\alpha)\, f(\mathbf{y}), \quad \forall \mathbf{x}, \mathbf{y};\ \mathbf{x} \ne \mathbf{y};\ 0 \le \alpha \le 1 \tag{A.214}$$

- The function $f(\mathbf{x})$ is *strongly convex* (with modulus m) if:

$$f(\alpha\,\mathbf{x} + (1-\alpha)\,\bar{y}) \le \alpha\, f(\mathbf{x}) + (1-\alpha)\, f(\mathbf{y}) - \tfrac{m}{2}(\alpha\,(1-\alpha))\, \|x-y\|_2^2, \quad \forall \mathbf{x}, \mathbf{y},\ 0 \le \alpha \le 1 \tag{A.215}$$

Again, if $f(\mathbf{x})$ is differentiable it is possible to consider gradients, obtaining some more inequalities corresponding to convexity.

- The function $f(\mathbf{x})$ is *convex* iff:

$$f(\mathbf{y}) \ge f(\mathbf{x}) + \nabla f(\mathbf{x})^T (\mathbf{y}-\mathbf{x}), \quad \forall \mathbf{x}, \mathbf{y} \tag{A.216}$$

- The function $f(\mathbf{x})$ is *strictly convex* if the equality holds only when $\mathbf{x} = \mathbf{y}$:

$$f(\mathbf{y}) > f(\mathbf{x}) + \nabla f(\mathbf{x})^T (\mathbf{y}-\mathbf{x}), \quad \forall \mathbf{x}, \mathbf{y};\ \mathbf{x} \ne \mathbf{y} \tag{A.217}$$

- The function $f(\mathbf{x})$ is *strongly convex* (with modulus m) if:

$$f(\mathbf{y}) \ge f(\mathbf{x}) + \nabla f(\mathbf{x})^T (\mathbf{y}-\mathbf{x}) + \frac{m}{2}\, \|\mathbf{y} - \mathbf{x}\|_2^2, \quad \forall \mathbf{x}, \mathbf{y} \tag{A.218}$$

For twice differentiable $f(\mathbf{x})$, the function is convex iff $\nabla^2 f(\mathbf{x}) \ge 0$, and strongly convex iff $\nabla^2 f(\mathbf{x}) \ge mI$. The condition $\nabla^2 f(\mathbf{x}) > 0$ is sufficient for $f(\mathbf{x})$ to be strictly convex, but it is not a necessary condition (for example, $f(\mathbf{x}) = x^4$).

Figure A.34 depicts the lower bound in case of $f(\mathbf{x})$ being differentiable and strongly convex.

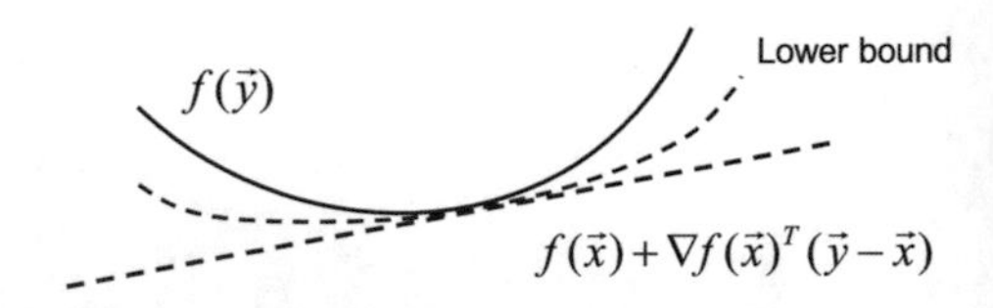

Fig. A.34 Lower bound corresponding to differentiable strongly convex $f(\mathbf{x})$

Notice that in the case of a differentiable and strongly convex $f(\mathbf{x})$ with the gradient being Lipschitz, one has the following "sandwich":

$$C(\mathbf{x}, \mathbf{y}) + \frac{m}{2} \|\mathbf{y} - \mathbf{x}\|_2^2 \leq f(\mathbf{y}) \leq C(\mathbf{x}, \mathbf{y}) + \frac{L}{2} \|\mathbf{y} - \mathbf{x}\|_2^2 \tag{A.219}$$

where:

$$C(\mathbf{x}, \mathbf{y}) = f(\mathbf{x}) + \nabla f(\mathbf{x})^T (\mathbf{y} - \mathbf{x}) \tag{A.220}$$

In order to complete our recital of inequalities, let us consider the *monotonicity of gradient*, which happens when:

$$(\nabla f(\mathbf{y}) - \nabla f(\mathbf{x}))^T (\mathbf{y} - \mathbf{x}) \geq 0 \; ; \; \forall \mathbf{x}, \mathbf{y} \tag{A.221}$$

A differentiable $f(\mathbf{x})$ is convex iff it enjoys monotonicity of gradient.

The gradient of a differentiable strictly convex function is strictly monotone:

$$(\nabla f(\mathbf{y}) - \nabla f(\mathbf{x}))^T (\mathbf{y} - \mathbf{x}) > 0 \; ; \; \forall \mathbf{x}, \mathbf{y} \tag{A.222}$$

And the gradient of a differentiable strongly convex function is strongly monotone or coercive:

$$(\nabla f(\mathbf{y}) - \nabla f(\mathbf{x}))^T (\mathbf{y} - \mathbf{x}) \geq m \; \|\mathbf{y} - \mathbf{x}\|_2^2 \; ; \; \forall \mathbf{x}, \mathbf{y} \tag{A.223}$$

A function $f(\mathbf{x})$ is said to be *coercive* if the limit of $f(\mathbf{x})$ when $\|\mathbf{x}\| \to \infty$ is ∞. A coercive function must increase without limit on any path on $\mathbf{x}$ going to infinity. For example, in 3D a coercive function looks like a bowl.

A function $f(\mathbf{x})$ is said to be *co-coercive* with modulus m, if:

$$(\mathbf{y} - \mathbf{x})^T (f(\mathbf{y}) - f(\mathbf{x})) \geq m \; \|f(\mathbf{y}) - f(\mathbf{x})\|_2^2 \; ; \; \forall \mathbf{x}, \mathbf{y} \tag{A.224}$$

In the case of differentiable $f(\mathbf{x})$ with gradient being Lipschitz:

$$(\nabla f(\mathbf{y}) - \nabla f(\mathbf{x}))^T (\mathbf{y} - \mathbf{x}) \geq \frac{1}{L} \|\nabla f(\mathbf{y}) - \nabla f(\mathbf{x})\|_2^2 \quad \forall \mathbf{x}, \mathbf{y} \tag{A.225}$$

and so, the gradient $\nabla f(\mathbf{x})$ is co-coercive with modulus $1/L$.

Strong convexity and Lipschitz gradient are related by Fenchel duality. According with the Lemma 5.10 in [114]:

1. If $f(.)$ is strongly convex with modulus m, the conjugate $f^*(.)$ has Lipschitz gradient with $L = 1/m$.
2. If $f(.)$ is convex and has Lipschitz gradient, the conjugate $f^*(.)$ is strongly convex with modulus $m = 1/L$.

A.7.1.2 Choosing the Step Size. Convergence

There are several ways of choosing the step size. For instance, you could try to determine the size which obtains the maximum descent; this is a line search optimization problem that must be solved for each step of the iteration. However, it has been recognized that this method is not the most convenient [121].

One could sometimes opt for a fixed step size. In this case, if $f(\mathbf{x})$ is differentiable and convex, and with Lipschitz gradient, then the gradient descent algorithm with $\alpha < (2/L)$ will converge to a stationary point. Let us use the lemma (A.211) to analyze this convergence:

$$\begin{aligned} & f(\mathbf{x}_{k+1}) \leq \; f(\mathbf{x}_k) \, + \, \nabla f(\mathbf{x}_k)^T (\mathbf{x}_{k+1} - \mathbf{x}_k)| \, + \tfrac{L}{2} \; \|\mathbf{x}_{k+1} - \mathbf{x}_k\|_2^2 = \\ & = f(\mathbf{x}_k) \, - \alpha \, \|\nabla f(\mathbf{x}_k)\|_2^2 + \tfrac{\alpha^2 L}{2} \, \|\nabla f(\mathbf{x}_k)\|_2^2 \, = \\ & = \; f(\mathbf{x}_k) \, - \alpha(1 - \alpha \tfrac{L}{2}) \; \|\nabla f(\mathbf{x}_k)\|_2^2 \end{aligned} \tag{A.226}$$

(recall that the gradient descent is: $\mathbf{x}_{k+1} \; = \mathbf{x}_k \; - \; \alpha \, \nabla f(\mathbf{x}_k)$)

The largest minimization of $f(\mathbf{x}_{k+1})$ in Eq. (A.226) is obtained by taking $\alpha = (1/L)$.

Based on the same line of reasoning, it can be shown that using $\alpha \leq (1/L)$ then:

$$f(\mathbf{x}_k) \; - \; f(\mathbf{x}^*) \; \leq \frac{\|\mathbf{x}_0 - \mathbf{x}^*\|_2^2}{2\,\alpha\,k} \tag{A.227}$$

Therefore, to get $f(\mathbf{x}_k) - f(\mathbf{x}^*) \; \leq \varepsilon$, one needs $O(1/\varepsilon)$ iterations.

If moreover $f(\mathbf{x})$ is strongly convex, better convergence can be obtained using $\alpha \leq (2/(m + L))$:

$$f(\mathbf{x}_k) \; - \; f(\mathbf{x}^*) \; \leq \; c^k \, \frac{L}{2} \, \left\|\mathbf{x}_0 - \mathbf{x}^*\right\|_2^2 \tag{A.228}$$

where $0 \; < c \; < 1$.

Hence, in this case, to get $f(\mathbf{x}_k) - f(\mathbf{x}^*) \; \leq \varepsilon$, one needs $O(\log(1/\varepsilon))$ iterations.

Instead of using a fixed step size, it seems advantageous to use an adaptive size, getting smaller as you approach the target. There exist exact line search procedures, that could optimize the size of each step, but they are cumbersome. The alternative could be an inexact procedure for deciding steps that "*sufficiently*" reduce $f(\mathbf{x})$. A first proposal in this sense was the Armijo's rule.

Figure A.35 illustrates the essence of the Armijo's rule. What one wants is a not too short nor large step. The Armijo's rule says that a step is too large if it goes beyond the line B.

In mathematical terms, a step is too large if:

$$f(x - \alpha \, g) \; > \; f(x) \; - \; \rho \alpha g \, \nabla f(x) \tag{A.229}$$

where the parameter ρ is typically chosen between 0.01 and 0.3.

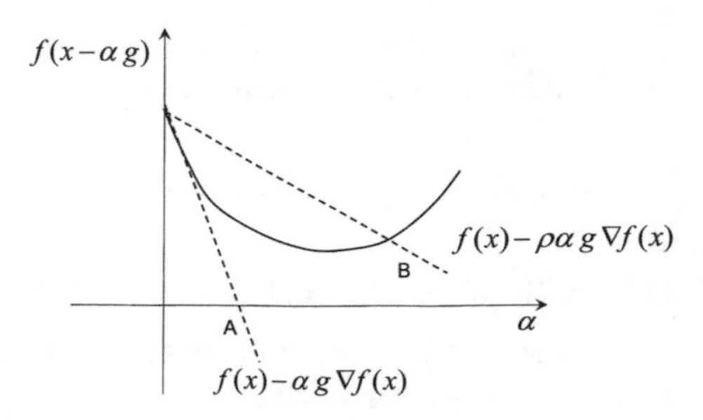

Fig. A.35 Illustration of Armijo's rule

A practical way for using the Armijo's rule is *'backtracking line search'*. It is a simple method that adapts the size of each step as follows:

- start with $t = 1$
- while: $f(\mathbf{x}_k - t\,\nabla f(\mathbf{x}_k)) > .f(\mathbf{x}_k) - \frac{t}{2}\,\|\nabla f(\mathbf{x}_k)\|_2^2$, update $t = \beta\, t$

where $0 < \beta < 1$ is a parameter you fix beforehand.

Like in the case of fixed step size, if $f(\mathbf{x})$ is differentiable and convex, and with Lipschitz gradient, then the convergence would be:

$$f(\mathbf{x}_k) \;-\; f(\mathbf{x}^*) \leq \frac{\|\mathbf{x}_0 - \mathbf{x}^*\|_2^2}{2\, t_{min}\, k} \tag{A.230}$$

where $t_{min} = \min\{1,\ \beta/L\}$.

In the case of $f(\mathbf{x})$ being also strongly convex, the convergence is the same as in the fixed step size (A.228).

A potential problem with the Armijo's rule is that it never declares a step size to be too small. Goldstein and Price proposed to consider a step to be too small if:

$$f(x - \alpha\, g) \;<\; f(x) \;-\; \mu\, \alpha g\, \nabla f(x) \tag{A.231}$$

where you have to specify a parameter $\mu \;>\; \rho$.

Another alternative, proposed by Wolfe, for judging if a step from x to $\tilde{x}$ is too small is to compare the derivative $\nabla f(\tilde{x})$ (the gradient at the new point) with $\eta\, \nabla f(x)$. If:

$$\nabla f(\tilde{x}) \;<\; \eta\, \nabla f(x) \tag{A.232}$$

then the step is too small (you have to specify a parameter $\rho < \eta < 1$).

A more detailed account of line searching methods can be found in [50].

A.7.1.3 The Quadratic Function Sase. Zig-Zag

In order to gain insight about gradient descent issues it is really opportune to study the quadratic function case, for several reasons.

One important reason is that, according with the matrix form of Taylor's theorem, one could devise the following approximation:

$$f(\mathbf{x}_{k+1}) \approx f(\mathbf{x}_k) + \nabla f(\mathbf{x}_k)^T (\mathbf{x}_{k+1} - \mathbf{x}_k) + \frac{1}{2}(\mathbf{x}_{k+1} - \mathbf{x}_k)^T H (\mathbf{x}_{k+1} - \mathbf{x}_k) \tag{A.233}$$

(H, the Hessian, can also be written as $\nabla^2 f(\mathbf{x})$).

This expression shows the interest of quadratic approximations for the study of gradient descent behaviour.

It is also opportune to mention that for a twice differentiable and strongly convex $f(\mathbf{x})$ with Lipschitz gradient, there is another "sandwich":

$$m\, I \leq \nabla^2 f(\mathbf{x}) \leq L\, I \tag{A.234}$$

the ratio $\kappa = L/m$ would be an upper bound on the condition number of the matrix $\nabla^2 f(\mathbf{x})$ (the condition number is the ratio of the largest eigenvalue to the smallest eigenvalue).

For the purposes of gradient descent convergence, it is good to have κ close to 1 (the best case is $\kappa = 1$).

By the way, it is said that an algorithm has *linear convergence* if its iterations satisfy:

$$\frac{f(\mathbf{x}_{k+1}) - f(\mathbf{x}^*)}{f(\mathbf{x}_k) - f(\mathbf{x}^*)} \leq \delta \tag{A.235}$$

where the constant $\delta < 1$ is the *convergence constant* (it is good to have a small δ)

Now, here is the quadratic function to be studied:

$$f(\mathbf{x}) = \frac{1}{2}\mathbf{x}^T Q\,\mathbf{x} + \mathbf{c}^T\mathbf{x} \tag{A.236}$$

where Q is a positive definite symmetric matrix, so all its eigenvalues are >0. Therefore, $f(\mathbf{x})$ is strongly convex.

The minimum of $f(\mathbf{x})$ is attained at:

$$\mathbf{x}^* = -Q^{-1}\,\mathbf{c} \tag{A.237}$$

And this minimum would be:

$$f(\mathbf{x}^*) = -\frac{1}{2}\,\mathbf{c}^T Q^{-1}\,\mathbf{c} \tag{A.238}$$

Given a certain step of the gradient descent, the current direction is:

$$\mathbf{d} = -\nabla f(\mathbf{x}_k) = -Q\,\mathbf{x}_k - \mathbf{c} \tag{A.239}$$

The next iterate would yield:

$$f(\mathbf{x}_k + \alpha\,\mathbf{d}) = \frac{1}{2}(\mathbf{x}_k + \alpha\,\mathbf{d})^T\,Q\,(\mathbf{x}_k + \alpha\,\mathbf{d}) \,+\, \mathbf{c}^T\,(\mathbf{x}_k + \alpha\,\mathbf{d}) \,= \\ = f(\mathbf{x}_k) \,-\, \alpha\,\mathbf{d}^T\,\mathbf{d} \,+\, \frac{1}{2}\,\alpha^2\mathbf{d}^T\,Q\,\mathbf{d} \tag{A.240}$$

On the basis of this equation, it is possible to determine the optimal step size, which is:

$$\alpha \,=\, \frac{\mathbf{d}^T d}{\mathbf{d}^T Q\,\mathbf{d}} \tag{A.241}$$

Using this step size, the convergence of the algorithm is:

$$\frac{f(\mathbf{x}_{k+1}) - f(\mathbf{x}^*)}{f(\mathbf{x}_k) - f(\mathbf{x}^*)} \,=\, 1 \,-\, \frac{1}{\beta} \tag{A.242}$$

with:

$$\beta \,=\, \frac{(\mathbf{d}^T\,Q\,\mathbf{d})(\mathbf{d}^T\,Q^{-1}\,\mathbf{d})}{(\mathbf{d}^T\mathbf{d})^2} \tag{A.243}$$

Kantorovitch [59] determined the following inequality, which is an upper bound on β:

$$\beta \leq \frac{(\lambda_M + \lambda_m)^2}{4\,\lambda_M \lambda_m} \tag{A.244}$$

where λ_M is the maximum eigenvalue of Q, and λ_m is the minimum eigenvalue of Q.The condition number of Q is: $\chi = \lambda_M/\lambda_m$.

In consequence:

$$\frac{f(\mathbf{x}_{k+1}) - f(\mathbf{x}^*)}{f(\mathbf{x}_k) - f(\mathbf{x}^*)} \,\leq\, \left(\frac{\chi - 1}{\chi + 1}\right)^2 \tag{A.245}$$

The reader is invited to take the following simple example:

$$Q = \begin{pmatrix} \lambda_1 & 0 \\ 0 & \lambda_2 \end{pmatrix} \;\; ; \;\; \mathbf{c} = \begin{pmatrix} 0 \\ 0 \end{pmatrix} \tag{A.246}$$

and study the evolution of the gradient descent for different eigenvalue choices.

Figure A.36 depicts a typical path of the gradient descent. Notice that each step is orthogonal to each other. Depending on the initial state, the directions of steps may point more or less precisely to the origin (which is the optimum). When the condition number increases the contours of the quadratic function become more elongated, and the zig-zags become more pronounced.

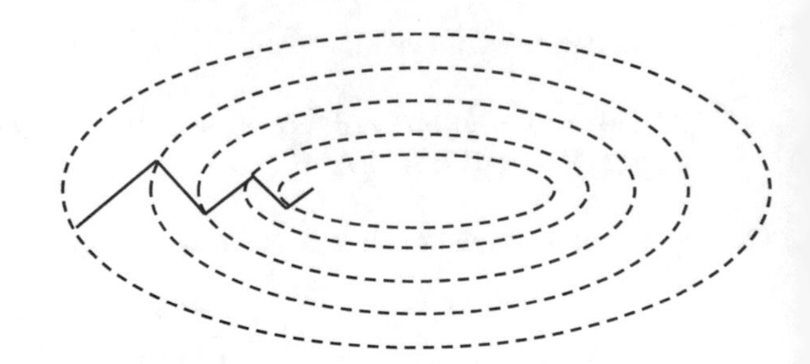

Fig. A.36 A typical gradient descent path

A.7.1.4 Acceleration

A simple idea for first-order methods to capture curvature information, is to use information from $\mathbf{x}_k$ and $\mathbf{x}_{k-1}$. The benefit of that is accelerated convergence.

For instance, the *'heavy-ball'* method [89] uses the following iteration:

$$\mathbf{x}_{k+1} = \mathbf{x}_k - \alpha \nabla f(\mathbf{x}_k) + \beta (\mathbf{x}_k - \mathbf{x}_{k-1}) \tag{A.247}$$

According with the heavy-ball metaphor, the state $\mathbf{x}_k$ is like a point mass with *momentum*, so it has a tendency to continue moving in the direction $\mathbf{x}_k - \mathbf{x}_{k-1}$. The constants are chosen as follows:

$$\alpha = \frac{4}{L} \frac{1}{(1 + 1/\sqrt{\kappa})^2} \; ; \; \beta = \left(1 - \frac{2}{1 + \sqrt{\kappa}}\right)^2 \tag{A.248}$$

Compared with steepest descent, it needs $q = \sqrt{\kappa}$ times fewer steps (for instance, if $\kappa = 144$, then $q = 12$ times less steps).

A popular method is the *conjugate gradient descent*. It can be represented as follows:

$$\mathbf{x}_{k+1} = \mathbf{x}_k + \alpha_k \, \mathbf{p}_k \tag{A.249}$$

$$\mathbf{p}_k = -\nabla f(\mathbf{x}_k) + \gamma_k \, \mathbf{p}_{k-1} \tag{A.250}$$

The parameter γ_k can be chosen in several ways. For example as the minimizer of $f(\mathbf{x})$ along $\mathbf{p}_k$. Another alternative is:

$$\gamma_k = \frac{\|\nabla f(\mathbf{x}_k)\|_2^2}{\|\nabla f(\mathbf{x}_{k-1})\|_2^2} \tag{A.251}$$

Three gradient acceleration methods have been introduced by Nesterov, in 1983, 1988, and 2007. The first method, also described in [83], can be found in the literature under several versions, like for instance [87]:

$$\mathbf{y}_k = \mathbf{x}_k + \beta_k (\mathbf{x}_k - \mathbf{x}_{k-1}) \tag{A.252}$$

$$\mathbf{x}_{k+1} = \mathbf{y}_k - \alpha \nabla f(\mathbf{y}_k) \tag{A.253}$$

with:

$$\beta_k = \frac{t_{k-1}}{t_{k+1}} ; \; t_{k+1} = \frac{1 + \sqrt{1 + 4\, t_k^2}}{2} ; \; \alpha = \frac{1}{L} \tag{A.254}$$

With this method, for differentiable convex $f(\mathbf{x})$ having Lipschitz gradient, Nesterov achieves:

$$f(\mathbf{x}_k) - f(\mathbf{x}^*) \leq \frac{4\, L\, \|\mathbf{x}_0 - \mathbf{x}^*\|_2^2}{(k+2)^2} \tag{A.255}$$

which is the optimal gradient descent convergence.

If $f(\mathbf{x})$ is in addition strongly convex:

$$f(\mathbf{x}_k) - f(\mathbf{x}^*) \leq L \left(1 - \sqrt{1/Q}\right)^k \|\mathbf{x}_0 - \mathbf{x}^*\|_2^2 \tag{A.256}$$

where $Q = L/m$.

In 1988 Barzilai and Borwein [9] introduced the following method:

$$\mathbf{x}_{k+1} = \mathbf{x}_k - \alpha_k \nabla f(\mathbf{x}_k) \tag{A.257}$$

$$\alpha_k = \arg \min_{\alpha} \|\mathbf{s}_k - \alpha\, \mathbf{z}_k\| \tag{A.258}$$

with:

$$\mathbf{s}_k = \mathbf{x}_k - \mathbf{x}_{k-1} ; \; \mathbf{z}_k = \nabla f(\mathbf{x}_k) - \nabla f(\mathbf{x}_{k-1}) \tag{A.259}$$

Frequently this method is cited as the BB method. According with [122] the BB method is in practice much better than the standard steepest descent. A more recent paper [92] contributes with an improved version of BB.

See [87] for more details on the accelerated gradient methods.

A.7.2 Trust Regions

As seen in the previous sub-section, gradient-based iterative methods for finding the maximum of a function proceed along a series of points step by step. Typically these methods first choose a step direction and then a step size. On the contrary, trust region (TR) methods first choose a step size, estimating a certain TR, and then choose a step direction. According with the review [120] of TR algorithms, they constitute a class of relatively new algorithms.

Before coming to details of TR algorithms, it seems opportune to include a summary of some traditional methods.

A.7.2.1 Line-Search Methods

The basic idea of line search methods is, for a given point $\mathbf{x}_k$ and a direction $\mathbf{p}_k$ to minimize:

$$f(\mathbf{x}_k + \alpha\,\mathbf{p}_k) \tag{A.260}$$

with respect to the parameter α.

In order to minimize (A.260), one has to solve:

$$\mathbf{p}_k^T\,\nabla f(\mathbf{x}_k + \alpha\,\mathbf{p}_k) = 0 \tag{A.261}$$

A.7.2.2 Newton Type Methods

As it was already introduced, the Newton's method is based on a second order Taylor expansion:

$$f(\mathbf{x}_k + \mathbf{p}_k) \approx f(\mathbf{x}_k) + \nabla f(\mathbf{x}_k)\,\mathbf{p}_k + \frac{1}{2}\mathbf{p}_k^T\,H_f(\mathbf{x}_k)\,\mathbf{p}_k \tag{A.262}$$

(where $H_f()$ is the Hessian)

To minimize it, the differentiation with respect to $\mathbf{p}_k$ is made equal to zero:

$$\nabla f(\mathbf{x}_k) + H_f(\mathbf{x}_k)\mathbf{p}_k = 0 \tag{A.263}$$

And therefore, considering that $\mathbf{p}_k = \mathbf{x}_{k+1} - \mathbf{x}_k$, one obtains:

$$\mathbf{x}_{k+1} = \mathbf{x}_k - \frac{\nabla f(\mathbf{x}_k)}{H_f(\mathbf{x}_k)} \tag{A.264}$$

Essentially, the idea of Newton is to use curvature information for a more direct path.

In practical terms, the use of the Hessian could be complicated. Moreover, matrix inversions may cause numerical difficulties. In consequence, some alternatives have been proposed to approximate the Hessian by a matrix, or even to directly approximate its inverse. This approximation is updated in each step, in order to improve it. The methods that put into practice these ideas are called *Quasi-Newton methods*.

A typical approximation would be:

$$f(\mathbf{x}_k + \mathbf{p}_k) \approx f(\mathbf{x}_k) + \nabla f(\mathbf{x}_k)\,\mathbf{p}_k + \frac{1}{2}\mathbf{p}_k^T\,B_k\,\mathbf{p}_k \tag{A.265}$$

where B_k is a positive definite symmetrical matrix.

Then:

$$\mathbf{x}_{k+1} = \mathbf{x}_k - \frac{\nabla f(\mathbf{x}_k)}{B_k} \tag{A.266}$$

The matrix B_k is chosen so that the direction of the step tends to approximate the Newton's step. In order to capture the curvature information, this is done according with the following equation:

$$\nabla f(\mathbf{x}_k + \Delta\mathbf{x}) = \nabla f(\mathbf{x}_k) + B_k\, \Delta\mathbf{x} \tag{A.267}$$

which is called the *secant equation*. Notice that it is a simple Taylor series approximation of the gradient.

The secant equation can also be expressed in the following way:

$$\mathbf{q}_k = B_k\, \mathbf{p}_k, \;\; with\; \mathbf{q}_k = \nabla f(\mathbf{x}_k + \Delta\mathbf{x}) - \nabla f(\mathbf{x}_k) \tag{A.268}$$

Except for the scalar case, the secant equation has many solutions. Taking advantage of the degrees of freedom, and assuming that the Hessian would not be wildly varying, the matrix B_k is chosen to be close to an initial matrix B_0 that is normally specified as a diagonal of suitable constants.

Suppose that the matrix B_{k+1} is obtained by adding a correction term to B_k; then the secant equation would be:

$$\mathbf{q}_{k+1} = (B_k + C_k)\, \mathbf{p}_{k+1} \tag{A.269}$$

Hence:

$$C_k\, \mathbf{p}_{k+1} = \mathbf{q}_{k+1} - B_k\, \mathbf{p}_{k+1} \tag{A.270}$$

The correction matrix could be built as follows:

$$C(\phi) = \frac{\mathbf{q}\,\mathbf{q}^T}{\mathbf{q}^T\mathbf{p}} - \frac{B\,\mathbf{p}\,\mathbf{p}^T B}{\mathbf{p}^T B\,\mathbf{p}} + \phi\, r\, \mathbf{v}\,\mathbf{v}^T \tag{A.271}$$

with:

$$\mathbf{v} = \frac{\mathbf{q}}{\mathbf{q}^T\mathbf{p}} - \frac{B\,\mathbf{p}}{r}\;;\; r = \mathbf{q}^T B\, \mathbf{q} \tag{A.272}$$

The famous *Broyden, Fletcher, Goldfarb and Shanno (BFGS) method* is obtained when taking $\phi = 0$.

If, instead of approximating the Hessian, one chooses to approximate its inverse, so the secant equation is written as:

$$D_k \mathbf{q}_k = \mathbf{p}_k \tag{A.273}$$

and so $D_{k+1} = D_k + G_k$ (where G_k is a correction term).

The following updating could be used:

$$G(\xi) = \frac{\mathbf{p}\,\mathbf{p}^T}{\mathbf{p}^T\mathbf{q}} - \frac{B\,\mathbf{q}\,\mathbf{q}^T B}{\mathbf{q}^T B\,\mathbf{q}} + \xi\, r\, \mathbf{w}\,\mathbf{w}^T \tag{A.274}$$

with:

$$\mathbf{w} = \frac{\mathbf{p}}{\mathbf{p}\mathbf{q}} - \frac{B\,\mathbf{q}}{r}\ ;\ r = \mathbf{q}^T B\,\mathbf{q} \tag{A.275}$$

By setting $\xi = 0$ one obtains the *Davidson, Fletcher and Powell (DFP) method*. And, setting $\xi = 1$ one obtains the BFGS method (which is superior to DFP). The general family of updates you can obtain with different values $0 \leq \xi \leq 1$ is the Broyden family.

When applied to quadratic functions, quasi-Newton methods result in conjugate direction methods.

Some of the MATLAB Optimization Toolbox functions use BFGS, or a low-memory variant called L-BFGS.

Reference [3] provides an overview of quasi-Newton methods with extensive bibliography. Likewise, it is easy to find on Internet several interesting dissertations on this topic. An alternative Newton-like step size selection has been proposed in [116].

A.7.2.3 Trust-Region Methods

Suppose you are at $\mathbf{x}_k$, the idea is to use an approximate model that tells you $f(\mathbf{x})$ for $\mathbf{x}$ near $\mathbf{x}_k$. This model is trusted in a certain region. If the model fits well, this region can be enlarged. Otherwise, if the model works not good enough, the TR should be reduced. The step direction is chosen inside the TR and it minimizes $f(\mathbf{x})$.

Frequently, a quadratic model (similar to the Taylor expansion) is used:

$$m(\mathbf{x}_k + \mathbf{p}) = f(\mathbf{x}_k) + \nabla f(\mathbf{x}_k)\,\mathbf{p} + \frac{1}{2}\mathbf{p}^T B_k\,\mathbf{p} \tag{A.276}$$

When the matrix B_k is chosen to be the Hessian, so $B_k = H_f(\mathbf{x}_k)$, the algorithm is called the *trust-region Newton method.*

In general, all that was assumed is that B_k is symmetric and uniformly bounded.

The locally constrained TR problem is:

$$\min_{\mathbf{p}\in T} m(\mathbf{x}_k + \mathbf{p}) = \min_{\mathbf{p}\in T} \left\{ f(\mathbf{x}_k) + \nabla f(\mathbf{x}_k)\,\mathbf{p}_k + \frac{1}{2}\mathbf{p}_k^T B_k\,\mathbf{p}_k \right\} \tag{A.277}$$

were T is the trust region. Usually this region is a ball with radius r:

$$T(r) = \{\mathbf{x} \mid \|\mathbf{x} - \mathbf{x}_k\| \leq r\} \tag{A.278}$$

The result of the minimization would be a certain new point $\mathbf{x}_+$. Now, some conditions are stated to accept a step or not. Define the actual reduction as:

$$R_a = f(\mathbf{x}_k) - f(\mathbf{x}_+) \tag{A.279}$$

and the predicted reduction as:

$$R_p = m(\mathbf{x}_k) - m(\mathbf{x}_+) \tag{A.280}$$

Now, one takes into account the ratio $\rho = R_a/R_p$ and three parameters: $\lambda_0 \leq \lambda_L < \lambda_H$. Based on these parameters:

1. If $\rho < \lambda_0$ then the step is rejected
2. If $\rho < \lambda_L$ then the radius of T should be decreased
3. If $\lambda_L \leq \rho \leq \lambda_H$ then accept the step
4. If $\rho > \lambda_H$ then the radius of T should be increased

Typically, the radius is divided by 2 when it should be decreased, or multiplied by 2 when it should be increased.

There are several alternatives for solving the minimization (A.277) [120]. One of them is the *trust-region-reflective* method, which uses as trust region a two-dimensional subspace V. A common choice of V is the linear space spanned by $\mathbf{v}_1$, the vector in the direction of the gradient $\mathbf{g}_k$, and $\mathbf{v}_2$, which is either the solution of:

$$H_k \mathbf{v}_2 = -\mathbf{g}_k \tag{A.281}$$

or a direction with:

$$\mathbf{v}_2^T H_k \mathbf{v}_2 < 0 \tag{A.282}$$

A.8 Cutting Plane in General Contexts

Cutting planes can be used in other contexts, not only in integer programming problems. Actually they have earned an important role in continuous variable optimization methods, including non-smooth scenarios.

This part of the Appendix is based in [19], which contains the mathematical details omitted in this succinct summary.

Given a feasible set F our aim would be to determine a (small) subset X such that the optimal solution belongs to X (called *the target set*). The idea is to apply a sequence of cuts to F, obtaining smaller and smaller regions that contain the optimum.

The cutting plane method uses an *oracle*. When you query the oracle about a certain point $\mathbf{x} \in \Re^n$, the oracle tells you that $\mathbf{x} \in X$, so a successful result has been reached, or it gives you the parameters $\mathbf{a}$, b corresponding to a separating hyperplane between $\mathbf{x}$ and X:

Fig. A.37 Example of cutting plane

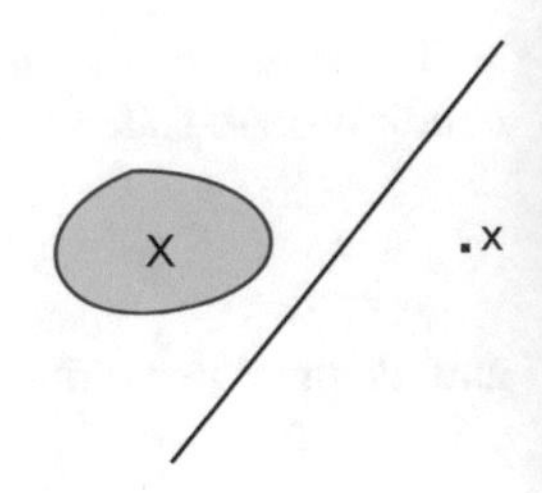

$$\mathbf{a}^T\mathbf{z} \leq b \ for \ z \in X \text{ and } \mathbf{a}^T\mathbf{x} \geq b$$

Figure A.37 illustrates the situation:

The hyperplane is called a *cutting plane* or *cut*. If it contains the point $\mathbf{x}$, it is called a *neutral cut*. The case depicted in Fig. A.37 corresponds to a *deep cut*, so the point $\mathbf{x}$ is inside the halfspace on the other side of X.

Suppose that the problem was to minimize a convex and differentiable $f_o(\mathbf{x})$. Given a certain point $\mathbf{x}$, a cutting plane containing $\mathbf{x}$ (a neutral cut) is the following:

$$\nabla f_o(\mathbf{x})^T(\mathbf{z} - \mathbf{x}) \leq 0 \tag{A.283}$$

(supposing that $\mathbf{x}$ is not the optimal and so $\nabla f_o(\mathbf{x}) \neq 0$).

Now, let us consider the constraints corresponding to F:

$$f_i(\mathbf{x}) \leq 0, \ i = 1, 2, \ldots m (\text{convex and differentiable}).$$

Given a not feasible point $\mathbf{x}$, this means that at least one of the constraints is violated; for instance: $f_j(\mathbf{x}) > 0$. Then, any feasible $\mathbf{z}$ satisfies:

$$f_j(\mathbf{x}) + \nabla f_j(\mathbf{x})^T(\mathbf{z} - \mathbf{x}) \leq 0 \tag{A.284}$$

Therefore, this is the equation of a deep cutting plane.

Let us look at the complete problem, for which one could combine the above results:

Minimize $f_o(\mathbf{x})$

Subject to: $f_i(\mathbf{x}) \leq 0, \ i = 1, 2, \ldots m$

If the query to the oracle concerns a certain point $\mathbf{x}$, this point may be not feasible, so a deep cut would be (A.284). It is called a *feasibility cut*, for one is cutting away a halfspace of infeasible points.

If the point is feasible and not optimal, then a neutral cut would be (A.283). It is called an *objective cut*, which cuts out a halfspace of points corresponding to too large objective values.

A.8.1 Basic Algorithm (Localization Polyhedron)

One could start the process choosing a ball large enough to contain X. Then, the oracle is queried at a series of points $\mathbf{x}_1, \ \mathbf{x}_2, \ \ldots, \ \mathbf{x}_k$. The sequence stops if a point belonging to X is found. Otherwise, what is most probable, one has obtained a series of cutting planes that can be written as:

$$\mathbf{a}_i^T \mathbf{z} \leq b_i \,, \ \ i = 1, \, 2, \, \ldots, k \tag{A.285}$$

These planes would constitute a localization polyhedron P_k, as depicted in Fig. A.38.

If one adds a new point, in the interior of P_k, not belonging to X, the oracle would return a new cutting plane (a new inequality), and this means a new polyhedron $P_{k+1} \subset P_k$.

The algorithm will iterate, obtaining better and better localization of X. The progress could be measured in several ways, like for instance with a ratio of volumes of P_{k+1} and P_k.

When trying a new query point, one wants to remove as much as possible from the current polyhedron P_k. Given the ignorance about where is X, the best choice on average would be a point near the "centre" of P_k, and to use a neutral cut.

Several alternatives have been proposed to implement this idea:

- Use the *center of gravity* of the polyhedron P_k, [110]
- Obtain the *maximum volume ellipsoid (MVE)* contained in P_k and then use its center, [110]
- Choose the *analytic center (ACCPM)*, [22]

Notice that the classical bisection method is a particular case of cutting-plane algorithm.

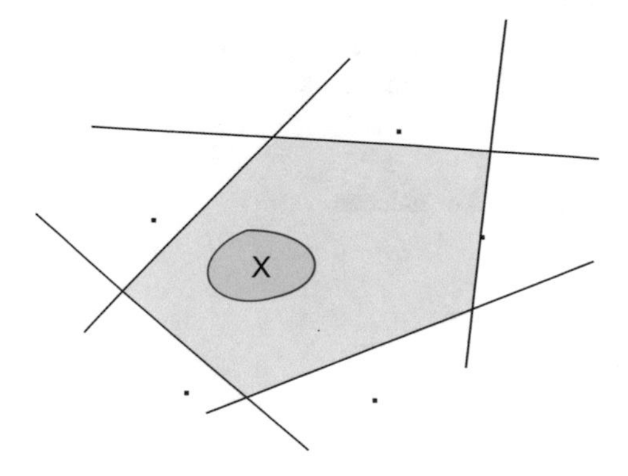

Fig. A.38 Example of localization polyhedron

A.9 Non-smooth Analysis and Optimization

As said before, in many cases the application of sparse representation leads to optimization issues. Typically optimization involves objective functions, to be maximised or minimised, and a set of constraints (see the appendix on optimization topics for fundamental concepts and methods).

It has been found that it is not always possible to have differentiable objective functions. The article [46] includes a long section describing several sources of non-differentiability in optimization problems; likewise, the book [103] includes, in Chap. 5, an extensive reference of this kind of problems. Standard analytical methods for optimization, which are based on derivatives, are difficult to apply in such cases.

This section is devoted to the treatment of non-differentiable optimization problems, [70]. Important concepts to be introduced are those of *subgradient* and *subdifferential*, which prolong to a certain extent the classical concepts of gradient and differential. The section focuses on minimization problems.

A.9.1 Non-smooth Analysis

Some initial examples would be helpful to illustrate why we are concerned with non-differentiability.

A typical case, which is found in image processing, involves an objective function like $z = |x|$, which is differentiable for all $x \neq 0$, but it is non-differentiable at $x = 0$. Figure A.39 shows a plot of this function. It happens that the minimum of the function takes place at $x = 0$.

A similar situation is depicted in Fig. A.40, where two curved components define a convex function that is non-differentiable at x_0. It also happens that this is the point where the function reaches its minimum.

Something has to be devised in order to keep, as much as possible, the traditional methodology, which determines the minimum by equalling derivatives to zero. With this in mind, a repertory of new concepts has been proposed. Let us now introduce some of these concepts.

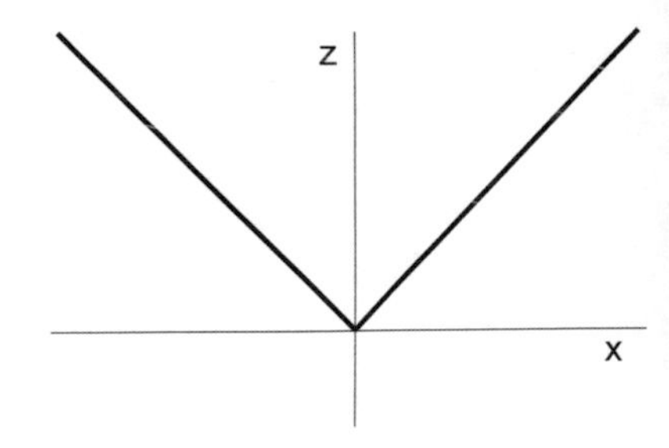

Fig. A.39 The function $z = |x|$

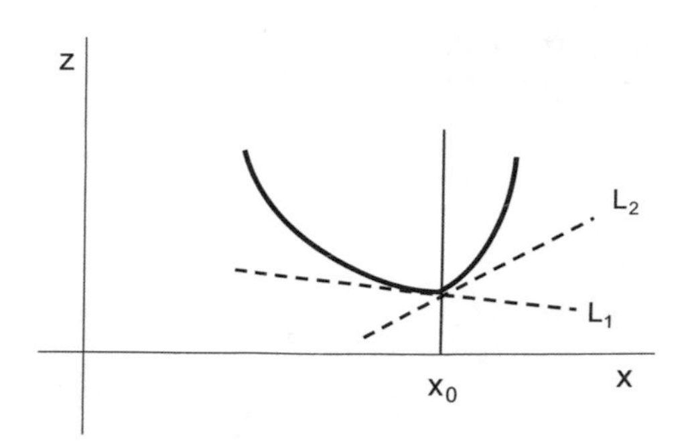

Fig. A.40 Two 'subtangent' lines

Considering in particular the Fig. A.40, there would be infinite lines that go through the point $(x_0,\ f(x_0))$ and which are either touching or below the graph of the function. The slope of any of these lines is a *'subderivative'*. The two lines, L_1 and L_2, on Fig. A.40 are two examples of these lines.

In more formal terms, a subderivative at x_0 is a real number c such that:

$$f(x) - f(x_0) \geq c\,(x - x_0) \tag{A.286}$$

The set $[a, b]$ of all subderivatives is called the *'subdifferential'* of the function $f()$ at x_0. The subdifferential is denoted as $\partial f(x_0)$.

These concepts can be generalized to functions of several variables. A vector **g** is called a *'subgradient'* of the function $f()$ at $\mathbf{x}_0$ if:

$$f(\mathbf{x}) - f(\mathbf{x}_0) \geq \mathbf{g}^T \cdot (\mathbf{x} - \mathbf{x}_0) \tag{A.287}$$

The set of all subgradients is called the *'subdifferential'* of the function $f()$ at $\mathbf{x}_0$. The subdifferential is denoted as $\partial f(\mathbf{x}_0)$.

Note that Eq. (A.287) can be written as:

$$f(\mathbf{x}) \geq f(\mathbf{x}_0) + \mathbf{g}^T \cdot (\mathbf{x} - \mathbf{x}_0) \tag{A.288}$$

A point $\mathbf{x}_0$ is a *global minimum* of a convex function if zero is contained in the subdifferential at $\mathbf{x}_0$.

Returning to the function $z = |x|$, Fig. A.41 shows its subdifferential $\partial f(x)$ as a function of x. The subdifferential at the origin is the interval $[-1, 1]$.

Notice that, according with Fig. A.41, $sign(x) \in \partial f(x)$.

Perhaps the reader could find (for instance in the literature about Filippov's differential inclusion), that some authors, like [33, 73], do use a *set-valued signum function*, that has values $[-1, 1]$ for $x = 0$ (so it is coincident with the subdifferential depicted in Fig. A.41).

See [21] for more details on subgradients and related concepts.

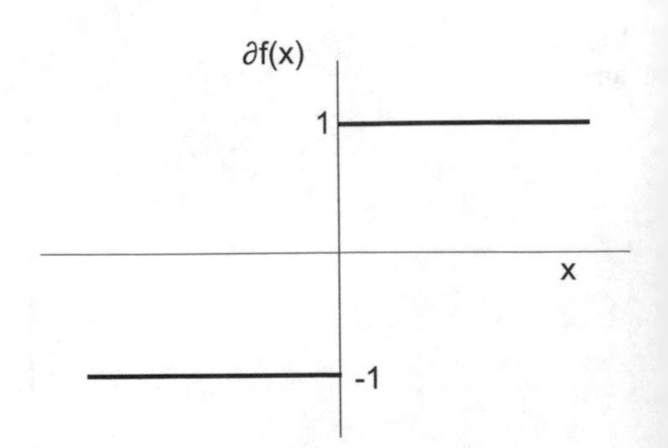

Fig. A.41 Subdifferential of the $z = |x|$ function

A.9.1.1 Brief Historical Sketch

The book of Rockafellar [95], in 1970 with the title *Convex Analysis*, provided momentum for new theory developments. It was realized that many convex optimization problems do not have a derivative at the optimum, and this motivated the introduction of subdifferentials. The term *nonsmooth analysis* was introduced by Clarke [30], 1983, who was the second student of Rockafellar and who extended the study to locally Lipschitz functions. New important contributions in order to further extensions have been done by Mordukhovich [79, 80] in 2006.

A.9.2 Optimization Methods for Non-smooth Problems

A basic methodology for non-smooth optimization is based on subgradients. Another type of methods is related to cutting planes.

A third type of methods, which is attracting a lot of attention, is based on the proximal mapping.

Of course, it is always possible to use direct searching methods [65], or heuristic methods. However, we prefer to put now the focus on the more specific methods already mentioned.

A.9.2.1 Subgradient Methods

The subgradient method uses the following simple iteration:

$$\mathbf{x}_{k+1} = \mathbf{x}_k - \alpha_k \, \mathbf{g}_k \tag{A.289}$$

where $\mathbf{g}_k$ is any subgradient of $f()$ at $\mathbf{x}_k$. The positive parameter α_k is the step size. When $f()$ is differentiable at $\mathbf{x}_k$ there is only one possible subgradient, which is the gradient. When it is not-differentiable the chosen subgradient would *not* necessarily lead to a descent of the function value; for this reason one usually retains a memory of the best point found along the iterations. The number of iterations could be high (perhaps millions).

The convergence of the method has been subject of much study. Some relevant results are summarized in [16], starting from the assumption that there is a bound G such that $\|\mathbf{g}_k\|_2 \leq G$, which is for instance the case when $f()$ satisfies the Lipschitz condition:

$$|f(u) - f(v)| \leq G \; \|u - v\|_2 \tag{A.290}$$

Considering in particular the difference $D = f_k^{best} - f^*$, then, under the previous assumption, it can be established that:

- for constant step size $\alpha_k = h$, then $D \rightarrow G^2 h/2$ as $k \rightarrow \infty$
- for constant step length $\alpha_k = h/\|\mathbf{g}_k\|_2$, then $D \rightarrow G\,h/2$ as $k \rightarrow \infty$

There are many other step size alternatives that have been proposed. For example the following nonsummable, i.e. $\sum_k \alpha_k = \infty$, step sizes:

- Square summable:

$$\sum_k \alpha_k^2 < \infty \tag{A.291}$$

(typical example: $\alpha_k = a/(b+k)$ with positive a and b).

- Diminishing step size:

$$\lim_{k \to \infty} \alpha_k = 0 \tag{A.292}$$

(typical example: $\alpha_k = a/\sqrt{k}$ with positive a).

- And also, diminishing step length:

$$\alpha_k = \gamma_k \,/\, \|\mathbf{g}_k\|_2\,; \;\; \lim_{k \to \infty} \gamma_k = 0; \;\; \sum_k \gamma_k = \infty \tag{A.293}$$

In these three alternatives the procedure converges to the optimum, that is: $D \rightarrow 0$ as $k \rightarrow \infty$.

When the optimal value f^* is known, it is convenient to use the Polyak's step size, which is given by:

$$\alpha_k = \gamma_k \, \frac{f(\mathbf{x}_k) - f^*}{\|\mathbf{g}_k\|_2^2} \tag{A.294}$$

where $0 < \gamma_k < 2$.

Indeed, there are some optimization problems for which f^* is known, like for example the *feasibility problem* of finding a point $\mathbf{x}_c$ satisfying a particular system of inequalities.

Some modifications of the Polyak's step size have been proposed to circumvent the use of f^*.

The *projected subgradient method* is a simple extension of the basic method that is suitable for the following convex constrained problem:

Minimize $f(x)$

Subject to $x \in C$

(C is a convex set)

The projected subgradient method is given by:

$$\mathbf{x}_{k+1} = P(\mathbf{x}_k - \alpha_k \, \mathbf{g}_k) \tag{A.295}$$

where P is a projection on C.

The step sizes can be chosen as in the basic method, with similar convergence properties.

For non-negative orthant, the projection would be: $P(\mathbf{x}_i) = \max(\mathbf{x}_i, 0)$

For a linear system such that: $C = \{\mathbf{x} : A\mathbf{x} = \mathbf{b}\}$, the projection would be:

$$P_C(\mathbf{x}_i) = \mathbf{x}_i - A^T (A\, A^T)^{-1} (A\mathbf{x}_i - \mathbf{b}) \tag{A.296}$$

In this last case, the subgradient update would be:

$$\mathbf{x}_{k+1} = \mathbf{x}_k - \alpha_k \, (I - A^T (A\, A^T)^{-1} A) \, \mathbf{g}_k \tag{A.297}$$

(note that $A\mathbf{x}_k = \mathbf{b}$)

Examples from literature show that the projected subgradient method usually has faster convergence.

The subgradient method was developed in 1962 by N.Z. Schor. Other authors, like Ermoliev and Polyak contributed to consolidate the method. See [45] for more technical and historical details, including main references. The method belongs to a large class of methods known as *first-order methods*, which are based on subgradients and the like; second-order methods usually employ Hessians.

Subgradient methods are very easy to implement, and can be applied to many different problems. The main drawback is that they it can be very slow, sometimes with a zig-zag behavior.

Some variants of the subgradient method

The subgradient method can be extended [117] to solve problems with inequalities:

Minimize $f_0(x)$

Subject to $f_i(\mathbf{x}) \leq 0, \; i = 1, 2, \ldots, m$

where $f_i(.)$ are convex.

In: $\mathbf{x}_{k+1} = \mathbf{x}_k - \alpha_k \, \mathbf{g}_k$, the subgradient is chosen as follows:

$$\mathbf{g}_k = \begin{cases} \partial f_0(\mathbf{x}_k) \; if \; f_i(\mathbf{x}_k) \leq 0, \; \forall i \\ \partial f_j(\mathbf{x}_k) \; for \; some \; j \; such \; that \; f_j(\mathbf{x}_k) > 0 \end{cases} \tag{A.298}$$

(if the current point is feasible one uses $\partial f_0(.)$, if not one uses a subgradient of any violated constraint).

As said before, subgradients are not guaranteed to be directions of descent. Some modifications have been proposed to detect good directions. One is the *conjugated subgradient method*, which includes a minimization task (to find the best direction) in each iteration. Another technique is called *ε-subgradient method*, which uses approximated subgradients. See [100] for details.

There are reasons, like for instance the vastness of the optimization problem at hand or its involved randomness, which can recommend the use of *stochastic subgradient methods* [17, 117]. This method is almost the same as the subgradient method, but using noisy subgradients and a reduced set of step size alternatives.

It is said that $\tilde{g}$ is a *noisy subgradient* of $f(.)$ at $\mathbf{x}$ (which is random) if $E(\tilde{g} \mid \mathbf{x}) \in \partial f(\mathbf{x})$ (where $E(.)$ is the expected value). One could consider the noisy subgradient as the sum of $\mathbf{g}$ and a zero-mean random variable.

The *stochastic subgradient iteration* is:

$$\mathbf{x}_{k+1} = \mathbf{x}_k - \alpha_k \, \tilde{g}_k \tag{A.299}$$

where $\tilde{g}_k$ is the noisy subgradient at $\mathbf{x}_k$.

A recently popularized idea [84, 94, 109] is the *random coordinate descent method.* A each step k of the iteration a coordinate index j is chosen at random (with uniform PDF), and then the update is done as follows:

$$(x_j)_{k+1} = (x_j)_k - \alpha_k \, (\mathbf{g}_j)_k \tag{A.300}$$

$$(x_i)_{k+1} = (x_i)_k \; for \; i \neq j \tag{A.301}$$

where (x_i) denotes the *i-component* (or coordinate) of $\mathbf{x}$.

Example of application (l_1 minimization):

Consider the following l_1 minimization problem:

Minimize $z = \|\mathbf{x}\|_1$

Subject to $A\mathbf{x} = \mathbf{b}$

This case corresponds to Figs. A.39 and A.41. A subgradient of z is given by $g = sgn(x)$ (the conventional signum function). Therefore, the optimization problem can be solved with the following projected subgradient iterations:

$$\mathbf{x}_{k+1} = \mathbf{x}_k - \alpha_k \, (I - A^T (A \, A^T)^{-1} A) \, sgn(\mathbf{x}_k) \tag{A.302}$$

A.9.2.2 Bundle Methods

In order to introduce the bundle methods it is important to make two observations.

The first is that for convex functions a first order Taylor approximation is a lower bound. The same can be said about subgradients, based on Eq. (A.288).

The second is that when running the subgradient methods described above, almost no record is kept on the exploration history. It would be more convenient to take into account the knowledge acquired about the objective function as the iterative optimization proceeds.

By using a combination of first order linear approximations it is possible to build a piecewise approximation of the objective function, which can be considered as model $f^M(.)$ of the objective function $f(.)$. This was the basis of the cutting-plane method proposed in [27, 62], which considers a series of subgradients $\mathbf{g}_1, \mathbf{g}_2, \ldots \mathbf{g}_n$ evaluated at points $\mathbf{x}_1, \mathbf{x}_2, \ldots \mathbf{x}_n$, and uses it for building the model:

$$f_n^M(\mathbf{x}) = \max_{1 \leq k \leq n} \{f(\mathbf{x}_k) + (\mathbf{g}_k^T(\mathbf{x} - \mathbf{x}_k)\} \tag{A.303}$$

This model is a lower bound of the objective function: $f_n^M(\mathbf{x}) \leq f(\mathbf{x})$.

The information on the current model is kept in a *bundle*:

$$B_n = \{(\mathbf{x}_k, \ f(\mathbf{x}_k), \ \mathbf{g}_k), \ k = 1, 2, \ldots, n\}$$

At iteration $n + 1$, a new point is added. This point is chosen as:

$$\mathbf{x}_{n+1} = \arg \min_{\mathbf{x}} f_n^M(\mathbf{x}) \tag{A.304}$$

Let us illustrate how the piecewise model is built. Figure A.42 shows a convex objective function $f(.) \geq 0$. The problem is to find its minimum. A first point $\mathbf{x}_1$ has been chosen, and a first model of the objective function has been found: the line L_1. The next point, $\mathbf{x}_2$, to be added to the exploration is simply found at the intersection of L_1 with the horizontal axis.

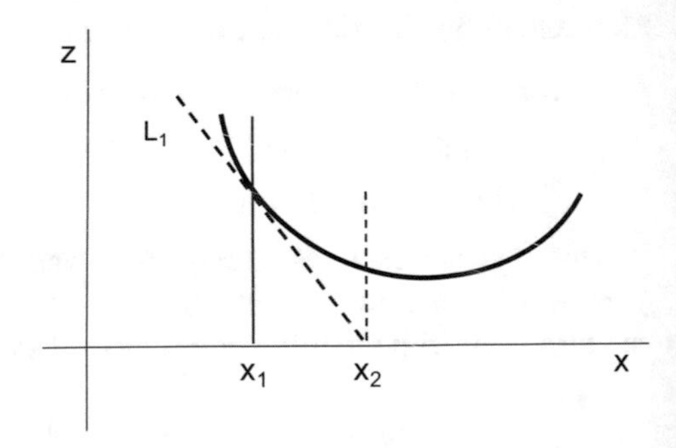

Fig. A.42 First step of the modelling process

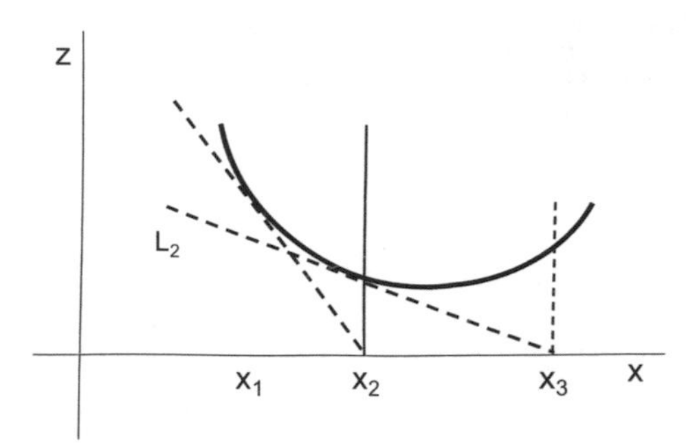

Fig. A.43 Second step of the modelling process

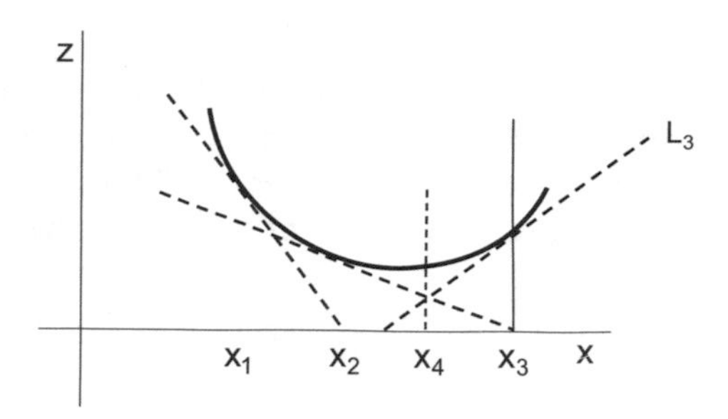

Fig. A.44 Third step of the modelling process

Taking $f(\mathbf{x}_2)$as new point for a linear approximation, which is plotted as the line L_2, it is possible to determine the next point to be added, $\mathbf{x}_3$. Now the model consists of two linear pieces (a "V"), taken from the two lines. Figure A.43 depicts this step.

Figure A.44 shows the result of a third step. Notice that now the next point to be added, $\mathbf{x}_4$, is found at the intersection of lines L_2 and L_3. The piecewise model, made of segments of L_1, L_2, and L_3 is getting better.

Along the modelling iterations, the cutting plane method monitors the following quantity, which monotonically decreases:

$$\varepsilon = \min_{0<k<n} \{f(\mathbf{x}_k) - f_n^M(\mathbf{x}_n)\} \tag{A.305}$$

Once this quantity becomes smaller than a certain specified threshold, the iterative process terminates.

A series of observed drawbacks of the cutting plane basic scheme has been reported by the scientific literature. Therefore, several improvements have been proposed, [69, 103]. The article [68] offers a modern perspective, and opportune references. A fairly recent survey can be found in [75].

A.9.2.3 Proximal Methods

Suppose that the function to be minimized has several points where it is not differentiable, and so the plot of the function has kinks (informally speaking) where

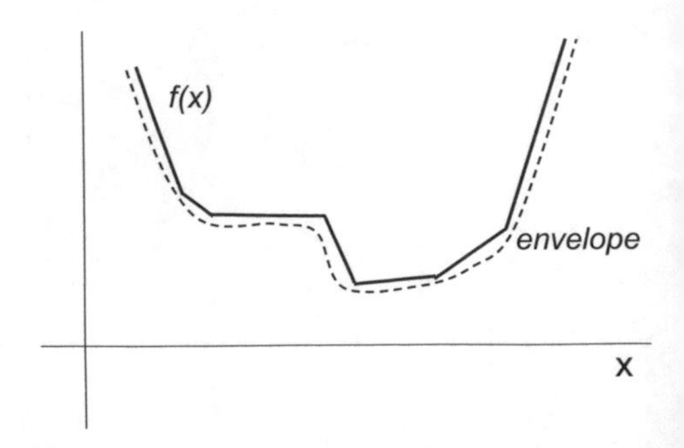

Fig. A.45 An example of envelope

smoothness is lost. An interesting idea to recover good mathematical conditions is to use *'envelopes'*. That is: to use a smooth approximation of the function so instead of kinks you have arcs. Figure A.45 tries to illustrate the idea:

An important example of envelope is the *Moreau's envelope*, defined by:

$$M_{\lambda f}(\mathbf{x}) = \inf_{\mathbf{u}} \left(f(\mathbf{u}) + \frac{1}{2\lambda} \|\mathbf{u} - \mathbf{x}\|_2^2 \right) \tag{A.306}$$

(Moreau employed $\lambda = 1$)

Directly related to this envelope, the '*proximal mapping*' is defined by:

$$\Pr ox_{\lambda f}(\mathbf{x}) = \arg \min_{\mathbf{u}} \left(f(\mathbf{u}) + \frac{1}{2\lambda} \|\mathbf{u} - \mathbf{x}\|_2^2 \right) \tag{A.307}$$

where $f(x)$ is a proper lower semi-continuous function (plsc).

In general, the Moreau's envelope approximates $f(\mathbf{x})$ from below, becoming a better approximation as λ is decreased. The envelope is a finite, continuous function. The lower bounds of $f(\mathbf{x})$ and $M_{\lambda f}(\mathbf{x})$ are equal:

$$\inf \ f(\mathbf{x}) = \inf \ M_{\lambda f}(\mathbf{x}) \tag{A.308}$$

Therefore, the original problem of minimizing a possibly non-differentiable $f(\mathbf{x})$ can be replaced with the minimization of the smooth function $M_{\lambda f}(\mathbf{x})$.

If $f(\mathbf{x})$ is convex and plsc, the gradient of the envelope would be:

$$\nabla M_{\lambda f}(\mathbf{x}) = \frac{1}{\lambda}(I - \Pr ox_{\lambda f}(\mathbf{x})) \tag{A.309}$$

Also, if $f(\mathbf{x})$is convex and plsc, the function inside the parenthesis in (A.307) is strongly convex because of the added Euclidean norm term.

Note the connection of (A.306) and (A.307) with aspects related to gradient descent in the appendix on optimization.

Moreau introduced the envelope and the proximal mapping in 1962 [81] as a way of regularizing and approximating a convex function. Actually it was called *prox-regularization*, and it can be regarded as analogue to the Tichonov regularization [54]. The book [98] places the envelope and the proximal mapping in a rigorous mathematical context.

The Moreau's envelope is an example of infimal convolution. The *infimal convolution* of closed proper convex functions $f(\mathbf{x})$ and $g(\mathbf{x})$ is defined as:

$$(f \otimes g)(\mathbf{v}) = \inf_{\mathbf{x}}(f(\mathbf{x}) + g(\mathbf{v} - \mathbf{x})) \tag{A.310}$$

There is a number of interpretations [88] of the approach initiated by Moreau. For instance, it can be considered as a penalty method for sequential unconstrained minimization [24]. In what follows some attention is paid to this aspect.

Penalty functions

The problem is to minimize a function $f(\mathbf{x})$. The *sequential unconstrained minimization* (SUM) algorithms use a set of auxiliary functions $p_k(\mathbf{x})$ so that at the kth step one minimizes:

$$P_k(\mathbf{x}) = f(\mathbf{x}) + g_k(\mathbf{x}) \tag{A.311}$$

to get a minimizer $\mathbf{x}^k$.

The functions $p_k(\mathbf{x})$ are chosen so that the infinite sequence $\{\mathbf{x}^k\}$ converges to a solution of the $f(\mathbf{x})$ minimization problem.

In the context of constrained minimization, where a feasible set exists, one could distinguish two approaches. Barrier-function methods (for instance, interior-point methods) require the minimizers $\{\mathbf{x}^k\}$ to belong to the feasible set. This is not the case with penalty-function methods, where constraint violations are discouraged but not prohibited (hence, sometimes they are called *exterior-point* methods).

Some examples of penalty functions, like the absolute-value, the Courant-Bertrami, the quadratic-loss, etc. can be found in [24]. Also, in [24], connections of penalty-function methods with the minimization of cross-entropy, the regularized least-squares, and the Lagrangian approach in optimization, are described.

As highlighted in [24], the proximal approach can be used for a sequential minimization algorithm (this will be introduced later on).

Some examples of envelopes and proximal maps

A first and important example is $f(x) = |x|$. Its Moreau's envelope is:

$$M_{\lambda f}(x) = \begin{cases} \frac{1}{2\lambda} x^2 & x \in [-\lambda, \lambda] \\ |x| - \frac{\lambda}{2} & otherwise \end{cases} \tag{A.312}$$

and the proximal mapping is:

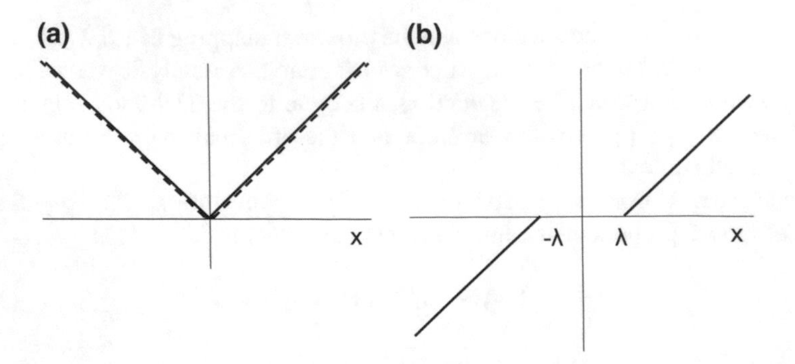

Fig. A.46 **a** Moreau's envelope (*dashes*), and **b** proximal mapping of $|x|$

$$Prox_{\lambda f}(x) = \begin{cases} 0 & x \in [-\lambda, \lambda] \\ x - \lambda & x > \lambda \\ x + \lambda & x < \lambda \end{cases} \tag{A.313}$$

Figure A.46 shows the Moreau's envelope and the proximal mapping of $|x|$.

In the case of the $l1$ norm of a vector, $\|\mathbf{x}\|_1 = \sum_i |x_i|$, the ith coordinate of $Prox_{\lambda f}(\mathbf{x})$ is given by (A.313) replacing x with x_i.

The proximal mapping of the $l1$ norm is also called the *soft thresholding operator*.

For $\lambda = 1$, the Moreau's envelope of $|x|$ would be a particular case of the *Huber penalty function*:

$$L_\lambda(x) = \begin{cases} \frac{1}{2}x^2 & x \in [-\lambda, \lambda] \\ \lambda\left(|x| - \frac{\lambda}{2}\right) & otherwise \end{cases} \tag{A.314}$$

Another example is the quadratic case: $f(\mathbf{x}) = \frac{1}{2}\mathbf{x}^T A\,\mathbf{x} + \mathbf{b}^T\mathbf{x} + c$. Its proximal mapping is:

$$Prox_{\lambda f}(\mathbf{x}) = (I + \lambda A)^{-1}(\mathbf{x} - \lambda\mathbf{b}) \tag{A.315}$$

There are three special cases of this mapping:

- If $f(\mathbf{x}) = \mathbf{b}^T\mathbf{x} + c$ then $Prox_{\lambda f}(\mathbf{x}) = \mathbf{x} - \lambda\mathbf{b}$.
- If $f(\mathbf{x}) = c$ then $Prox_{\lambda f}(\mathbf{x}) = \mathbf{x}$ (the proximal operator is the identity)
- If $f(\mathbf{x}) = \frac{1}{2}\|\cdot\|_2^2$ then $Prox_{\lambda f}(\mathbf{x}) = (\frac{1}{1+\lambda})\,\mathbf{x}$.

Other example of interest, in connection with logarithmic barriers, is the case of $f(x) = -\log x$. Its proximal mapping is:

$$Prox_{\lambda f}(\mathbf{x}) = \frac{x + \sqrt{x^2 + 4\lambda}}{2} \tag{A.316}$$

The next case is of mathematical relevance, for minimization problems where solutions are constrained to belong to a certain set.

If $f(\mathbf{u}) = i_C(\mathbf{u})$, where $i_C(\mathbf{u})$ is the indicator function of the set C:

$$i_C(\mathbf{u}) = \begin{cases} 0 & x \in C \\ \infty & x \notin C \end{cases} \tag{A.317}$$

then: $Prox_{\lambda f}(\mathbf{x}) = \arg\min_{\mathbf{u}} \|\mathbf{u} - \mathbf{x}\|_2^2 = P_C(\mathbf{x})$, which is the projection over set C.

And, finally, the case of the Euclidean norm $f(x) = \|\mathbf{x}\|_2$. Its proximal mapping is

$$Prox_{\lambda f}(\mathbf{x}) = \begin{cases} \left(\frac{1-\lambda}{\|\mathbf{x}\|_2}\right) \mathbf{x} & \|\mathbf{x}\|_2 \geq \lambda \\ 0 & otherwise \end{cases}. \tag{A.318}$$

See [88] for more details.

Characterization and properties of the proximal mapping

It can be easily shown, [88], that if the point $\mathbf{x}^*$ minimizes $f(\mathbf{x})$, then $Prox_{\lambda f}(\mathbf{x}^*) = \mathbf{x}^*$.

Also, the point $\mathbf{x}^*$ minimizes:

$$\left(f(\mathbf{x}) + \frac{1}{2\lambda} \|\mathbf{x} - \mathbf{v}\|_2^2 \right) \tag{A.319}$$

if and only if:

$$0 \in \partial f(\mathbf{x}^*) + \frac{1}{\lambda} (\mathbf{x}^* - \mathbf{v}) \tag{A.320}$$

Taking $\mathbf{x}^* = \mathbf{v}$, it follows that $0 \in \partial f(\mathbf{x}^*)$, so $\mathbf{x}^*$ minimizes $f(\mathbf{x})$.

Therefore, from the above results, the point $\mathbf{x}^*$ minimizes $f(\mathbf{x})$ iff:

$$\mathbf{x}^* = Prox_{\lambda f}(\mathbf{x}^*) \tag{A.321}$$

Then, the point $\mathbf{x}^*$ is a fixed point of the operator $Prox_{\lambda f}$.

The expression (A.320) is the *subgradient characterization* of the proximal mapping.

The proximal mapping of a separable function such that $f(x, y) = \alpha(x) + \beta(y)$, would be:

$$Prox_{\lambda f}(v, w) = (Prox_{\lambda \alpha}(v), Prox_{\lambda \beta}(w)) \tag{A.322}$$

Thus, the proximal mapping of $f(x, y)$ is done with the proximal mapping of each of the separable parts. In the general case of a fully separable $f(\mathbf{x})$, the proximal mapping would consist of proximal mappings of scalar functions.

The proximal point algorithm

As announced before, when dealing with penalty-functions, there is a sequential minimization algorithm, based on the proximal mapping. The algorithm is simply expressed a follows:

$$\mathbf{x}^{k+1} = Prox_{\lambda f}(\mathbf{x}^k) \quad (A.323)$$

Covergence of the sequence $\{\mathbf{x}^k\}$ to the set of minimizers of $f(\mathbf{x})$ is guaranteed for values of λ that satisfy $\lambda^k > 0$ and $\sum_{i=1}^{\infty} \lambda^k = \infty$. The convergence of the algorithm has been studied in detail in [97, 51]. It may be quite slow. In [104] a modification of the algorithm was proposed in order to obtain strong convergence.

Part of the literature uses the term proximal fixed-point algorithm, emphasizing that the solution would be a fixed point of the proximal operator. The article [88] remarks that if $Prox_{\lambda f}$ was a contraction (Lipschitz with $\lambda < 1$), the repeated application of the proximal mapping would find a fixed point. However, $Prox_{\lambda f}$ needs not to be a contraction. It turns out that it has a different property which is sufficient for fixed point iteration. This property is called *firm nonexpansiveness*:

$$\| Prox_{\lambda f}(\mathbf{x}) - Prox_{\lambda f}(\mathbf{y}) \|_2^2 \leq (\mathbf{x} - \mathbf{y})^T (Prox_{\lambda f}(\mathbf{x}) - Prox_{\lambda f}(\mathbf{y})) \quad (A.324)$$

Now, a short disgression seems opportune. Firmly nonexpansive operators are a particular case of nonexpansive operators, which are Lipschitz with $\lambda = 1$. Iterations of a nonexpansive operator need not converge, like for instance in the case of rotation operators. Anyhow, given a nonexpansive operator F, the sequence:

$$\mathbf{x}^{k+1} = (1 - \alpha)\, \mathbf{x}^k + \alpha\, F(\mathbf{x}^k) \quad (A.325)$$

with $0 < \alpha < 1$, will converge to one of the fixed points of F.

Operators like T are called α-averaged operators. The iteration $\mathbf{x}^{k+1} = T(\mathbf{x}^k)$ converges to a fixed point of T.

Firmly nonexpansive operators are averaged. In particular, the proximal operator is an averaged operator, and so the proximal point algorithm converges to a fixed point.

By the way, given a set of closed proper convex functions, $f_1, f_2, \ldots, f_n$, one has:

$$\frac{1}{m} \sum_{i=1}^{n} Prox_{\lambda f_i} = Prox_{\lambda g} \quad (A.326)$$

The function g is called the *proximal average* of the set of functions. Not to be confused with the concept of averaged operator.

Let us enter into another aspect. The already cited article [97] connects *monotone operators* and the proximal mapping. In the case of functions, the concept of monotonicity is clear: a monotonically increasing function $f(x)$ would never decrease, or vice-versa. In the case of operators, it is said that a set-valued operator M (mapping x to a set $M(x)$) is *monotone* if:

$$(\mathbf{u} - \mathbf{v})^T (\mathbf{x} - \mathbf{y}) \geq 0\,, \quad \forall \mathbf{x},\, \mathbf{y},\; \mathbf{u} \in M(\mathbf{x}),\; \mathbf{v} \in M(\mathbf{y}) \quad (A.327)$$

The subdifferential of a closed convex function is a monotone operator.

The *resolvent* of a set-valued mapping M is the mapping:

$$R_\lambda = (I + \lambda M)^{-1} \tag{A.328}$$

If M is monotone then R_λ is firmly nonexpansive, so R_λ is single valued and Lipschitz.

The resolvent of ∂f is the proximal mapping:

$$(I + \lambda \partial f)^{-1}(\mathbf{x}) = Prox_{\lambda f}(\mathbf{x}) \tag{A.329}$$

Now, consider the following case:

$$M(\mathbf{x}) = A(\mathbf{x}) + B(\mathbf{x}) \tag{A.330}$$

with A and B monotone, $A(\mathbf{x})$ single valued, and B with easy to compute resolvent. An important algorithm is the following:

$$\mathbf{x}^{k+1} = (I + \lambda_k B)^{-1} (I - \lambda_k A) (\mathbf{x}^k) \tag{A.331}$$

It is called the *'forward-backward'* algorithm, since it combines a "forward operator" $(I - \lambda_k A)$ and a "backward operator" $(I + \lambda_k B)^{-1}$.

Particular cases of this algorithm are:

- The *proximal gradient method*:

$$\mathbf{x}^{k+1} = Prox_{\lambda f}(\mathbf{x}^k - \lambda_k \nabla g(\mathbf{x}^k)) \tag{A.332}$$

with $A(\mathbf{x}) = \nabla g(\mathbf{x})$ and $B(\mathbf{x}) = \partial f(\mathbf{x})$

- The *projection method for variational inequality*:

$$\mathbf{x}^{k+1} = P_C(\mathbf{x}^k - \lambda_k F(\mathbf{x}^k)) \tag{A.333}$$

with $A(\mathbf{x}) = F(\mathbf{x})$ and $B(\mathbf{x}) = N_C$ (the normal cone operator, which is 0 for $\mathbf{x} \notin C$, and $\{\mathbf{w} \mid \mathbf{w}^T(\mathbf{z} - \mathbf{x}) \leq 0,\ \forall \mathbf{z} \in C\}$ for $\mathbf{x} \in C$)

Let us study next, in more detail, the proximal gradient method.

Composite functions and the proximal gradient method

Consider the minimization of:

$$f(\mathbf{x}) = g(\mathbf{x}) + h(\mathbf{x}) \tag{A.334}$$

where $g(\mathbf{x})$ and $h(\mathbf{x})$ are closed, proper and convex, and $g(\mathbf{x})$ is differentiable. The function $h(\mathbf{x})$ can be extended-valued, so $h(\mathbf{x}) : \Re^n \rightarrow \Re \cup \{+\infty\}$.

The proximal gradient method for the minimization problem is:

$$\mathbf{x}^{k+1} = Prox_{\lambda h}(\mathbf{x}^k - \lambda_k \nabla g(\mathbf{x}^k)) \tag{A.335}$$

When $\nabla g(\mathbf{x})$ is Lipschitz with constant L, the method converges with rate $O(1/k)$ if $\lambda_k \in (0, 1/L)$. If L is not known, the step size λ_k can be found by line search [88].

The iterative method can be interpreted as a fixed point iteration. A point $\mathbf{x}^*$ is a solution of the minimization problem iff:

$$0 \in \nabla g(\mathbf{x}^*) + \partial h(\mathbf{x}^*) \tag{A.336}$$

This condition is satisfied if and only:

$$\mathbf{x}^* = (I + \lambda \partial h)^{-1} (I - \lambda \nabla g) (\mathbf{x}^*) \tag{A.337}$$

or equivalently:

$$\mathbf{x}^* = Prox_{\lambda h}(\mathbf{x}^* - \lambda \nabla g(\mathbf{x}^*) \tag{A.338}$$

The point $\mathbf{x}^*$ would be a fixed point of the forward-backward operator $(I + \lambda \partial h)^{-1} (I - \lambda \nabla g)$.

Another interpretation is that the method is an example of *majorization-minimization algorithm* [88]. This class of algorithms embodies many important instances, like the gradient descent, the Newton's method, the EM algorithm, etc.

Given a function $g(\mathbf{x})$, its minimization could be done with the following iterative algorithm:

$$\mathbf{x}^{k+1} = \arg\min_{\mathbf{x}} \hat{g}(\mathbf{x}, \mathbf{x}^k) \tag{A.339}$$

where $\hat{g}(\mathbf{x})$ is a convex upper bound to $g(\mathbf{x})$ satisfying that $\hat{g}(\mathbf{x}, \mathbf{x}^k) \geq g(\mathbf{x})$ and $\hat{g}(\mathbf{x}, \mathbf{x}) = g(\mathbf{x})$.

This generic algorithm iteratively exerts a majorization, with the upper bound, and then minimizes the majorization.

An example of upper bound could be the following:

$$\hat{g}(\mathbf{x}, \mathbf{y}) = g(\mathbf{y}) + \nabla g(\mathbf{y})^T (\mathbf{x} - \mathbf{y}) + \frac{1}{2\lambda} \|\mathbf{x} - \mathbf{y}\|_2^2 \tag{A.340}$$

Consider now the function:

$$q(\mathbf{x}, \mathbf{y}) = \hat{g}(\mathbf{x}, \mathbf{y}) + h(\mathbf{x}) \tag{A.341}$$

This function is a surrogate for $g(\mathbf{x}) + h(\mathbf{x})$, with fixed $\mathbf{y}$. It can be shown that the majorization-minimization iteration:

$$\mathbf{x}^{k+1} = \arg\min_{\mathbf{x}} q(\mathbf{x}, \mathbf{x}^k) \tag{A.342}$$

is equivalent to the proximal gradient iteration.

The minimization of $q(\mathbf{x}, \mathbf{x}^k)$ can be expressed as the minimization of:

$$\frac{1}{2} \parallel \mathbf{x} - (\mathbf{x}^k - \lambda \nabla g(\mathbf{x}^k)) \parallel_2^2 + \lambda\, h(\mathbf{x}) \tag{A.343}$$

Therefore, the solution tries to balance the gradient step with the weighted contribution of $h(\mathbf{x})$.

Nesterov [82] introduced and accelerated proximal gradient method, with the following scheme:

$$\mathbf{y}^{k+1} = \mathbf{x}^k + w_k\, (\mathbf{x}^k - \mathbf{x}^{k-1}) \tag{A.344}$$

$$\mathbf{x}^{k+1} = Prox_{\lambda h}(\mathbf{y}^{k+1} - \lambda_k \nabla g(\mathbf{y}^{k+1})) \tag{A.345}$$

where $w_k \in [0, 1)$ is an extrapolation parameter. A choice of this parameter could be:

$$w_k = \frac{k}{k+3} \tag{A.346}$$

When $\nabla g(\mathbf{x})$ is Lipschitz with constant L, the accelerated method converges with rate $O(1/k^2)$ if $\lambda_k \in (0, 1/L)$. Notice the difference with the standard method, which converges with rate $O(1/k)$. Also like before, if L is not known, the step size λ_k can be found by line search

An extensive treatise on monotone operators and proximal methods can be found in [10].

For more material on non-smooth optimization, it would be recommended to visit the web page of N. Karmitsa (see the Resources section).

A.10 The MATLAB Optimization Toolbox

The MATLAB Optimization Toolbox (OpT) provides functions for most of the optimization aspects described in this Appendix.

By using the *optimtool( )* function, MATLAB offers access via a GUI to four general categories of optimization solvers:

- Minimizers, including:
  - Unconstrained optimization
  - Linear programming
  - Quadratic programming
  - Nonlinear programming

- Multiobjective minimizers, with:
  - Minimizing the maximum of a set of functions
  - Finding a location where a set of functions is below some bounds
- Least-squares (curve-fitting), for different models
- Equation solving, linear or non-linear

In accord with the topics considered in this Appendix, our first comments are devoted to linear and quadratic programming, and then to other optimization problems. See [41] for a MATLAB OpT tutorial.

A.10.1 Linear Programming

The function for linear programming is *linprog( )*, which implements three types of algorithms:

- A simplex algorithm
- An active-set algorithm
- A primal-dual interior point method

A.10.2 Quadratic Programming

Quadratic programming problems can be tackled with the function *quadprog( )*. The function includes two types of algorithms:

- For medium scale problems: active-set
- For large scale problems: interior reflective Newton method coupled with a trust region methods

Large scale algorithms use sparse representations, so they do not need large computational resources. On the contrary, medium scale algorithms use full matrices and may require large computational efforts, taking perhaps a long time. You can use large scale algorithms on a small problem. However, the medium scale algorithm offer extra functionality, such as more types of constraints, and maybe better performance.

A.10.3 Unconstrained Minimization

Many unconstrained minimization problems are of the form:

$$\min_{\mathbf{x}} f(\mathbf{x}), \quad x \in R \tag{A.347}$$

MATLAB offers two alternatives for solving this kind of problem: to use *fminsearch( )* or to use *fminunc( )*.

The function *fminsearch( )* is based on the Nelder-Mead simplex algorithm. Here the word *'simplex'* denotes a polytope with $n + 1$ vertices (so it has nothing to do with the simplex method employed for linear programming). The algorithm starts with an initial simplex, and then modifies the simplex repeatedly, moving it and getting it smaller and smaller until convergence to the solution.

The function *fminunc( )* has two algorithm options. If you have information on the gradient, use the *'trust-region'* algorithm; otherwise, use the *'quasi-newton'* algorithm.

Before using any of these functions for unconstrained optimization, you can use the function *optimset( )* to specify options, like for instance to specify the chosen algorithm when using *fminunc( )*.

A.10.4 Constrained Minimization

In case of single-variable bounded non-linear function $f(x)$, so $l < x < m$, one could use the function *fminbnd( )*. This function combines a golden-section search and a parabolic interpolation.

When there is a set of constraints, including equalities and inequalities, the function to be used is *fmincon( )*. This function has four algorithm options: *'trust-region-reflective'*, *'interior-point'*, *'sqp'*, *'active-set'*. The recommendations from MATLAB is to use the interior-point first, and then try SQP next and active-set last, to get more speed; if your problem includes gradient information, only bounds *xor* only equality constraints, use trust-region-reflective.

One of the advices given in the OpT documentation concerns potential inaccuracy of interior-point algorithms (this kind of algorithm is one of the options that can be used in *linprog( )*, *quadprog( )* and *fmincon( )*). In order to reduce the inaccuracy it is recommended to modify some tolerances that can be specified, and to run a different algorithm starting from the interior-point solution (although this could fail).

A.10.5 Multiobjective Minimizers

In words of MATLAB documentation, the function *fminimax( )* minimizes the worst-case (largest) value of a set of multivariable functions, starting at an initial estimate.

Another multiobjective function is *fgoalattain( )*, which tries to make the objective function to attain specified goals. When using this function, it also returns the amount of over- or under-achievement of the goals.

A.10.6 General GUI

Figure A.47 shows the screen deployed by *optimtool()*.

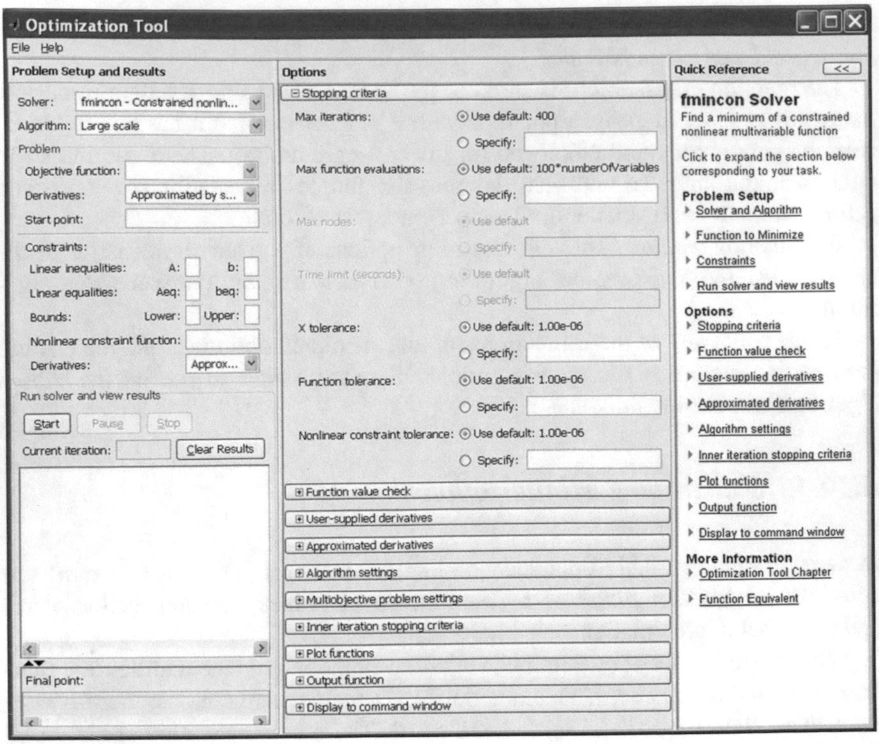

Fig. A.47 The optimtool ( ) screen

If one opens the solver menu, a series of functions can be accessed, as shown in Fig. A.48. Some of them, those beginning with *lsq*, are devoted to least squares. Most of the others have been already described.

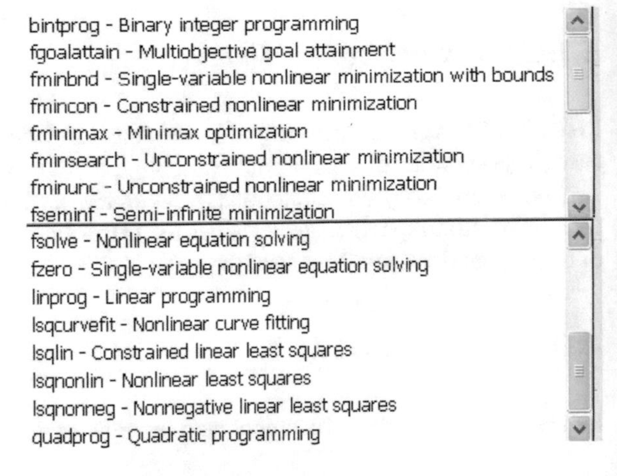

Fig. A.48 The solver functions

A.11 Resources

A.11.1 MATLAB

A.11.1.1 Toolboxes

- CVX:
  http://www.stanford.edu/~boyd/software.html
- YALMIP:
  http://control.ee.ethz.ch/index.cgi?action=details\&id=2088\&
- MOSEK:
  http://docs.mosek.com/7.0/toolbox/
- UNLocBoX:
  https://lts2.epfl.ch/unlocbox/
- OPTI Toolbox:
  http://www.i2c2.aut.ac.nz/Wiki/OPTI/index.php/Main/WhatIsOPTI?
- TOMLAB:
  http://tomopt.com/tomlab/about/

A.11.1.2 Matlab Code

- Stephen P. Boyd:
  http://www.stanford.edu/~boyd/software.html
- LIPSOL:
  www.caam.rice.edu/~zhang/lipsol/
- Guanghui Lan:
  http://www.ise.ufl.edu/glan/computer-codes/
- Stanford.edu:
  http://web.stanford.edu/~yyye/matlab.html
- NCSU.edu:
  http://www4.ncsu.edu/~ctk/matlab_darts.html
- SDPT3:
  http://www.math.nus.edu.sg/~mattohkc/sdpt3.html
- Mark Schmidt:
  https://www.cs.ubc.ca/~schmidtm/Software/code.html
- Optimization in Practice with MATLAB (book):
  http://www.cambridge.org/us/academic/subjects/engineering/control-systems
  -and-optimization/

A.11.2 Internet

A.11.2.1 Web sites

- Linear Programming on the Web:
  http://www.economicsnetwork.ac.uk/cheer/ch11_1/ch11_1p6.htm
- Peter Carbonetto:
  http://www.cs.ubc.ca/~pcarbo/#pubs
- Gould, N., Toint, P.: A Quadratic Programming Page,
  www.numerical.rl.ac.uk/people/nimg/qp/qp.html
- Michael A. Trick: A Tutorial on Integer Programming,
  http://mat.gsia.cmu.edu/orclass/integer/integer.html
- J.E. Beasley: OR Notes (Integer Programming),
  http://people.brunel.ac.uk/~mastjjb/jeb/or/ip.html
- L. El Ghaoui: Hyper-textbook on Optimization Models and Applications,
  https://inst.eecs.berkeley.edu/~ee127a/book/login/index.html
- N. Karmitsa: Non-smooth optimization,
  http://napsu.karmitsa.fi/

A.11.2.2 Link Lists

- Local Optimization Software:
  http://www.mat.univie.ac.at/~neum/glopt/software_l.html
- Decision Tree for Optimization Software:
  http://plato.asu.edu/sub/nlores.html
- Optimization solvers (CRAN page):
  https://cran.r-roject.org/web/views/Optimization.html

References

1. I. Adler, The equivalence of linear programs and zero-sum games. Int. J. Game Theory **42**(1), 165–177 (2013)
2. A.A. Ahmadi, A. Olshevsky, P.A. Parrillo, J.N. Tsitsiklis, NP-hardness of deciding convexity of quartic polynomials and related problems. Math. Programm. **137**(1–2), 453–476 (2013) series A
3. M. Al-Baali, H. Khalfan, An overview of some practical quasi-Newton methods for unconstrained optimization. SQU J. Sci. **12**(2), 199–209 (2007)
4. F. Alizadeh, Optimization over the positive-definite cone: interior point methods and combinatorial applications, eds. by P.M. Pardalos. *Advances in Optimization and Parallel Computing* (North-Holland, 1992)
5. F. Alizadeh, D. Goldfarb, Second-order cone programming. Math. Programm. **95**(1), 3–51 (2003)

6. E.D. Andersen, How to use Farkas' lemma to say something important about linear infeasible problems. Technical report, MOSEK Technical Report, 2011. TR-2011–1.
7. Y. Bai, M.E. Ghami, C. Roos, A comparative study of kernel functions for primal-dual interior-point algorithms in linear optimization. SIAM J. Optim. **15**(1), 101–128 (2004)
8. T. Baran, D. Wei, A.V. Oppenheim, Linear programming algorithms for sparse filter design. IEEE T. Signal Process. **58**(3), 1605–1617 (2010)
9. J. Barzilai, J.M. Borwein, Two point step size gradient methods. IMA J. Numer. Anal. **8**, 141–148 (1988)
10. H.H. Bauschke, P.L. Combettes, *Convex Analysis and Monotone Operator Theory* (Springer Verlag, 2010)
11. A. Beck, M. Teboulle, Gradient-based algorithms with applications to signal recovery, in *Convex Optimization in Signal Processing and Communications*, pp. 42–88 (2009)
12. R. Bellman, K. Fan, On systems of linear inequalities in hermitian matrix variables. Proc. Symp. Pure Math. **VII**, 1–11. Amer. Math. Soc., 1963. Providence
13. P. Belotti, *Optimization Models and Applications* Clemson University, IE 426 Lecture Notes, Lecture 20, 2009. www.myweb.clemson.edu/~pbelott/bulk/teaching/lehigh/ie426-f09/lecture20.pdf
14. D.P. Bertsekas, *Convex Optimization Theory* (Athena Scientific, 2014) supplement. http://www-mit.mit.edu/dimitrib/www/convexdualitychapter.pdf
15. K.C. Border, *Separating Hyperplane Theorems* (Caltech, 2010). http://www.hss.caltech.edu/~kcb/Notes/SeparatingHyperplane.pdf
16. S. Boyd, A. Mutapcic, *Subgradient Methods* (Stanford University, Lecture Notes for EE364b, 2007). https://web.stanford.edu/class/ee364b/lectures/subgrad_method_notes.pdf
17. S. Boyd, A. Mutapcic, *Stochastic Subgradient Methods* (Stanford University, Lecture Notes for EE364b, 2008). https://web.stanford.edu/class/ee364b/lectures/stoch_subgrad_notes.pdf.
18. S. Boyd, L. Vandenberghe, *Convex Optimization* (Cambridge University Press, 2004)
19. S. Boyd, L. Vandenberghe, *Localization and Cutting Plane Methods* (Stanford University, Lecture Notes, 2007). www.stanford.edu/class/ee392o/localization-methods.
20. S. Boyd, L. Vandenberghe, *Interior-point Methods* (2008). https://web.stanford.edu/class/ee364a/lectures/barrier.pdf
21. S. Boyd, L. Vandenberghe, *Subgradients* (Stanford University, Lecture Notes for EE364b, 2008). https://web.stanford.edu/class/ee364b/lectures/subgradients_notes.pdf
22. S. Boyd, L. Vandenberghe, J. Skaf, *Analytic Center Cutting-plane Method* (Stanford University, 2008). http://112.112.8.207/resource/data/20100601/U/stanford201001010/06-accpm_notes.pdf
23. S. Burer, Copositive programming, in *Handbook on Semidefinite, Conic and Polynomial Optimization*, pp. 201–218 (Springer USA, 2012.)

24. C.L. Byrne, *Continuous Optimization* (Department of Mathematical Sciences University of Massachusetts, Lowell, 2013). http://faculty.uml.edu/cbyrne/optfirst.pdf
25. C. Caramanis, S. Sanghavi, *Large Scale Optimization* (The University of Texas at Austin, Lecture 21 of EE381V Course, 2012). http://users.ece.utexas.edu/~cmcaram/EE381V_2012F/Lecture_21_Scribe_Notes.final.pdf
26. V. Chandru, M.R. Rao, 175 years of linear programming. *Resonance*, pp. 4–13 (1999)
27. E.W. Cheney, A.A. Goldstein, Newton's method for convex programming and Tchebycheff approximation. Num. Math. **1**(1), 253–268 (1959)
28. A.P. Chottera, G.A. Jullien, A linear programming approach to recursive digital filter design with linear phase. IEEE T. Circuits Syst. **29**(3), 139–149 (1982)
29. V. Chvatal, *Linear Programming* (W.H. Freeman, 1983)
30. F.H. Clarke, *Optimization and Nonsmooth Analysis* (Wiley-Interscience, 1983)
31. V. Conitzer, *Solving (mixed) Integer Programs Using Branch and Bound* (Duke University, Durham, USA, Lecture Notes, 2008). https://www.cs.duke.edu/courses/spring08/cps296.2/branch_and_bound.pdf.
32. S. Cui, *Lecture 4: LP, QP, SOCP, SDP, and GP* (Texas A&M University, Lecture Presentation, 2015). http://ece.tamu.edu/~cui/ECEN629/lecture4.pdf
33. M.F. Danca, S. Codreanu, On a possible approximation of discontinuous dynamical systems. Chaos, Solitons & Fractals **13**(4), 681–691 (2002)
34. F. Eisenbrand, *Integer Programming and Branch and Bound* (Ecole Polytechnique Federale de Lausanne, Switzerland, Lecture 11 Notes, 2010). http://disopt.epfl.ch/files/content/sites/disopt/files/shared/OptInFinance10/scribes11_Boris_Oumow.pdf
35. M.A. Epelman, C.-W. Kuo, *Continuous Optimization Methods, Section 1* (University of Michigan, Lecture Notes, course IOE 511/ math 562, 2007). http://www.cse.iitd.ernet.in/~naveen/courses/CSL866/511notes.pdf
36. J. Fairbrother, A. Letchford, *Mathematical Programming Through Cones* (STOR-I, Lancaster University, 2011). http://www.lancaster.ac.uk/pg/fairbrot/files/cones.pdf
37. A.V. Fiacco, G.P. McCormick, *Nonlinear Programming: Sequential Unconstrained Minimization Techniques* (John Wiley, 1968)
38. M.L. Fisher, The lagrangian relaxation method for solving integer programming problems. Manag. Sci. **50**(12), 1861–1871 (2004) supplement.
39. J.B.G. Frenk, G. Kassay, J. Kolumbán, On equivalent results in minimax theory. Eur. J. Oper. Res. **157**, 46–58 (2004)
40. S.I. Gass, *Linear Programming: Methods and Applications* (Dover, 2010)
41. A. Geletu, *Solving Optimization Problems Using the Matlab Optimization Toolbox-A Tutorial* (TU-Ilmenau, Fakultät für Mathematik und Naturwissenschaften, 2007). http://www.tu-ilmenau.de/fileadmin/media/simulation/Lehre/Vorlesungsskripte/Lecture_materials_Abebe/Optimiz atioWithMatlab.pdf
42. A. Geletu, *Quadratic programming problems-a review on algorithms and applications (Active-set and interior point methods)* (TU-Ilmenau, Fakultät

für Mathematik und Naturwissenschaften, 2012). https://www.tu-ilmenau.de/fileadmin/media/simulation/Lehre/Vorlesungsskripte/Lecture_materials_Abebe/QPs_with_IPM_and_ASM.pdf
43. A.M. Geoffrion, R.E. Marsten, Integer programming algorithms: A framework and state-of-the-art survey. Manag. Sci. **18**(9), 465–491 (1972)
44. F. Glineur, Topics in Convex Optimization: Interior-Point Methods, Conic Duality and Approximations. Ph.D. thesis, Fac. Polytechnique de Mons, Belgium, 2001
45. J.L. Goffin, Subgradient optimization in nonsmooth optimization (including the Soviet revolution). Doc. Math. extra vol. ISMP, 277–290 (2012)
46. J.L. Goffin, J.P. Vial, Convex nondifferentiable optimization: A survey focused on the analytic center cutting plane method. Optim. Meth. Softw. **17**(5), 805–867 (2002)
47. R.E. Gomory, Outline of an algorithm for integer solutions to linear programs. Bull. Am. Math. Soc. **64**, 275–278 (1958)
48. C.C. Gonzaga, E.W. Karas, Optimal steepest descent algorithms for unconstrained convex problems: Fine tuning Nesterov's method. Technical report, Department of Mathematics, Federal University of Santa Catarina, Florianópolis, 2008
49. V. Gopal, L.T. Biegler, Large scale inequality constrained optimization and control. IEEE Control Syst. Mgz. **18**(6), 59–68 (1998)
50. M. Grasmair, *Continuous Optimisation* (University of Vienna, Lecture Notes, 2011). http://www.csc.univie.ac.at/files/Continuous_Optimisation.pdf
51. O. Güler, On the convergence of the proximal point algorithm for convex optimization. SIAM J. Control Optim. **29**(2), 403–419 (1991)
52. D. Henrion, *What Is an LMI?* (LAAS, France, Lecture Notes, 2010). http://www.eeci-institute.eu/pdf/M1-Textes/hycon-henrion-1.pdf
53. H. Hindi, A tutorial on convex optimization. Proc. Am. Control Conf. **4**, 3252–3265 (2004)
54. J.-B. Hiriart-Urruty, H.Y. Le, From Eckart and Young approximation to Moreau envelopes and vice versa. RAIRO - Oper. Res. **47**(3), 299–310 (2013)
55. M. Jacimovic, Farkas' lemma of alternative. Teach. Math. **14**(2), 77–86 (2011)
56. P.A. Jensen, J.F. Bard, Nonlinear programming methods.s2 quadratic programming, eds. by Jensen, Bard. *Operations Research Models and Methods* (John Wiley, 2003)
57. A. Kamath, N. Karmarkar, A continuous approach to compute upper bounds in quadratic maximization problems with integer constraints, eds. by Floudas and Pardalos *Recent Advances in Global Optimization*, pp. 125–140 (Princeton University Press, 1992)
58. A. Kamath, N. Karmarkar, An O(nL) iteration algorithm for computing bounds in quadratic optimization problems, ed. by P.M. Pardalos. *Complexity in Numerical Optimization*, pp. 254–268 (World Scientific Publishing, 1992)
59. L.V. Kantorovich, G.P. Akilov, *Functional Analysis* (Pergamon Press, 1982)
60. N. Karmarkar, A new polynomial-time algorithm for linear programming. Combinatorica **4**(4), 373–395 (1984)

61. C.T. Kelley, *Iterative Methods for Optimization* (SIAM, 1999)
62. J.E. Jr. Kelley, The cutting-plane method for solving convex programs. J. Soc. Indus. Appl. Math. **8**(4), 703–712 (1960)
63. S.J. Kim, K. Koh, M. Lustig, S. Boyd, D. Gorinevsky, An interior-point method for large-scale L1-regularized least squares. IEEE J. Sel. Top. Sign. Process. **1**(4), 606–617 (2007)
64. K. Koh, S.J. Kim, S.P. Boyd, An interior-point method for large-scale L1-regularized logistic regression. J. Mach. Learn. Res. **8**(8), 1519–1555 (2007)
65. T.G. Kolda, R.M. Lewis, V. Torczon, Optimization by direct search: New perspectives on some classical and modern methods. SIAM Rev. **45**(3), 385–482 (2003)
66. H.E. Krogstad, *Quadratic Programming Basics* (Norwegian University of Science and Technology, Trondheim, Norway, Lecture Notes, 2005). https://www.math.ntnu.no/~hek/Optimering2012/QPbasicsOH2012.pdf
67. S. Kumar, M.K. Luhandjula, E. Munapo, B.C. Jones, Fifty years of integer programming: A review of the solution approaches. Asia Pacific Bus. Rev. **6**(3), 5–15 (2010)
68. G. Lan, *Bundle-type Methods Uniformly Optimal for Smooth and Non-smooth Convex Optimization* (Department of Industrial and Systems Engineering, University of Florida, 2010). http://www.optimization-online.org/DB_FILE//11/2796.pdf
69. C-R Lee, *Numerical Optimization-Unit 8* (National Tsing Hua University, Taiwan, Lecture Notes, 2013). http://www.cs.nthu.edu.tw/~cherung/teaching/2011cs5321/handout8.pdf
70. C. Lemaréchal, Nondifferentiable optimization, ed. by Nemhauser. *Handbook of OR & MS*, vol. 1, pp. 529–572 (Elsevier, 1989)
71. J. Linderoth, *IE418: Integer Programming* (Lehigh University, Lecture Presentation, 2005). http://homepages.cae.wisc.edu/~linderot/classes/ie418/lecture2.pdf
72. M.S. Lobo, L. Vandenberghe, S. Boyd, H. Lebret, Applications of second-order cone programming. Linear Algebra Appl. **284**(1), 193–228 (1998)
73. D.A. Lorenz, S. Schiffler, D. Trede, Beyond convergence rates: Exact recovery with the Tikhonov regularization with sparsity constraints. Inverse Probl. **27**(8), 085009 (2011)
74. Z.Q. Luo, W. Yu, An introduction to convex optimization for communications and signal processing. IEEE J. Sel. Areas Commun. **24**(8), 1426–1438 (2006)
75. M. Mäkelä, Survey of bundle methods for nonsmooth optimization. Optim. Meth. Softw. **17**(1), 1–29 (2002)
76. R. Marsten, R. Subramanian, M. Saltzman, I. Lustig, D. Shanno, Interior point methods for linear programming: Just call Newton, Lagrange, and Fiacco and McCormick. Interfaces **20**(4), 105–116 (1990)
77. L. McLinden, An extension of Fenchel's duality theorem to saddle functions and dual minimax problems. Pacific J. Math. **50**(1), 135–158 (1974)

78. H.D. Mittelmann, The state-of-the-art in conic optimization software, in *Handbook on Semidefinite, Conic and Polynomial Optimization*, pp. 671–686 (Springer USA, 2012)
79. B.S. Mordukhovich, *Variational Analysis and Generalized Differentiation I: Basic Theory* (Springer Verlag, 2006)
80. B.S. Mordukhovich, *Variational Analysis and Generalized Differentiation II: Applications* (Springer Verlag, 2006)
81. J.-J. Moreau, Fonctions convexes duales et points proximaux dans un espace Hilbertien. C.R. Acad. Sci. **255**, 2897–2899 (1962)
82. Y. Nesterov, A method of solving a convex programming problem with convergence rate O(1/k2). Sov. Math. Doklady **27**, 372–376 (1983)
83. Y. Nesterov, *Introductory Lectures on Convex Optimization: A Basic Course* (Kluwer, 2003)
84. Y. Nesterov, Efficiency of coordinate descent methods on huge-scale optimization problems. SIAM J. Optim. **22**(2), 341–362 (2012)
85. Y. Nesterov, A.S. Nemirovsky, A general approach to polynomial-time algorithms design for convex programming. Technical report, Centr. Econ. Math. Inst., USSR Acad. Sci., 1988
86. Y. Nesterov, A.S. Nemirovsky, *Interior-point Polynomial Algorithms in Convex Programming*, vol. 13 (SIAM, 1994)
87. J. Nocedal, S. Wright, *Numerical Optimization* (Springer Verlag, 2006)
88. D.P. Palomar, Y.C. Eldar (Eds.), *Convex Optimization in Signal Processing and Communications* (Cambridge University Press, 2009)
89. B.T. Polyak, *Introduction to Optimization* (Optimization Software Inc., 1987)
90. L.R. Rabiner, The design of finite impulse response digital filters using linear programming techniques. The Bell System Technical J. **51**(6), 1177–1198 (1972)
91. L.R. Rabiner, N.Y. Graham, H.D. Helms, Linear programming design of FIR digital filters with arbitrary magnitude function. IEEE T. Acoust. Speech Signal Process. **22**(2) 117–123 (1974)
92. M. Raydan, B.F. Svaiter, Relaxed steepest descent and Cauchy-Barzilai-Borwein method. Comput. Optim. Appl. **21**(2), 155–167 (2002)
93. J. Renegar, *A Mathematical View of Interior-Point Methods in Convex Optimization* (SIAM, 2001)
94. P. Richtárik, M. Takáè, Iteration complexity of randomized block-coordinate descent methods for minimizing a composite function. Math. Programm. **144**(1–2), 1–38 (2014)
95. R.T. Rockafellar, *Convex Analysis* (Princeton University Press, 1970)
96. R.T. Rockafellar, *Conjugate Duality and Optimization* (SIAM, 1974)
97. R.T. Rockafellar, Monotone operators and the proximal point algorithm. SIAM J. Control Optim. **14**(5), 877–898 (1976)
98. R.T. Rockafellar, R.J.-B. Wets, M. Wets, *Variational Analysis* (Springer Verlag, 2009)
99. M. Rudnev, *Farkas Alternative and Duality Theorem* (University of Bristol, 2011). http://www.maths.bris.ac.uk/~maxmr/opt/farkas.pdf

100. R. Säljö, Implementation of a bundle algorithm for convex optimization. Master's thesis, Goteborg University, 2004
101. M. Searcóid, *Metric Spaces* (Springer, 2007)
102. I. Selesnick, *Linear-phase FIR Filter Design by Linear Programming* (Polytechnic Institute of NYU, EL 713 Lecture Notes, 2003). www.eeweb.poly.edu/iselesni/EL713/linprog/linprog.pdf
103. N.Z. Shor, *Minimization Methods for Non-Differentiable Functions* (Springer Verlag, 1985)
104. M.V. Solodov, B.F. Svaiter, Forcing strong convergence of proximal point iterations in a Hilbert space. Math. Programm. Ser. A **87**, 189–202 (2000)
105. Pearson staff, *The Simplex Method of Linear Programming* (2009) Online Tutorial 3. http://wps.prenhall.com/wps/media/objects/9434/9660836/online_tutorials/heizer10e_tut3.pdf
106. K. Tanabe, Centered Newton method for mathematical programming, eds. by M. Iri, K. Yajima. *Lecture Notes in Control and Information Sciences*, pp. 197–206 (Springer-Verlag, Berlin, 1988)
107. M.J. Todd, Y.Ye, A centered projective algorithm for linear programming. Math. Oper. Res. **15**(3), 508–529 (1990)
108. H. Touchette, *Legendre-Fenchel transforms in a nutshell* (School of Mathematical Sciences, Queen Mary, University of London, 2005). www.maths.qmul.ac.uk/~ht/archive/lfth2.pdf
109. P. Tseng, S. Yun, A coordinate gradient descent method for nonsmooth separable minimization. Math. Programm. **117**(1–2), 387–423 (2009)
110. L. Vandenberghe, *Cutting-plane Methods* (UCLA, Lecture Presentation, 2003). http://ee.ucla.edu/~vandenbe/236C/lectures/localization.pdf
111. L. Vandenberghe, S. Boyd, Semidefinite programming. SIAM Rev. **38**(1), 49–95 (1996)
112. R.J. Vanderbel, *Linear Programming: Foundations and Extensions* (Springer Verlag, 2008)
113. G. Vatchtsevanos, N. Papamarkos, B. Mertzios, Design of two-dimensional IIR digital filters via linear programming. Sign. Process. **12**, 17–26 (1987)
114. S. V.N. Vishwanathan, *Introduction to Machine Learning, Chapter 5: Optimization* (Purdue University, Lecture Notes, CS590, 2010). http://www.stat.purdue.edu/~vishy/introml/introml.html.
115. J. von Neumann, O. Morgenstern, *Theory of Games and Economic Behavior* (Princeton University Press, 1944)
116. G.K. Wen, M. Mamat, I. Mohd, Y. Dasril, A novel of step size selection procedures for steepest descent method. Appl. Math. Sci. **6**(51), 2507–2518 (2012)
117. L. Weng, Y. Chen, *Stochastic subgradient methods* (Bren School of Information and Computer Sci., University of California at Irvine, 2012). http://www.ics.uci.edu/~xhx/courses/ConvexOpt/projects/stochastic_subgradient_methods_report.pdf
118. T.U. Wien, *Convex Optimization for Signal Processing and Communications* (TU Wien, Lecture Notes, Lecture 7, 2012). http://www.nt.tuwien.ac.at/uploads/media/lecture7.pdf

119. W. Wiesemann, *Integer Programming* (Imperial College, London, Lecture Presentation, 2009). http://www.doc.ic.ac.uk/~br/berc/IPlecture2.pdf
120. Y.-X. Yuan, A review of trust region algorithms for optimization, in *Proceedings 4th International Congress on Industrial and Applied Mathematics*, pp. 271–282 (2000)
121. Y.X. Yuan, Step-sizes for the gradient method. AMS IP Stud. Adv. Math. **42**(2), 785–798 (2008)
122. B. Zhou, L. Gao, Y.H. Dai, Gradient methods with adaptive step-sizes. Comput. Optim. Appl. **35**(1), 69–86 (2006)

Appendix B
Long Programs

B.1 Chapter 1: Kalman Filter

B.1.1 Evolution of Filter Variables (8.3.2.)

The example of the two-tank system was chosen. Figure B.1 depicts the outputs of the system, which are the measurements of tank heights

Figure B.2 compares the evolution of the 2-variable system, in continuous curves, and the state estimation yield by the Kalman filter, depicted by x marks.

Figure B.3 shows the evolution of the error.

Figure B.4 shows the evolution of the Kalman filter gains.

Figure B.5 shows the evolution of the a priori state covariance.

Figure B.6 shows the evolution of the estimated state covariance.

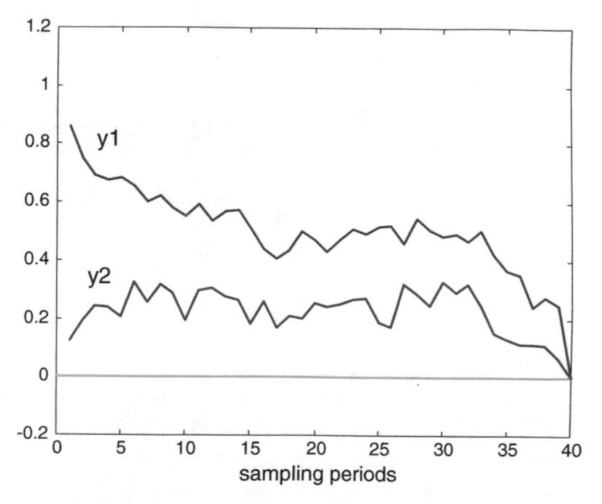

Fig. B.1 System outputs (measurements) (Fig. 1.6)

J.M. Giron-Sierra, *Digital Signal Processing with Matlab Examples, Volume 3*, Signals and Communication Technology, DOI 10.1007/978-981-10-2540-2

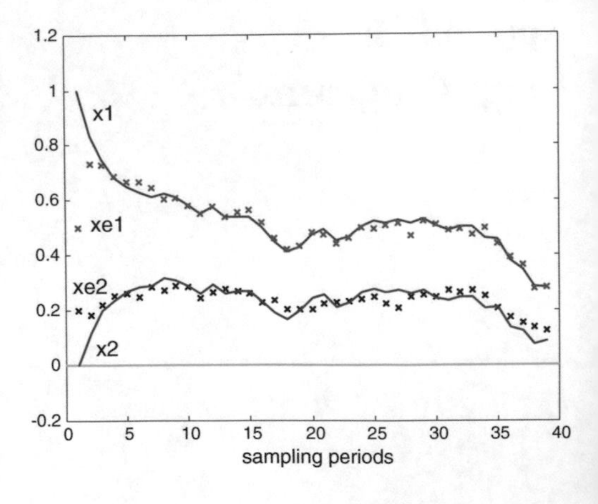

Fig. B.2 System states, and states estimated by the Kalman filter (Fig. 1.7)

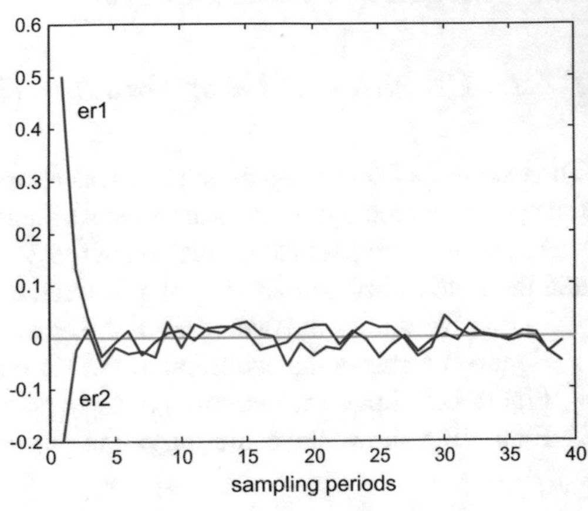

Fig. B.3 Error evolution (Fig. 1.8)

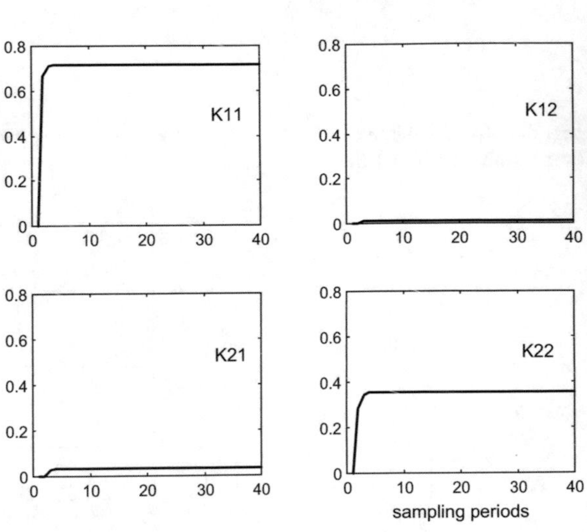

Fig. B.4 Evolution of the Kalman gains (Fig. 1.9)

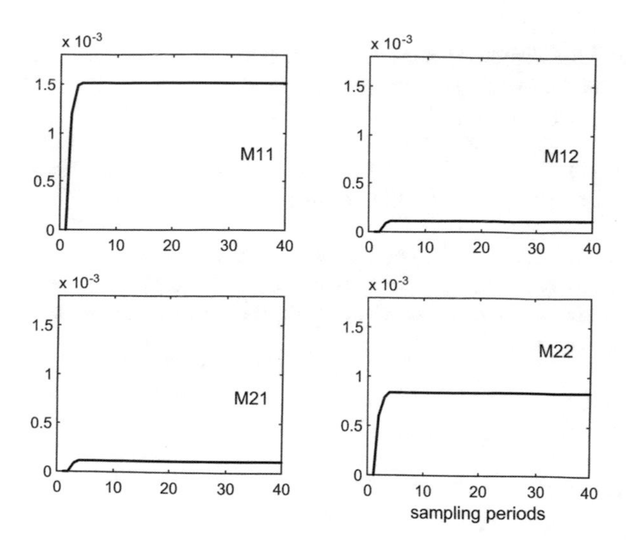

Fig. B.5 Evolution of the a priori state covariance

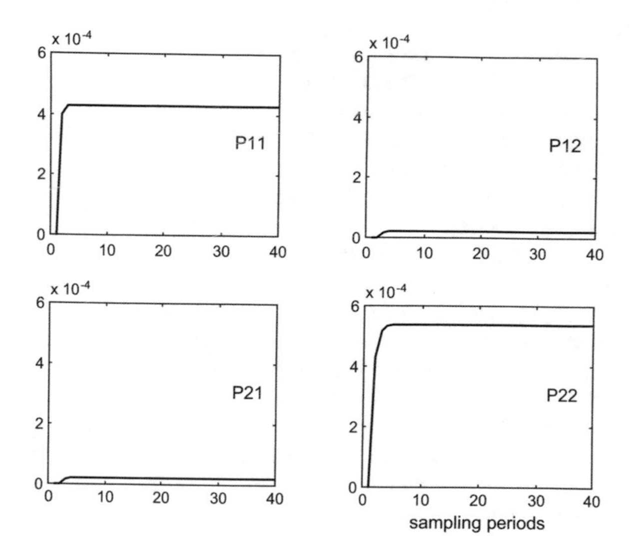

Fig. B.6 Evolution of the estimated state covariance (Fig. 1.10)

Program B.1 Kalman filter example, in noisy conditions

```
%Kalman filter example, in noisy conditions
%state space system model (2 tank system):
A1=1; A2=1; R1=0.5; R2=0.4;
cA=[-1/(R1*A1) 1/(R1*A1); 1/(R1*A2) -(1/A2)*((1/R1)+(1/R2))];
cB=[1/A1; 0]; cC=[1 0; 0 1]; cD=0;
Ts=0.1; %sampling period
csys=ss(cA,cB,cC,cD); %setting the continuous time model
dsys=c2d(csys,Ts,'zoh'); %getting the discrete-time model
[A,B,C,D]=ssdata(dsys); %retrieves discrete-time model matrices
%simulation horizon
Nf=40;
%process noise
Sw=[12e-4 0; 0 6e-4]; %cov
sn=zeros(2,Nf);
sn(1,:)=sqrt(Sw(1,1))*randn(1,Nf);
sn(2,:)=sqrt(Sw(2,2))*randn(1,Nf);
%observation noise
Sv=[6e-4 0; 0 15e-4]; %cov.
on=zeros(2,Nf);
on(1,:)=sqrt(Sv(1,1))*randn(1,Nf);
on(2,:)=sqrt(Sv(2,2))*randn(1,Nf);
% system simulation preparation
%space for recording x1(n), x2(n)
x1=zeros(1,Nf-1); x2=zeros(1,Nf-1);
x=[1;0]; % state vector with initial tank levels
u=0.4; %constant input
% Kalman filter simulation preparation
%space for matrices
K=zeros(2,2,Nf); M=zeros(2,2,Nf); P=zeros(2,2,Nf);
%space for recording er(n), xe1(n), xe2(n),ym(n)
er=zeros(2,Nf-1); xe1=zeros(1,Nf-1);
xe2=zeros(1,Nf-1);rym=zeros(2,Nf);
xe=[0.5; 0.2]; % filter state vector with initial values
%behaviour of the system and the Kalman
% filter after initial state
% with constant input u
for nn=1:Nf-1,
  x1(nn)=x(1); x2(nn)=x(2); %recording the system state
  xe1(nn)=xe(1); xe2(nn)=xe(2); %recording the observer state
  er(:,nn)=x-xe; %recording the error
  %
  %system simulation
  xn=(A*x)+(B*u)+sn(nn); %next system state
  x=xn; %system state actualization
  ym=(C*x)+on(:,nn); %output measurement
  rym(:,nn)=ym;
  %
  %Prediction
```

```
    xa=(A*xe)+(B*u); %a priori state
    M(:,:,nn+1)=(A*P(:,:,nn)*A')+ Sw;
    %Update
    K(:,:,nn+1)=(M(:,:,nn+1)*C')*inv((C*M(:,:,nn+1)*C')+Sv);
    P(:,:,nn+1)=M(:,:,nn+1)-(K(:,:,nn+1)*C*M(:,:,nn+1));
    %estimated (a posteriori) state :
    xe=xa+(K(:,:,nn+1)*(ym-(C*xa)));
end;
%------------------------------------------------------------
% display of system outputs
figure(1)
plot([0 Nf],[0 0],'g'); hold on; %horizontal axis
plot([0 0],[-0.2 1.2],'k'); %vertical axis
plot(rym(1,:),'r'); %plots y1
plot(rym(2,:),'b'); %plots y2
xlabel('sampling periods');
title('system outputs');
% display of state evolution
figure(2)
plot([0 Nf],[0 0],'g'); hold on; %horizontal axis
plot([0 0],[-0.2 1.2],'k'); %vertical axis
plot(x1,'r'); %plots x1
plot(x2,'b'); %plots x2
plot(xe1,'mx'); %plots xe1
plot(xe2,'kx'); %plots xe2
xlabel('sampling periods');
title('system and Kalman filter states');
% display of error evolution
figure(3)
plot([0 Nf],[0 0],'g'); hold on; %horizontal axis
plot([0 0],[-0.2 0.6],'k'); %vertical axis
plot(er(1,:),'r'); %plots x1 error
plot(er(2,:),'b'); %plots x2 error
xlabel('sampling periods');
title('estimation error');
% display of K(n) evolution
figure(4)
aa=[0 40 0 0.8];
subplot(2,2,1); plot(shiftdim(K(1,1,:)),'k');
axis(aa);title('K11');%plots M11
subplot(2,2,2); plot(shiftdim(K(1,2,:)),'k');
axis(aa);title('K12');%plots M12
subplot(2,2,3); plot(shiftdim(K(2,1,:)),'k');
axis(aa);title('K21');%plots M21
subplot(2,2,4); plot(shiftdim(K(2,2,:)),'k');
axis(aa);title('K22'); %plots M22
xlabel('sampling periods');
% display of M(n) evolution
figure(5)
```

```
aa=[0 40 0 0.0018];
subplot(2,2,1); plot(shiftdim(M(1,1,:)),'k');
axis(aa);title('M11');%plots M11
subplot(2,2,2); plot(shiftdim(M(1,2,:)),'k');
axis(aa);title('M12');%plots M12
subplot(2,2,3); plot(shiftdim(M(2,1,:)),'k');
axis(aa);title('M21');%plots M21
subplot(2,2,4); plot(shiftdim(M(2,2,:)),'k');
axis(aa);title('M22'); %plots M22
xlabel('sampling periods');
% display of P(n) evolution
figure(6)
aa=[0 40 0 0.0006];
subplot(2,2,1); plot(shiftdim(P(1,1,:)),'k');
axis(aa);title('P11');%plots P11
subplot(2,2,2); plot(shiftdim(P(1,2,:)),'k');
axis(aa);title('P12');%plots P12
subplot(2,2,3); plot(shiftdim(P(2,1,:)),'k');
axis(aa);title('P21');%plots P21
subplot(2,2,4); plot(shiftdim(P(2,2,:)),'k');
axis(aa);title('P22'); %plots P22
xlabel('sampling periods');
```

B.1.2 Uncertainties (8.3.2.)

It is interesting to visualize regions of uncertainty in the scenario proposed by the Kalman filter. Figure B.7 corresponds to the prediction step.

Figure B.8 corresponds to the output measurement equation.

Figure B.9 contains, from left to right, the two things to combine in the update step, prediction and measurement, and the final result.

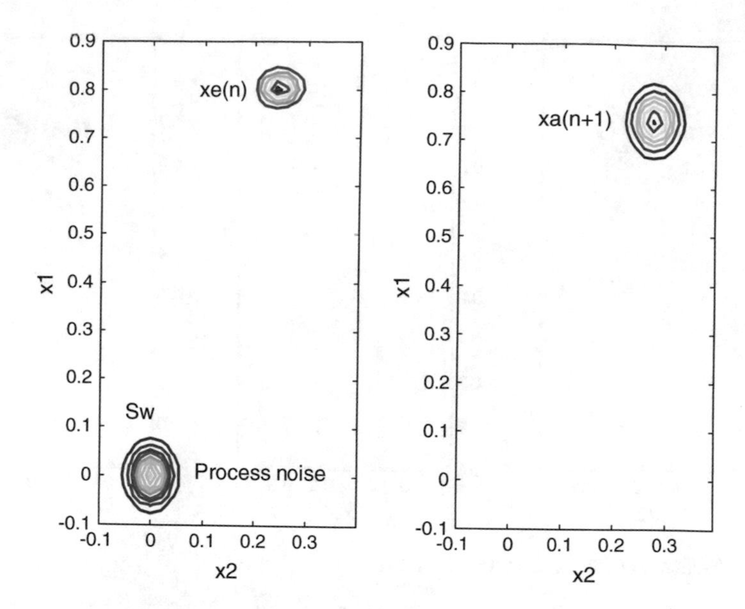

Fig. B.7 The prediction step, from *left* to *right* (Fig. 1.11)

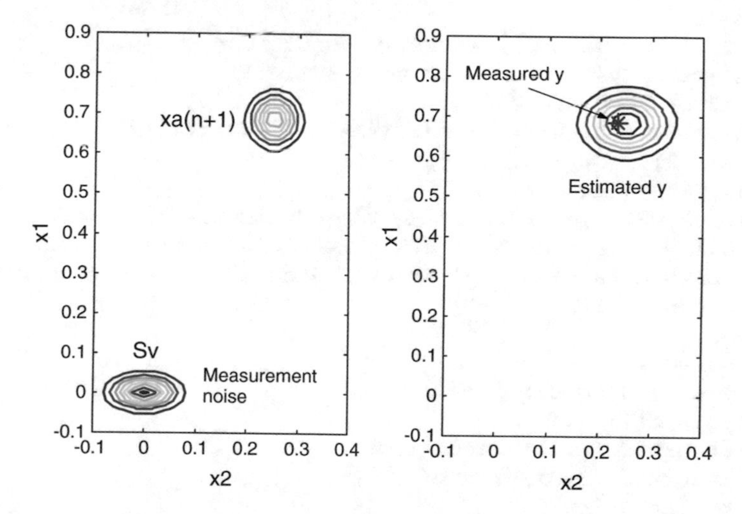

Fig. B.8 The measurement (Fig. 1.12)

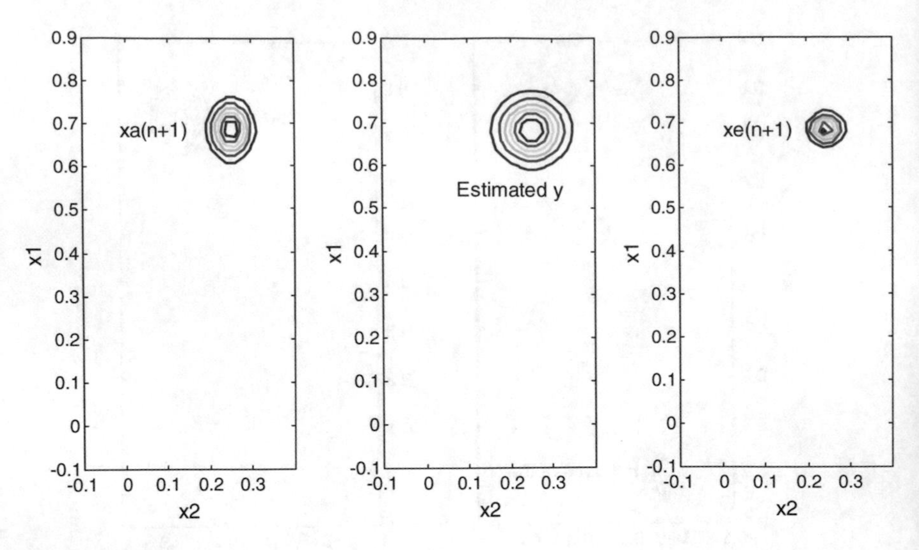

Fig. B.9 Estimation of the next state (Fig. 1.13)

Program B.2 Kalman filter ellipsoids

```
%Ellipsoids
%Kalman filter example, in noisy conditions
%state space system model (2 tank system):
A1=1; A2=1; R1=0.5; R2=0.4;
cA=[-1/(R1*A1) 1/(R1*A1); 1/(R1*A2) -(1/A2)*((1/R1)+(1/R2))];
cB=[1/A1; 0]; cC=[1 0; 0 1]; cD=0;
Ts=0.1; %sampling period
csys=ss(cA,cB,cC,cD); %setting the continuous time model
dsys=c2d(csys,Ts,'zoh'); %getting the discrete-time model
[A,B,C,D]=ssdata(dsys); %retrieves discrete-time model matrices
%simulation horizon
Nf=10;
%process noise
Sw=[12e-4 0; 0 6e-4]; %cov
sn=zeros(2,Nf);
sn(1,:)=sqrt(Sw(1,1))*randn(1,Nf);
sn(2,:)=sqrt(Sw(2,2))*randn(1,Nf);
%observation noise
Sv=[6e-4 0; 0 15e-4]; %cov.
on=zeros(2,Nf);
on(1,:)=sqrt(Sv(1,1))*randn(1,Nf);
on(2,:)=sqrt(Sv(2,2))*randn(1,Nf);
% system simulation preparation
%space for recording x1(n), x2(n)
x1=zeros(1,Nf-1); x2=zeros(1,Nf-1);
x=[1;0]; % state vector with initial tank levels
u=0.4; %constant input
% Kalman filter simulation preparation
```

```
%space for matrices
K=zeros(2,2,Nf); M=zeros(2,2,Nf); P=zeros(2,2,Nf);
aM=zeros(2,2,Nf); cY=zeros(2,2,Nf);
rxa=zeros(2,Nf); rya=zeros(2,Nf); rym=zeros(2,Nf);
%space for recording er(n), xe1(n), xe2(n)
er=zeros(2,Nf-1); xe1=zeros(1,Nf-1); xe2=zeros(1,Nf-1);
xe=[0.5; 0.2]; % filter state vector with initial values
%behaviour of the system and the Kalman
% filter after initial state
% with constant input u
for nn=1:Nf-1,
  x1(nn)=x(1); x2(nn)=x(2); %recording the system state
  xe1(nn)=xe(1); xe2(nn)=xe(2); %recording the observer state
  er(:,nn)=x-xe; %recording the error
  %
  %system simulation
  xn=(A*x)+(B*u)+sn(nn); %next system state
  x=xn; %system state actualization
  ym=(C*x)+on(:,nn); %output measurement
  rym(:,nn)=ym;
  %
  %Prediction
  xa=(A*xe)+(B*u); %a priori state
  M(:,:,nn+1)=(A*P(:,:,nn)*A')+ Sw;
  aM(:,:,nn)=(A*P(:,:,nn)*A');
  rxa(:,nn+1)=xa;
  %Update
  K(:,:,nn+1)=(M(:,:,nn+1)*C')*inv((C*M(:,:,nn+1)*C')+Sv);
  cY(:,:,nn+1)=((C*M(:,:,nn+1)*C')+Sv);
  P(:,:,nn+1)=M(:,:,nn+1)-(K(:,:,nn+1)*C*M(:,:,nn+1));
  %estimated (a posteriori) state:
  xe=xa+(K(:,:,nn+1)*(ym-(C*xa)));
  rya(:,nn+1)=C*xa;
end;
%-------------------------------------------------------------
ns1=3; ns2=4; %state number
%Distribution ellipsoids
px1=-0.1:0.02:0.9;
px2=-0.1:0.02:0.4;
pN1=length(px1);
pN2=length(px2);
%the Sw PDF
mu1=0; mu2=0;
C=Sw; D=det(C);
K=1/(2*pi*sqrt(D)); Q=(C(1,1)*C(2,2))/(2*D);
Swpdf=zeros(pN1,pN2); %space for the PDF
for ni=1:pN1,
  for nj=1:pN2,
    aux1=(((px1(ni)-mu1)^2)/C(1,1))...
```

```
    +(((px2(nj)-mu2).^2)/C(2,2));
    Swpdf(ni,nj)= K*exp(-Q*aux1);
  end;
end;
%the aP PDF (P(n))
mu1=xe1(ns1); mu2=xe2(ns1);
C=P(:,:,ns1); D=det(C);
K=1/(2*pi*sqrt(D)); Q=(C(1,1)*C(2,2))/(2*D);
aPpdf=zeros(pN1,pN2); %space for the PDF
for ni=1:pN1,
  for nj=1:pN2,
    aux1=(((px1(ni)-mu1)^2)/C(1,1))...
    +(((px2(nj)-mu2).^2)/C(2,2))...
    -(((px1(ni)-mu1).*(px2(nj)-mu2)/C(1,2)*C(2,1)));
    aPpdf(ni,nj)= K*exp(-Q*aux1);
  end;
end;
%the M PDF
mu1=rxa(1,ns2); mu2=rxa(2,ns2);
C=M(:,:,ns2); D=det(C);
K=1/(2*pi*sqrt(D)); Q=(C(1,1)*C(2,2))/(2*D);
Mpdf=zeros(pN1,pN2); %space for the PDF
for ni=1:pN1,
  for nj=1:pN2,
    aux1=(((px1(ni)-mu1)^2)/C(1,1))...
    +(((px2(nj)-mu2).^2)/C(2,2))...
    -(((px1(ni)-mu1).*(px2(nj)-mu2)/C(1,2)*C(2,1)));
    Mpdf(ni,nj)= K*exp(-Q*aux1);
  end;
end;
%the Sv PDF
mu1=0; mu2=0;
C=Sv; D=det(C);
K=1/(2*pi*sqrt(D)); Q=(C(1,1)*C(2,2))/(2*D);
Svpdf=zeros(pN1,pN2); %space for the PDF
for ni=1:pN1,
  for nj=1:pN2,
    aux1=(((px1(ni)-mu1)^2)/C(1,1))...
    +(((px2(nj)-mu2).^2)/C(2,2));
    Svpdf(ni,nj)= K*exp(-Q*aux1);
  end;
end;
%the eY PDF (estimated y)
ya=rya(:,ns2); %estimated y
mu1=ya(1); mu2=ya(2);
C=cY(:,:,ns2); D=det(C);
K=1/(2*pi*sqrt(D)); Q=(C(1,1)*C(2,2))/(2*D);
eYpdf=zeros(pN1,pN2); %space for the PDF
for ni=1:pN1,
```

```
   for nj=1:pN2,
    aux1=(((px1(ni)-mu1)^2)/C(1,1))+(((px2(nj)-mu2).^2)/C(2,2));
    eYpdf(ni,nj)= K*exp(-Q*aux1);
   end;
end;
%the P PDF (P(n+1))
mu1=xe1(ns2); mu2=xe2(ns2);
C=P(:,:,ns2); D=det(C);
K=1/(2*pi*sqrt(D)); Q=(C(1,1)*C(2,2))/(2*D);
Ppdf=zeros(pN1,pN2); %space for the PDF
for ni=1:pN1,
  for nj=1:pN2,
    aux1=(((px1(ni)-mu1)^2)/C(1,1))...
    +(((px2(nj)-mu2).^2)/C(2,2))...
    -(((px1(ni)-mu1).*(px2(nj)-mu2)/C(1,2)*C(2,1)));
    Ppdf(ni,nj)= K*exp(-Q*aux1);
  end;
end;
%--------------------------------------------------
%display
figure(1)
subplot(1,2,1)
contour(px2,px1,Swpdf); hold on;
contour(px2,px1,aPpdf);
xlabel('x2'); ylabel('x1');
title('<process noise>, <xe(n)>');
subplot(1,2,2)
contour(px2,px1,Mpdf); hold on;
xlabel('x2'); ylabel('x1');
title('<xa(n+1)>');
figure(2)
subplot(1,2,1)
contour(px2,px1,Svpdf); hold on;
contour(px2,px1,Mpdf);
xlabel('x2'); ylabel('x1');
title('<measurement noise>, xa(n+1)>');
subplot(1,2,2)
contour(px2,px1,eYpdf); hold on;
plot(rym(2,ns1),rym(1,ns1),'r*','MarkerSize',12);
xlabel('x2'); ylabel('x1');
title('<estimated y>');
figure(3)
subplot(1,3,1)
contour(px2,px1,Mpdf);
xlabel('x2'); ylabel('x1');
title('<xa(n+1)>');
subplot(1,3,2)
contour(px2,px1,eYpdf);
xlabel('x2'); ylabel('x1');
```

```
title('<estimated y>');
subplot(1,3,3)
contour(px2,px1,Ppdf);
xlabel('x2'); ylabel('x1');
title('<xe(n+1)>');
```

B.1.3 Propagation and Nonlinearity (8.4.1.)

Figure B.10 shows a possible situation,when the original PDF, with $\sigma = 2$, is shifted 0.8 to the right.

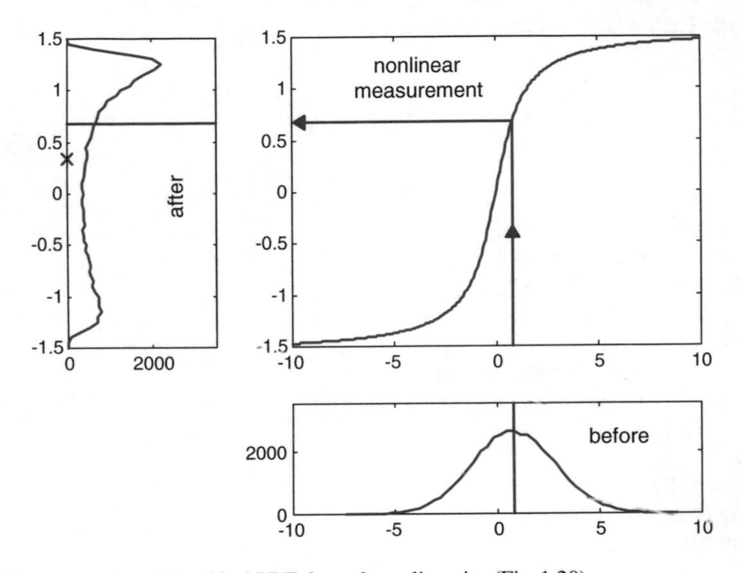

Fig. B.10 Propagation of a shifted PDF through nonlinearity (Fig. 1.20)

Program B.3 Propagation through nonlinearity

```
%Propagation through nonlinearity
% Asymmetrical result
%
%nonlinear function
vx=-10:0.1:10;
vy=atan(vx);
%gaussian random data
Nd=40000;
bx=-1.5:0.05:1.5; %location of histogram bins
sig=2;
hsf=0.8;
```

```
adat=hsf+sig*randn(1,Nd); %with horizontal shift
%histogram of a priori data
Nbins=50;
[ha,hax]=hist(adat,Nbins);
%propagated random data
pdat=atan(adat);
%histogram of posterior data
hpt=hist(pdat,bx);
mup=mean(pdat); %mean of posterior data
figure(1)
pl1=0.1; pb1=0.35; pw1=0.2; ph1=0.55;
pl2=0.4; pb2=0.05; pw2=0.55; ph2=0.2;
cnl=atan(hsf); %shifted arrow position
subplot('position',[pl1 pb1 pw1 ph1]) %left plot
plot(hpt,bx); hold on;
plot([0 3500],[cnl cnl],'r--');
plot(0,mup,'rx','Markersize',10);
axis([0 3500 -1.5 1.5]);
ylabel('after');
subplot('position',[pl2 pb1 pw2 ph1]) %central plot
plot(vx,vy); hold on;
plot([-10 hsf],[cnl cnl],'r--',-9.5,cnl,'r<');
plot([hsf hsf],[-1.5 cnl],'r--',hsf,-0.4,'r^');
title('nonlinear measurement');
subplot('position',[pl2 pb2 pw2 ph2]) %bottom plot
plot(hax,ha); hold on;
plot([hsf hsf],[0 3500],'r--');
axis([-10 10 0 3500]);
title('before');
```

B.1.4 Example of a Body Falling Towards Earth (8.4.4.)

Figures B.11 and B.12 show the evolution of state variables, measurements, and air drag during the fall under noisy conditions.

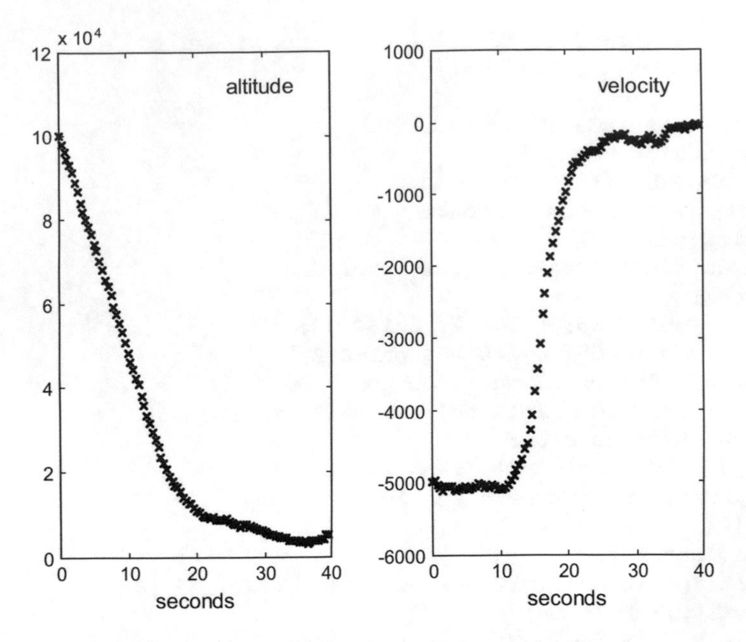

Fig. B.11 System states (*cross marks*) under noisy conditions (Fig. 1.27)

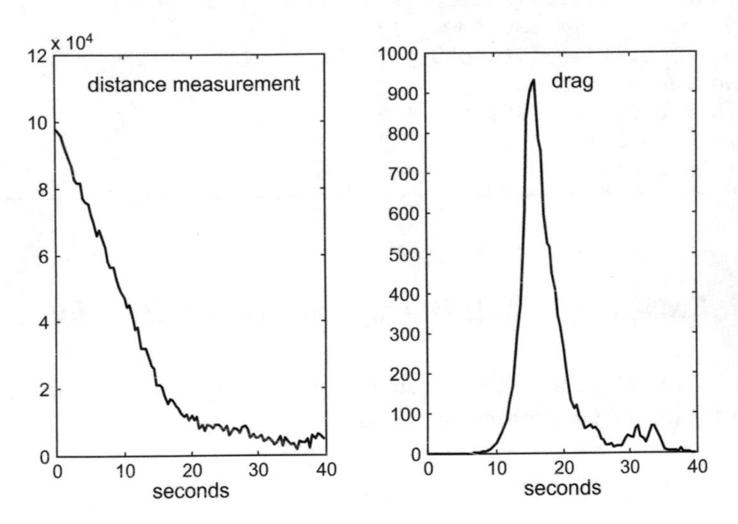

Fig. B.12 Distance measurement and drag (Fig. 1.28)

Program B.4 Influence of perturbations and noise

```
%Influence of perturbations and noise
%Radar monitoring of falling body
%--------------------------------------------------------
%Prepare for the simulation of the falling body
T=0.4; %sampling period
g=-9.81;
rho0=1.225; %air density, sea level
k=6705.6; %density vs. altitude constant
L=100; %horizontal distance radar$<->$object
L2=L^2;
Nf=100; %maximum number of samples
rx=zeros(3,Nf); %space for state record
rd=zeros(1,Nf); %space for drag record
ry=zeros(1,Nf); %space for measurement record
tim=0:T:(Nf-1)*T; %time
%process noise
Sw=[10^5 0 0; 0 10^3 0; 0 0 10^2]; %cov
bn=randn(3,Nf); sn=zeros(3,Nf);
sn(1,:)=sqrt(Sw(1,1))*bn(1,:); %state noise along simulation
sn(2,:)=sqrt(Sw(2,2))*bn(2,:); %" " "
sn(3,:)=sqrt(Sw(3,3))*bn(3,:); %" " "
%observation noise
Sv=10^6; %cov.
on=sqrt(Sv)*randn(1,Nf); %observation noise along simulation
x=[10^5; -5000; 400]; %initial state
%--------------------------------------------------------
%simulation
nn=1;
while nn<Nf+1,
%system
rx(:,nn)=x; %state recording
rho=rho0*exp(-x(1)/k); %air density
d=(rho*(x(2)^2))/(2*x(3)); %drag
rd(nn)=d; %drag recording
%next system state
x(1)=x(1)+(x(2)*T)+sn(1,nn);
x(2)=x(2)+((g+d)*T)+sn(2,nn);
x(3)=x(3)+sn(3,nn);
%system output
ym=on(nn)+sqrt(L2+(x(1)^2)); %measurement
ry(nn)=ym; %measurement recording
nn=nn+1;
end;
%--------------------------------------------------------
%display
figure(1)
subplot(1,2,1)
plot(tim,rx(1,1:Nf),'kx');
```

```
title('altitude'); xlabel('seconds')
axis([0 Nf*T 0 12*10^4]);
subplot(1,2,2)
plot(tim,rx(2,1:Nf),'kx');
title('velocity'); xlabel('seconds');
axis([0 Nf*T -6000 1000]);
figure(2)
subplot(1,2,1)
plot(tim,ry(1:Nf),'k');
title('distance measurement');
xlabel('seconds');
axis([0 Nf*T 0 12*10^4]);
subplot(1,2,2)
plot(tim,rd(1:Nf),'k');
title('drag');
xlabel('seconds');
axis([0 Nf*T 0 1000]);
```

B.1.5 Extended Kalman Filter (8.5.)

Next figures show the EKF results for the falling body in noisy conditions (Figs. B.13, B.14, B.15, B.16, B.17 and B.18).

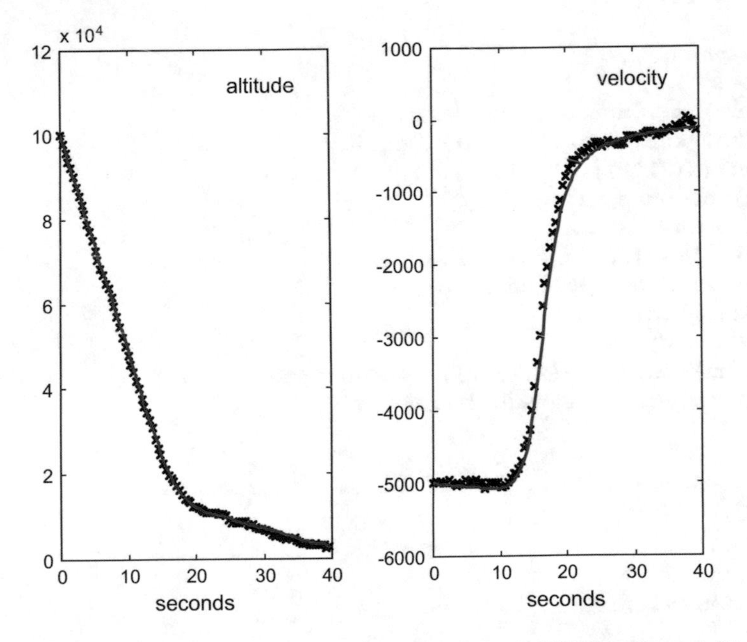

Fig. B.13 System states (*cross marks*), and states estimated by the EKF (*continuous*) (Fig. 1.29)

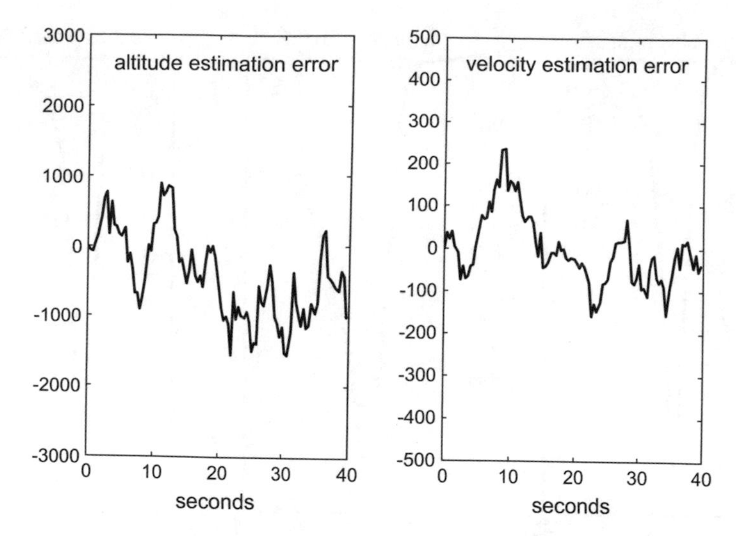

Fig. B.14 Error evolution (Fig. 1.30)

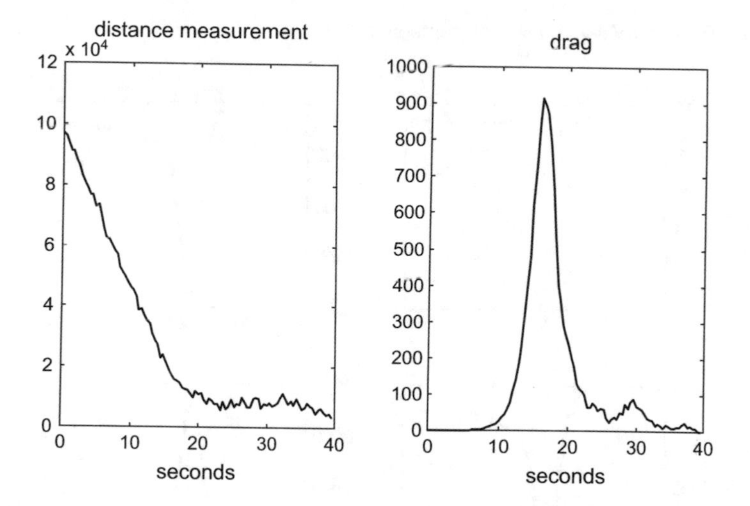

Fig. B.15 Distance measurement and drag

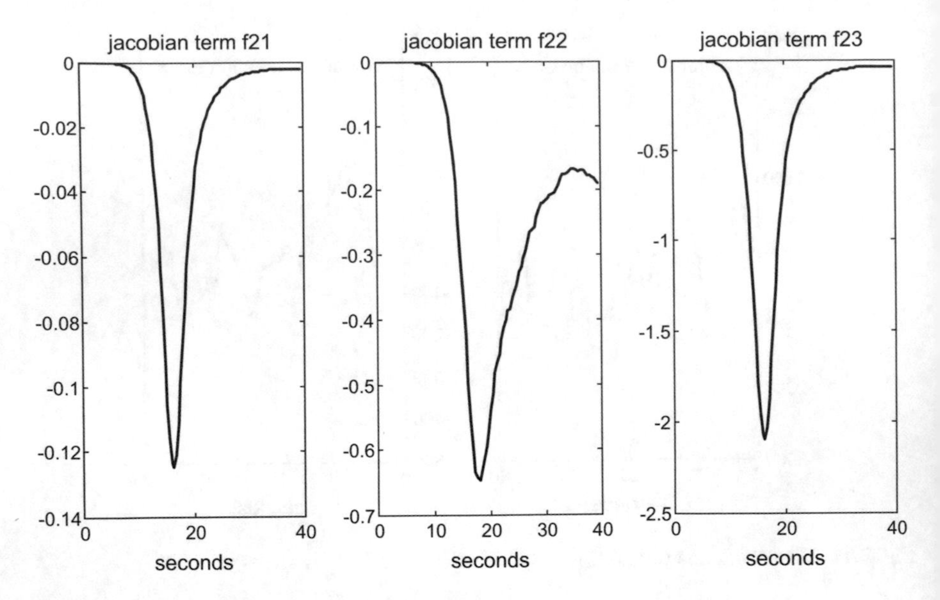

Fig. B.16 The three non-zero jacobian components

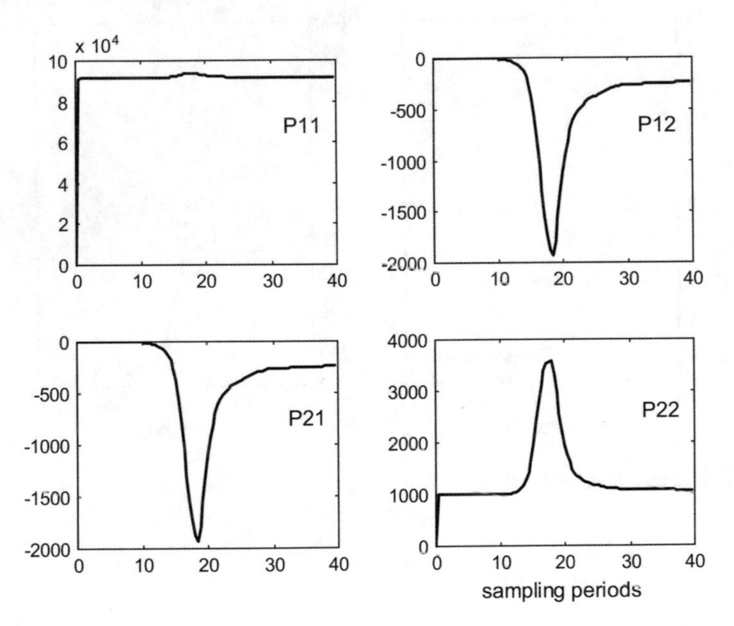

Fig. B.17 Evolution of matrix P (Fig. 1.31)

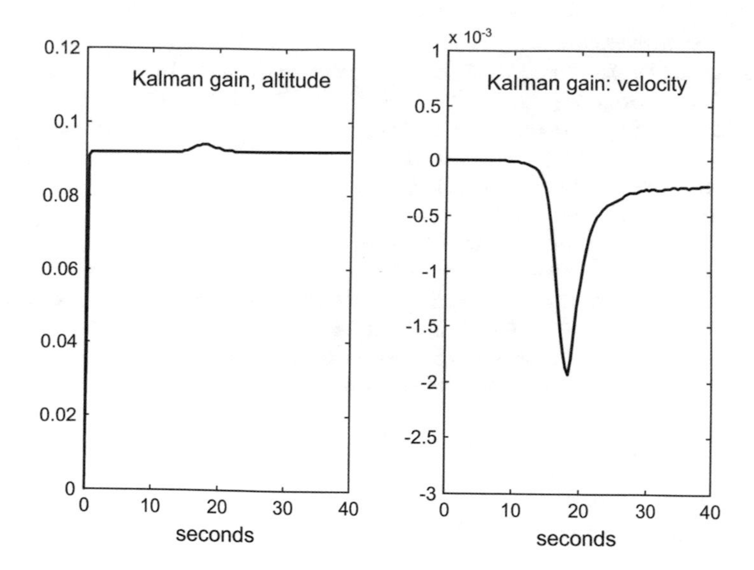

Fig. B.18 Evolution of the altitude and velocity Kalman gains (Fig. 1.32)

Program B.5 Extended Kalman filter example

```
%Extended Kalman filter example
%Radar monitoring of falling body
%-----------------------------------------------------
%Prepare for the simulation of the falling body
T=0.4; %sampling period
g=-9.81;
rho0=1.225; %air density, sea level
k=6705.6; %density vs. altitude constant
L=100; %horizontal distance radar$<->$object
L2=L^2;
Nf=100; %maximum number of samples
rx=zeros(3,Nf); %space for state record
rd=zeros(1,Nf); %space for drag record
ry=zeros(1,Nf); %space for measurement record
tim=0:T:(Nf-1)*T; %time
%process noise
Sw=[10^5 0 0; 0 10^3 0; 0 0 10^2]; %cov
bn=randn(3,Nf); sn=zeros(3,Nf);
sn(1,:)=sqrt(Sw(1,1))*bn(1,:); %state noise along simulation
sn(2,:)=sqrt(Sw(2,2))*bn(2,:); %" " "
sn(3,:)=sqrt(Sw(3,3))*bn(3,:); %" " "
%observation noise
Sv=10^6; %cov.
on=sqrt(Sv)*randn(1,Nf); %observation noise along simulation
%-----------------------------------------------------
%Prepare for filtering
```

```
%space for matrices
K=zeros(3,Nf); M=zeros(3,3,Nf); P=zeros(3,3,Nf);
%space for recording er(n), xe(n)
rer=zeros(3,Nf); rxe=zeros(3,Nf);
%space for recording jacobians
rf2=zeros(3,Nf); rh1=zeros(1,Nf);
W=eye(3,3); %process noise jacobian
V=1; %observation noise jacobian
%------------------------------------------------------
%Behaviour of the system and the filter after initial state
x=[10^5; -5000; 400]; %initial state
xe=x; % initial values of filter state
xa=xe; %initial intermediate state
nn=1;
while nn<Nf+1,
%estimation recording
rxe(:,nn)=xe; %state
rer(:,nn)=x-xe; %error
%system
rx(:,nn)=x; %state recording
rho=rho0*exp(-x(1)/k); %air density
d=(rho*(x(2)^2))/(2*x(3)); %drag
rd(nn)=d; %drag recording
%next system state
x(1)=x(1)+(x(2)*T)+sn(1,nn);
x(2)=x(2)+((g+d)*T)+sn(2,nn);
x(3)=x(3)+sn(3,nn);
%system output
y=on(nn)+sqrt(L2+(x(1)^2));
ym=y; %measurement
ry(nn)=ym; %measurement recording
%Prediction
%a priori state
rho=rho0*exp(-xe(1)/k); %air density
d=(rho*(xe(2)^2))/(2*xe(3)); %drag
xa(1)=xe(1)+(xe(2)*T);
xa(2)=xe(2)+((g+d)*T);
xa(3)=xe(3);
%a priori cov.
f21=-d/k; f22=(rho*xe(2)/xe(3)); f23=-(d/xe(3));
F=[0 1 0; f21 f22 f23; 0 0 0]; %state jacobian
M(:,:,nn+1)=(F*P(:,:,nn)*F')+ (W*Sw*W');
%
%Update
ya=sqrt(L2+xa(1)^2);
h1=xa(1)/ya;
H=[h1 0 0]; %measurement jacobian
K(:,nn+1)=(M(:,:,nn+1)*H')*inv((H*M(:,:,nn+1)*H')+(V*Sv*V'));
P(:,:,nn+1)=M(:,:,nn+1)-(K(:,nn+1)*H*M(:,:,nn+1));
```

```
xe=xa+(K(:,nn+1)*(ym-ya)); %estimated (a posteriori) state
%jacobian recording
rf2(:,nn)=[f21; f22; f23]; rh1(nn)=h1;
nn=nn+1;
end;
%-------------------------------------------------------
%display
figure(1)
subplot(1,2,1)
plot(tim,rx(1,1:Nf),'kx'); hold on;
plot(tim,rxe(1,1:Nf),'r');
title('altitude'); xlabel('seconds')
axis([0 Nf*T 0 12*10^4]);
subplot(1,2,2)
plot(tim,rx(2,1:Nf),'kx'); hold on;
plot(tim,rxe(2,1:Nf),'r');
title('velocity'); xlabel('seconds');
axis([0 Nf*T -6000 1000]);
figure(2)
subplot(1,2,1)
plot(tim,rer(1,1:Nf),'k');
title('altitude estimation error');
xlabel('seconds');
axis([0 Nf*T -3000 3000]);
subplot(1,2,2)
plot(tim,rer(2,1:Nf),'k');
title('velocity estimation error');
xlabel('seconds');
axis([0 Nf*T -500 500]);
figure(3)
subplot(1,2,1)
plot(tim,ry(1:Nf),'k');
title('distance measurement');
xlabel('seconds');
axis([0 Nf*T 0 12*10^4]);
subplot(1,2,2)
plot(tim,rd(1:Nf),'k');
title('drag');
xlabel('seconds');
axis([0 Nf*T 0 1000]);
figure(4)
subplot(1,3,1)
plot(tim,rf2(1,1:Nf),'k');
title('jacobian term f21');
xlabel('seconds');
subplot(1,3,2)
plot(tim,rf2(2,1:Nf),'k');
title('jacobian term f22');
xlabel('seconds');
```

```
subplot(1,3,3)
plot(tim,rf2(3,1:Nf),'k');
title('jacobian term f23');
xlabel('seconds');
figure(5)
subplot(1,2,1)
plot(tim,K(1,1:Nf),'k');
title('Kalman gain, altitude');
xlabel('seconds');
axis([0 Nf*T 0 0.12]);
subplot(1,2,2)
plot(tim,K(2,1:Nf),'k');
title('Kalman gain: velocity');
xlabel('seconds');
axis([0 Nf*T -0.003 0.001]);
figure(6)
% display of P(n) evolution
subplot(2,2,1); plot(tim,shiftdim(P(1,1,1:Nf)),'k');
title('P11');%plots P11
subplot(2,2,2); plot(tim,shiftdim(P(1,2,1:Nf)),'k');
title('P12');%plots P12
subplot(2,2,3); plot(tim,shiftdim(P(2,1,1:Nf)),'k');
title('P21');%plots P21
subplot(2,2,4); plot(tim,shiftdim(P(2,2,1:Nf)),'k');
title('P22'); %plots P22
xlabel('sampling periods');
```

B.1.6 The Unscented Transform (8.6.1.)

With the propagated sigma points it is possible to compute the mean and variance of the propagated data (Fig. B.19).

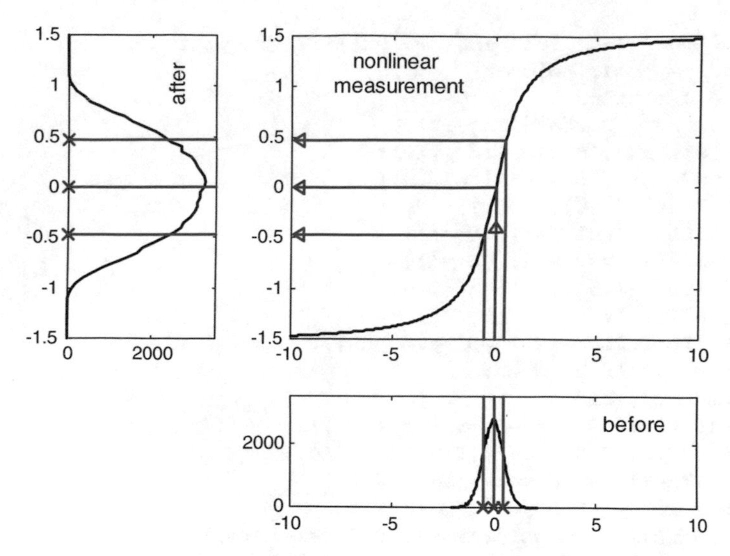

Fig. B.19 Propagation of sigma points (Fig. 1.38)

Program B.6 Propagation of sigma points through nonlinearity

```
%Propagation of sigma points through nonlinearity
%
%nonlinear function
vx=-10:0.1:10;
vy=atan(vx);
%gaussian random data
Nd=80000;
bx=-1.5:0.05:1.5; %location of histogram bins
sig=0.5;
adat=sig*randn(1,Nd);
%histogram of a priori data
Nbins=100;
[ha,hax]=hist(adat,Nbins);
%sigma points:
xs0=0;
xs1=xs0+sig;
xs2=xs0-sig;
%propagated random data
pdat=atan(adat);
%histogram of posterior data
hpt=hist(pdat,bx);
%measured sigma points
ys0=atan(xs0);
ys1=atan(xs1);
ys2=atan(xs2);
figure(1)
pl1=0.1; pb1=0.35; pw1=0.2; ph1=0.55;
pl2=0.4; pb2=0.05; pw2=0.55; ph2=0.2;
```

```
subplot('position',[pl1 pb1 pw1 ph1]) %left plot
plot(hpt,bx,'k'); hold on;
%sigma projections
plot(0,0,'rx','MarkerSize',10);
plot(0,ys1,'rx','MarkerSize',10);
plot(0,ys2,'rx','MarkerSize',10);
plot([0 3500],[0 0],'r--');
plot([0 3500],[ys1 ys1],'b--');
plot([0 3500],[ys2 ys2],'b--');
axis([0 3500 -1.5 1.5]);
ylabel('after');
subplot('position',[pl2 pb1 pw2 ph1]) %central plot
plot(vx,vy,'k'); hold on;
%sigma projections
plot([-10 0],[0 0],'r--',-9.5,0,'r<');
plot([0 0],[-1.5 0],'r--',0,-0.4,'r^');
plot([xs1 xs1],[-1.5 ys1],'b--');
plot([xs2 xs2],[-1.5 ys2],'b--');
plot([-10 xs1],[ys1 ys1],'b--',-9.5,ys1,'b<');
plot([-10 xs2],[ys2 ys2],'b--',-9.5,ys2,'b<');
title('Propagation of sigma points through nonlinear measurement');
subplot('position',[pl2 pb2 pw2 ph2]) %bottom plot
plot(hax,ha,'k'); hold on;
%sigma points
plot(0,0,'rx','MarkerSize',10);
plot(xs1,0,'rx','MarkerSize',10);
plot(xs2,0,'rx','MarkerSize',10);
plot([0 0],[0 3500],'r--');
plot([xs1 xs1],[0 3500],'b--');
plot([xs2 xs2],[0 3500],'b--');
axis([-10 10 0 3500]);
title('before');
```

In order to illustrate the UT steps, the example of satellite tracking is considered. A set of five sigma points have been selected. They are shown with cross marks (Figs. B.20, B.21 and B.22).

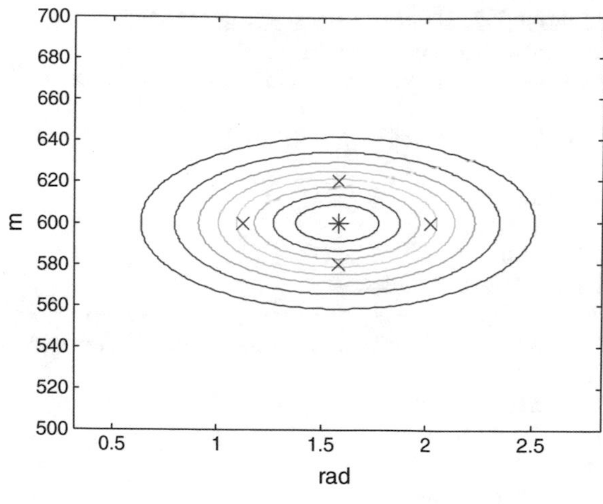

Fig. B.20 Uncertainty on the angle-radius plane (satellite example), and sigma points (Fig. 1.39)

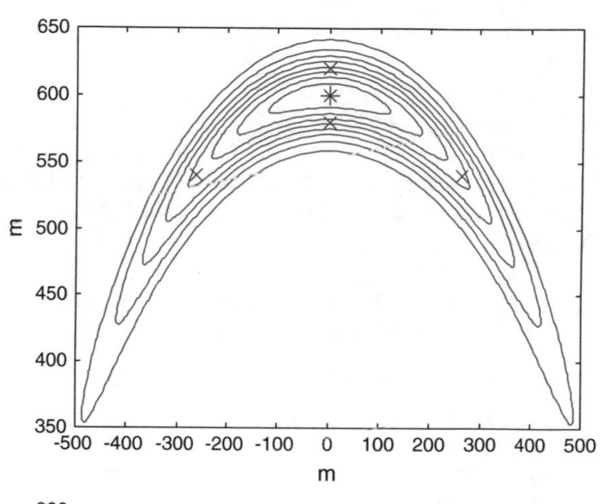

Fig. B.21 The uncertainty on the Cartesian plane, and sigma points (Fig. 1.40)

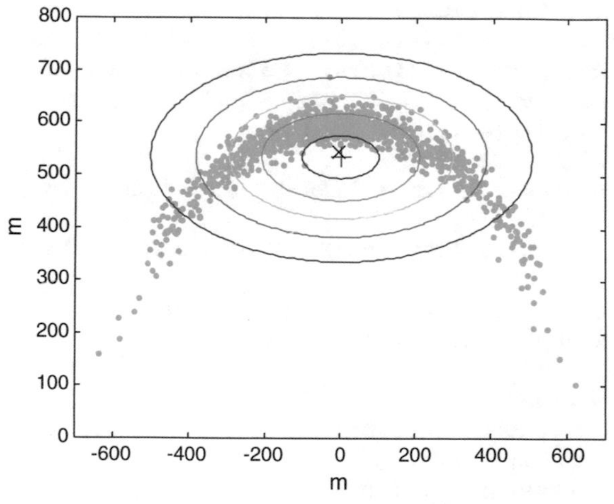

Fig. B.22 The UT approximation and propagated data points (Fig. 1.41)

Program B.7 Unscented Transformation example

```
%Unscented Transformation example
% from polar to Cartesian coordinates
%
%reference position at T0
r0=600;
alpha0=pi/2;
%variances
sigr=20;
siga=0.45; %radians
%-------------------------------------------------------
%bivariate Gaussian PDF on the alpha-r plane
al=0.4*pi;
da=-al:(al/50):al;
pN1=length(da);
dr=-100:1:100;
pN2=length(dr);
r=r0+dr;
a=alpha0+da;
C=[siga^2 0; 0 sigr^2];
mu1=alpha0; mu2=r0;
D=det(C);
K=1/(2*pi*sqrt(D)); Q=(C(1,1)*C(2,2))/(2*D);
p=zeros(pN1,pN2); %space for the PDF
for ni=1:pN1,
  for nj=1:pN2,
    aux1=(((a(ni)-mu1)^2)/C(1,1))+(((r(nj)-mu2).^2)/C(2,2));
    p(ni,nj)= K*exp(-Q*aux1);
  end;
end;
%sigma points
mu1=alpha0; mu2=r0;
xs=zeros(2,6); %reserve space
xs0=[mu1; mu2];
xs(:,1)=xs0+[siga;0]; xs(:,2)=xs0+[0;sigr];
xs(:,3)=xs0-[siga;0]; xs(:,4)=xs0-[0;sigr];
figure(1)
contour(a,r,p'); hold on;
%sigma points
plot(xs0(1),xs0(2),'r*','MarkerSize',12);
for nn=1:4,
  plot(xs(1,nn),xs(2,nn),'rx','MarkerSize',12);
end;
title('Original sigma points and PDF');
xlabel('rad'); ylabel('m');
%-------------------------------------------------------
%Transform PDF from alpha-r plane to x-y plane
MC=contourc(a,r,p');
auxA=1; aux0=0;
ncs=zeros(1,8); csa=zeros(8,500); csr=zeros(8,500);
for nn=1:8,
```

```
    aux0=MC(2,auxA); %number of segments in contour leveln
    ncs(nn)=aux0;
    %select the leveln segments:
    csa(nn,1:aux0)=MC(1,(auxA+1):auxA+aux0);
    csr(nn,1:aux0)=MC(2,(auxA+1):auxA+aux0);
    auxA=auxA+aux0+1;
end;
%transform sigma points
muy1=mu2*cos(mu1);
muy2=mu2*sin(mu1);
ys=zeros(2,6); %reserve space
ys0=[muy1; muy2];
for nn=1:4,
   ys(1,nn)=xs(2,nn)*cos(xs(1,nn));
   ys(2,nn)=xs(2,nn)*sin(xs(1,nn));
end;
figure(2)
for m=1:8,
   L=ncs(m);
   sm=zeros(2,L);
   for nn=1:L,
      axa=csa(m,nn); axr=csr(m,nn);
      sm(1,nn)=axr*cos(axa);
      sm(2,nn)=axr*sin(axa);
   end;
   plot(sm(1,:),sm(2,:),'b'); hold on;
end;
%transformed sigma points
plot(ys0(1),ys0(2),'r*','MarkerSize',12); hold on;
for nn=1:4,
   plot(ys(1,nn),ys(2,nn),'rx','MarkerSize',12);
end;
title('Propagated sigma points and PDF');
xlabel('m'); ylabel('m');
%------------------------------------------------------
% Use sigma points
%UKF parameters (to be edited here)
N=2; %space dimension
alpha=0.55; kappa=1; beta=2;
%pre-computation of constants
lambda= ((alpha^2)*(N+kappa))-N;
aab=(1-(alpha^2)+beta);
lN=lambda+N; LaN=lambda/lN; aaN=aab+LaN;
%mean: weighted average
ya=[0;0];
for nn=1:4,
   ya=ya+ys(:,nn);
end;
ya=ya/(2*lN);
ya=ya+(LaN*ys0);
%covariance: weighted average
```

```
aux=zeros(2,2); aux1=zeros(2,2);
for nn=1:4,
  aux=aux+(ys(:,nn)-ya(:))*(ys(:,nn)-ya(:))';
end;
aux=aux/(2*lN);
aux1=(ys0-ya)*(ys0-ya)';
Py=aux+(aaN*aux1);
%the Gaussian PDF approximation
C=Py;
mu1=ya(1); mu2=ya(2);
D=det(C);
K=1/(2*pi*sqrt(D)); Q=(C(1,1)*C(2,2))/(2*D);
x1=-700:10:700;
x2=0:10:900;
pN1=length(x1); pN2=length(x2);
yp=zeros(pN1,pN2); %space for the PDF
for ni=1:pN1,
  for nj=1:pN2,
    aux1=(((x1(ni)-mu1)^2)/C(1,1))+(((x2(nj)-mu2).^2)/C(2,2));
    yp(ni,nj)= K*exp(-Q*aux1);
  end;
end;
%Set of random measurements
Np=1000; %number of random points
px=zeros(1,Np);
py=zeros(1,Np);
nr=sigr*randn(1,Np);
na=siga*randn(1,Np);
for nn=1:Np,
  r=r0+nr(nn);
  a=alpha0+na(nn);
  px(nn)=r*cos(a);
  py(nn)=r*sin(a);
end;
xmean=sum(px/Np);
ymean=sum(py/Np);
figure(3)
plot(px,py,'g.'); hold on; %the points
contour(x1,x2,yp'); %the UT PDF approximation
plot(ya(1),ya(2),'b+', 'MarkerSize',12); %the PDF center
plot(xmean,ymean,'kx', 'MarkerSize',12); %the data mean
title('Some propagated data points, and the PDF approximation by
UT');
xlabel('m'); ylabel('m'); axis([-700 700 0 800]);
```

B.1.7 *The Unscented Kalman Filter (UKF) (8.6.2.)*

The example of the falling body is used to illustrate the UKF algorithm (Figs. B.23, B.24, B.25, B.26 and B.27).

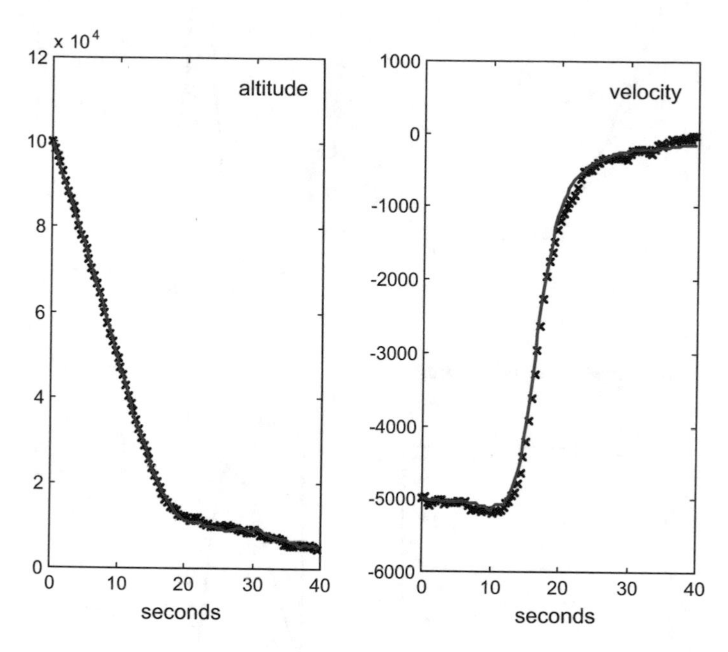

Fig. B.23 System states (*cross marks*), and states estimated by the UKF (*continuous*) (Fig. 1.42)

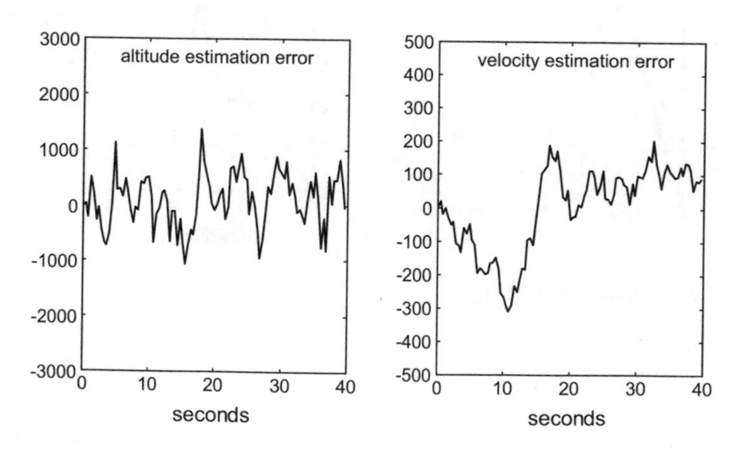

Fig. B.24 Error evolution (Fig. 1.43)

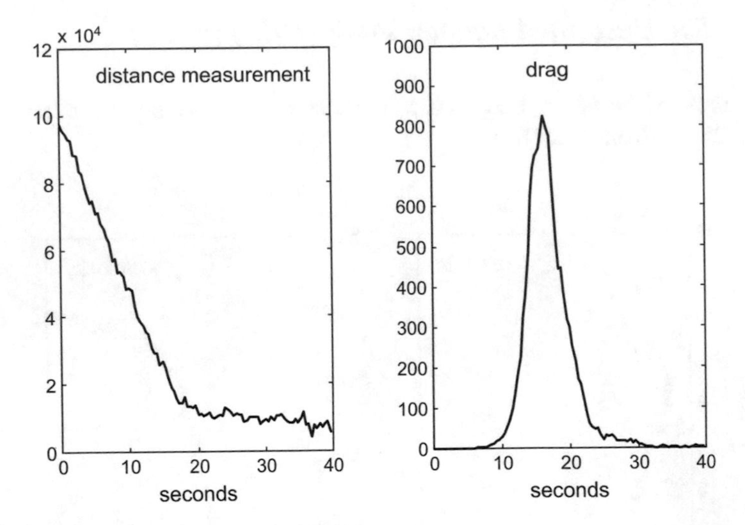

Fig. B.25 Distance measurement and drag

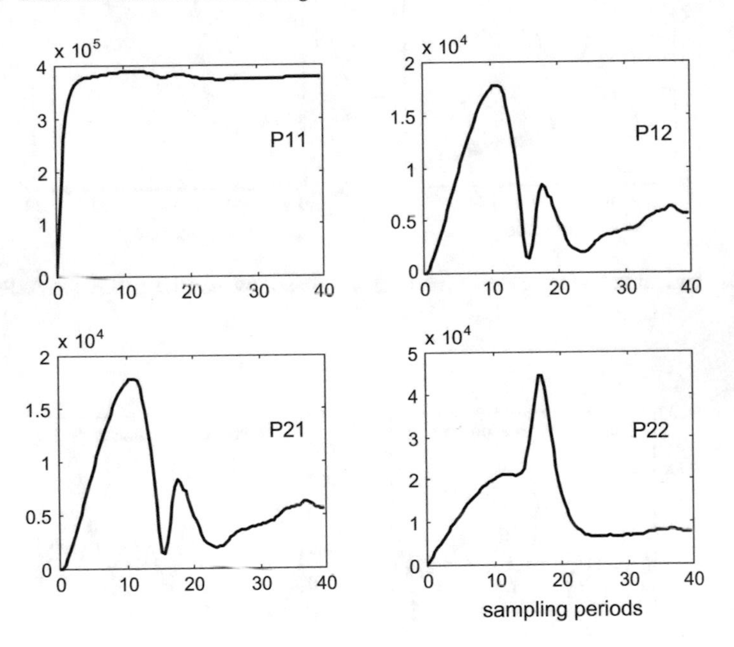

Fig. B.26 Evolution of matrix P (Fig. 1.44)

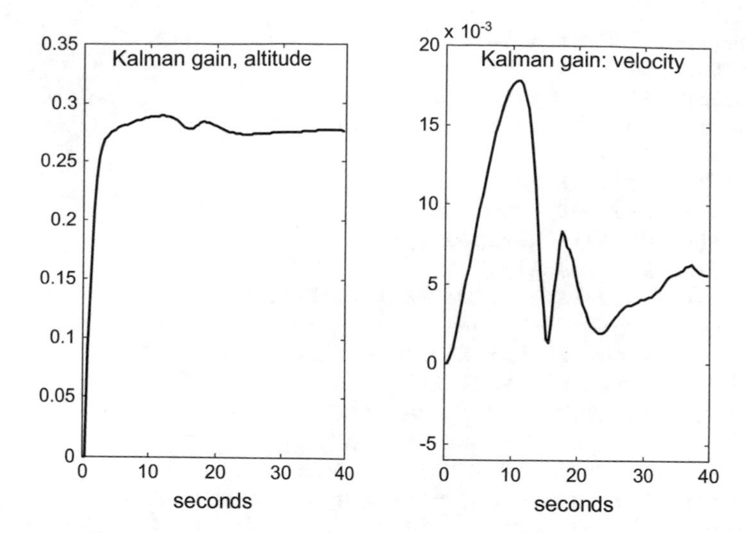

Fig. B.27 Evolution of the altitude and velocity Kalman gains (Fig. 1.45)

Program B.8 Unscented Kalman filter example

```
%Unscented Kalman filter example
%Radar monitoring of falling body
%--------------------------------------------------
%Prepare for the simulation of the falling body
T=0.4; %sampling period
g=-9.81;
rho0=1.225; %air density, sea level
k=6705.6; %density vs. altitude constant
L=100; %horizontal distance radar$<->$object
L2=L^2;
Nf=100; %maximum number of samples
rx=zeros(3,Nf); %space for state record
rd=zeros(1,Nf); %space for drag record
ry=zeros(1,Nf); %space for measurement record
tim=0:T:(Nf-1)*T; %time
%process noise
Sw=[10^5 0 0; 0 10^3 0; 0 0 10^2]; %cov
bn=randn(3,Nf); sn=zeros(3,Nf);
sn(1,:)=sqrt(Sw(1,1))*bn(1,:); %state noise along simulation
sn(2,:)=sqrt(Sw(2,2))*bn(2,:); %" " "
sn(3,:)=sqrt(Sw(3,3))*bn(3,:); %" " "
%observation noise
Sv=10^6; %cov.
on=sqrt(Sv)*randn(1,Nf); %observation noise along simulation
%--------------------------------------------------
%Prepare for filtering
%space for matrices
```

```
K=zeros(3,Nf); M=zeros(3,3,Nf); P=zeros(3,3,Nf);
%space for recording er(n), xe(n)
rer=zeros(3,Nf); rxe=zeros(3,Nf);
%UKF parameters (to be edited here)
N=3; %space dimension
alpha=0.7; kappa=0; beta=2;
%pre-computation of constants
lambda= ((alpha^2)*(N+kappa))-N;
aab=(1-(alpha^2)+beta);
lN=lambda+N; LaN=lambda/lN; aaN=aab+LaN;
%------------------------------------------------------
%Behaviour of the system and the filter after initial state
x=[10^5; -5000; 400]; %initial state
xe=x; % initial value of filter state
xa=xe; %initial intermediate state
xs=zeros(3,7); %space for sigma points
xas=zeros(3,7); %space for propagated sigma points
yas=zeros(1,7); %" " "
P(:,:,1)=0.001*eye(3,3); %cov. non-zero init.
nn=1;
while nn<Nf+1,
  %estimation recording
  rxe(:,nn)=xe; %state
  rer(:,nn)=x-xe; %error
  %system
  rx(:,nn)=x; %state recording
  rho=rho0*exp(-x(1)/k); %air density
  d=(rho*(x(2)^2))/(2*x(3)); %drag
  rd(nn)=d; %drag recording
  %next system state
  x(1)=x(1)+(x(2)*T)+sn(1,nn);
  x(2)=x(2)+((g+d)*T)+sn(2,nn);
  x(3)=x(3)+sn(3,nn);
  %system output
  y=on(nn)+sqrt(L2+(x(1)^2));
  ym=y; %measurement
  ry(nn)=ym; %measurement recording
  %Prediction
  %sigma points
  sqP=chol(lN*P(:,:,nn)); %matrix square root
  xs(:,7)=xe;
  xs(:,1)=xe+sqP(1,:)'; xs(:,2)=xe+sqP(2,:)';
  xs(:,3)=xe+sqP(3,:)';
  xs(:,4)=xe-sqP(1,:)'; xs(:,5)=xe-sqP(2,:)';
  xs(:,6)=xe-sqP(3,:)';
  %a priori state
  %propagation of sigma points (state transition)
  for m=1:7,
    rho=rho0*exp(-xs(1,m)/k); %air density
```

```
    d=(rho*(xs(2,m)^2))/(2*xs(3,m)); %drag
    xas(1,m)=xs(1,m)+(xs(2,m)*T);
    xas(2,m)=xs(2,m)+((g+d)*T);
    xas(3,m)=xs(3,m);
  end;
  %a priori state mean (a weighted sum)
  xa=0;
  for m=1:6,
    xa=xa+(xas(:,m));
  end;
  xa=xa/(2*lN);
  xa=xa+(LaN*xas(:,7));
  %a priori cov.
  aux=zeros(3,3); aux1=zeros(3,3);
  for m=1:6,
    aux=aux+((xas(:,m)-xa(:))*(xas(:,m)-xa(:))');
  end;
  aux=aux/(2*lN);
  aux1=((xas(:,7)-xa(:))*(xas(:,7)-xa(:))');
  aux=aux+(aaN*aux1);
  M(:,:,nn+1)=aux+Sw;
  %Update
  %propagation of sigma points (measurement)
  for m=1:7,
    yas(m)=sqrt(L2+(xas(1,m)^2));
  end;
  %measurement mean
  ya=0;
  for m=1:6,
    ya=ya+yas(m);
  end;
  ya=ya/(2*lN);
  ya=ya+(LaN*yas(7));
  %measurement cov.
  aux2=0;
  for m=1:6,
    aux2=aux2+((yas(m)-ya)^2);
  end;
  aux2=aux2/(2*lN);
  aux2=aux2+(aaN*((yas(7)-ya)^2));
  Syy=aux2+Sv;
  %cross cov
  aux2=0;
  for m=1:6,
    aux2=aux2+((xas(:,m)-xa(:))*(yas(m)-ya));
  end;
  aux2=aux2/(2*lN);
  aux2=aux2+(aaN*((xas(:,7)-xa(:))*(yas(7)-ya)));
  Sxy=aux2;
```

```
  %Kalman gain, etc.
  K(:,nn+1)=Sxy*inv(Syy);
  P(:,:,nn+1)=M(:,:,nn+1)-(K(:,nn+1)*Syy*K(:,nn+1)');
  xe=xa+(K(:,nn+1)*(ym-ya)); %estimated (a posteriori) state
  nn=nn+1;
end;
%-------------------------------------------------
%display
figure(1)
subplot(1,2,1)
plot(tim,rx(1,1:Nf),'kx'); hold on;
plot(tim,rxe(1,1:Nf),'r');
title('altitude'); xlabel('seconds')
axis([0 Nf*T 0 12*10^4]);
subplot(1,2,2)
plot(tim,rx(2,1:Nf),'kx'); hold on;
plot(tim,rxe(2,1:Nf),'r');
title('velocity'); xlabel('seconds');
axis([0 Nf*T -6000 1000]);
figure(2)
subplot(1,2,1)
plot(tim,rer(1,1:Nf),'k');
title('altitude estimation error');
xlabel('seconds');
axis([0 Nf*T -3000 3000]);
subplot(1,2,2)
plot(tim,rer(2,1:Nf),'k');
title('velocity estimation error');
xlabel('seconds');
axis([0 Nf*T -500 500]);
figure(3)
subplot(1,2,1)
plot(tim,ry(1:Nf),'k');
title('distance measurement');
xlabel('seconds');
axis([0 Nf*T 0 12*10^4]);
subplot(1,2,2)
plot(tim,rd(1:Nf),'k');
title('drag');
xlabel('seconds');
axis([0 Nf*T 0 1000]);
figure(4)
subplot(1,2,1)
plot(tim,K(1,1:Nf),'k');
title('Kalman gain, altitude');
xlabel('seconds');
axis([0 Nf*T 0 0.35]);
subplot(1,2,2)
plot(tim,K(2,1:Nf),'k');
```

```
title('Kalman gain: velocity');
xlabel('seconds');
axis([0 Nf*T -0.006 0.02]);
figure(5)
% display of P(n) evolution
subplot(2,2,1); plot(tim,shiftdim(P(1,1,1:Nf)),'k');
title('P11');%plots P11
subplot(2,2,2); plot(tim,shiftdim(P(1,2,1:Nf)),'k');
title('P12');%plots P12
subplot(2,2,3); plot(tim,shiftdim(P(2,1,1:Nf)),'k');
title('P21');%plots P21
subplot(2,2,4); plot(tim,shiftdim(P(2,2,1:Nf)),'k');
title('P22'); %plots P22
xlabel('sampling periods');
```

B.1.8 Particle Filters (8.7.)

Using a particle filter for the falling body example (Figs. B.28, B.29 and B.30).

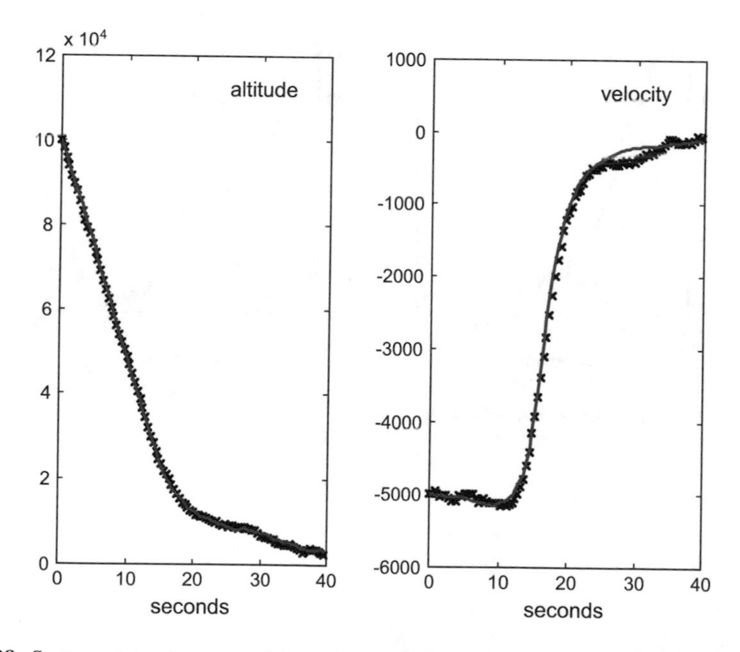

Fig. B.28 System states (*cross marks*), and states estimated by the particle filter (*continuous*) (Fig. 1.46)

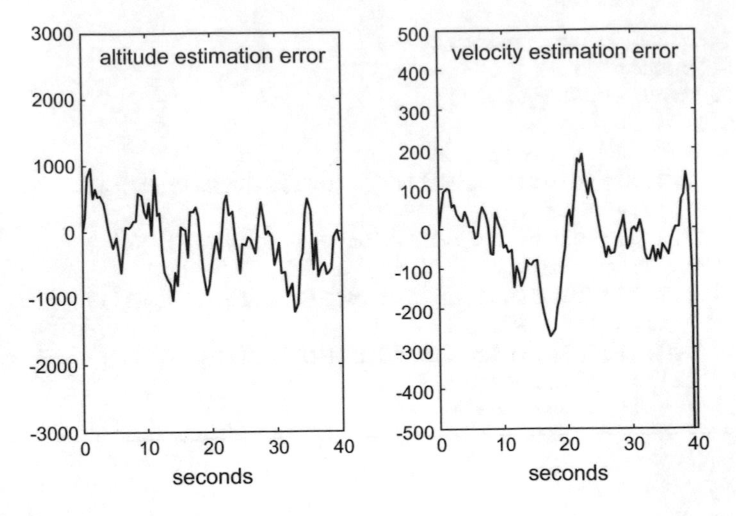

Fig. B.29 Error evolution (Fig. 1.47)

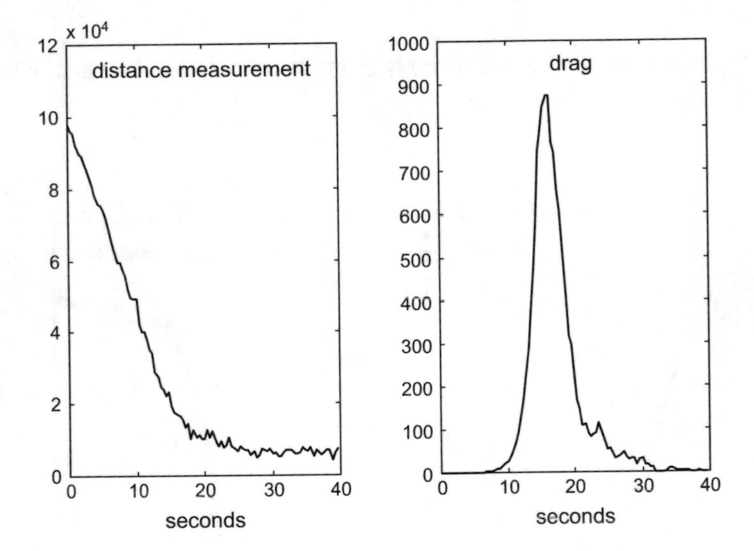

Fig. B.30 Distance measurement and drag

Program B.9 Particle filter example

```
%Particle filter example
%Radar monitoring of falling body
disp('please wait a bit');
%---------------------------------------------------------
%Prepare for the simulation of the falling body
T=0.4; %sampling period
g=-9.81;
rho0=1.225; %air density, sea level
k=6705.6; %density vs. altitude constant
L=100; %horizontal distance radar$<->$object
L2=L^2;
Nf=100; %maximum number of steps
rx=zeros(3,Nf); %space for state record
rd=zeros(1,Nf); %space for drag record
ry=zeros(1,Nf); %space for measurement record
tim=0:T:(Nf-1)*T; %time
%process noise
Sw=[10^5 0 0; 0 10^3 0; 0 0 10^2]; %cov
w11=sqrt(Sw(1,1)); w22=sqrt(Sw(2,2)); w33=sqrt(Sw(3,3));
w=[w11; w22; w33];
%observation noise
Sv=10^6; %cov.
v11=sqrt(Sv);
%---------------------------------------------------------
%Prepare for filtering
%space for recording er(n), xe(n)
rer=zeros(3,Nf); rxe=zeros(3,Nf);
%---------------------------------------------------------
%Behaviour of the system and the filter after initial state
x=[10^5; -5000; 400]; %initial state
xe=x; %initial estimation
%prepare particles
Np=1000; %number of particles
%reserve space
px=zeros(3,Np); %particles
apx=zeros(3,Np); %a priori particles
ny=zeros(1,Np); %particle measurements
vy=zeros(1,Np); %meas. dif.
pq=zeros(1,Np); %particle likelihoods
%particle generation
wnp=randn(3,Np); %noise (initial particles)
for ip=1:Np,
  px(:,ip)=x+(w.*wnp(:,ip)); %initial particles
end;
%system noises
wx=randn(3,Nf); %process
wy=randn(1,Nf); %output
nn=1;
```

```
while nn<Nf+1,
  %estimation recording
  rxe(:,nn)=xe; %state
  rer(:,nn)=x-xe; %error
  %Simulation of the system
  %system
  rx(:,nn)=x; %state recording
  rho=rho0*exp(-x(1)/k); %air density
  d=(rho*(x(2)^2))/(2*x(3)); %drag
  rd(nn)=d; %drag recording
  %next system state
  x(1)=x(1)+(x(2)*T);
  x(2)=x(2)+((g+d)*T);
  x(3)=x(3);
  x=x+(w.*wx(:,nn)); %additive noise
  %system output
  y=sqrt(L2+(x(1)^2))+(v11*wy(nn)); %additive noise
  ym=y; %measurement
  ry(nn)=ym; %measurement recording
  %Particle propagation
  wp=randn(3,Np); %noise (process)
  vm=randn(1,Np); %noise (measurement)
  for ip=1:Np,
    rho=rho0*exp(-px(1,ip)/k); %air density
    d=(rho*(px(2,ip)^2))/(2*px(3,ip)); %drag
    %next state
    apx(1,ip)=px(1,ip)+(px(2,ip)*T);
    apx(2,ip)=px(2,ip)+((g+d)*T);
    apx(3,ip)=px(3,ip);
    apx(:,ip)=apx(:,ip)+(w.*wp(:,ip)); %additive noise
    %measurement (for next state)
    ny(ip)=sqrt(L2+(apx(1,ip)^2))+(v11*vm(ip)); %additive noise
    vy(ip)=ym-ny(ip);
  end;
  %Likelihood
  %(vectorized part)
  %scaling
  vs=max(abs(vy))/4;
  ip=1:Np;
  pq(ip)=exp(-((vy(ip)/vs).^2));
  spq=sum(pq);
  %normalization
  pq(ip)=pq(ip)/spq;
  %Prepare for roughening
  A=(max(apx')-min(apx'))';
  sig=0.2*A*Np^(-1/3);
  rn=randn(3,Np); %random numbers
  %==========================================================
  %Resampling (systematic)
```

```
    acq=cumsum(pq);
    cmb=linspace(0,1-(1/Np),Np)+(rand(1)/Np); %the "comb"
    cmb(Np+1)=1;
    ip=1; mm=1;
    while(ip<=Np),
      if (cmb(ip)<acq(mm)),
        aux=apx(:,mm);
        px(:,ip)=aux+(sig.*rn(:,ip)); %roughening
        ip=ip+1;
      else
        mm=mm+1;
      end;
    end;
    %=============================================================
    %Results
    %estimated state (the particle mean)
    xe=sum(px,2)/Np;
    nn=nn+1;
end;
%-----------------------------------------------------
%display
figure(1)
subplot(1,2,1)
plot(tim,rx(1,1:Nf),'kx'); hold on;
plot(tim,rxe(1,1:Nf),'r');
title('altitude'); xlabel('seconds')
axis([0 Nf*T 0 12*10^4]);
subplot(1,2,2)
plot(tim,rx(2,1:Nf),'kx'); hold on;
plot(tim,rxe(2,1:Nf),'r');
title('velocity'); xlabel('seconds');
axis([0 Nf*T -6000 1000]);
figure(2)
subplot(1,2,1)
plot(tim,rer(1,1:Nf),'k');
title('altitude estimation error');
xlabel('seconds');
axis([0 Nf*T -3000 3000]);
subplot(1,2,2)
plot(tim,rer(2,1:Nf),'k');
title('velocity estimation error');
xlabel('seconds');
axis([0 Nf*T -500 500]);
figure(3)
subplot(1,2,1)
plot(tim,ry(1:Nf),'k');
title('distance measurement');
xlabel('seconds');
axis([0 Nf*T 0 12*10^4]);
```

```
subplot(1,2,2)
plot(tim,rd(1:Nf),'k');
title('drag');
xlabel('seconds');
axis([0 Nf*T 0 1000]);
```

B.1.9 Resampling Schemes. Algorithm Modifications (8.7.2.)

A typical histogram of weights, obtained in a experiment. Resampling is used to eliminate the particles with low importance weights and multiply particles with high importance weights (Figs. B.31, B.32, B.33 and B.34).

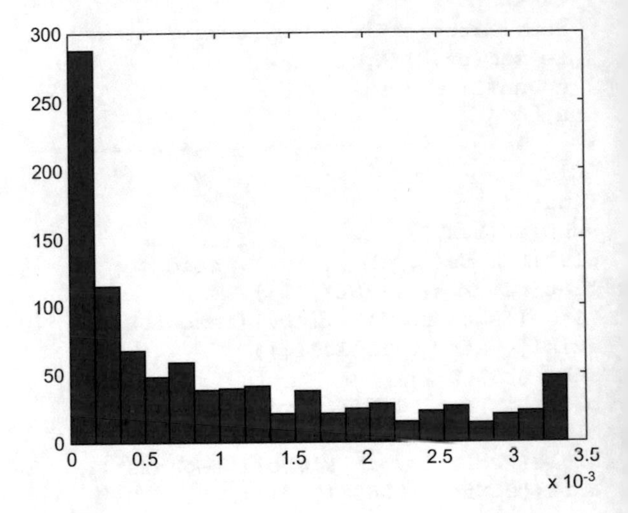

Fig. B.31 Histogram of weights, resampling example (Fig. 1.48)

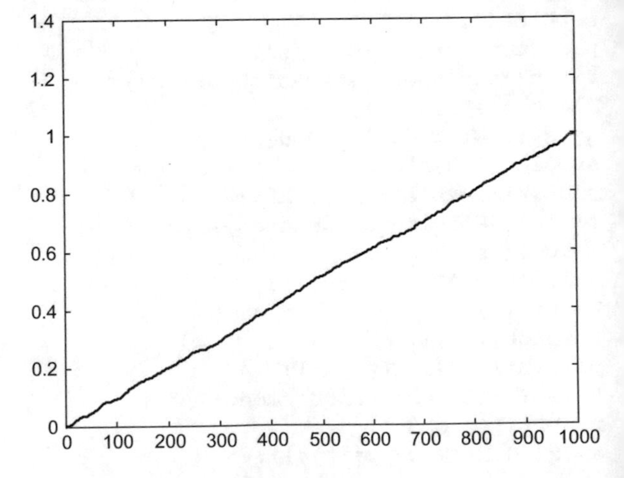

Fig. B.32 Cumsum() of weights, resampling example (Fig. 1.49)

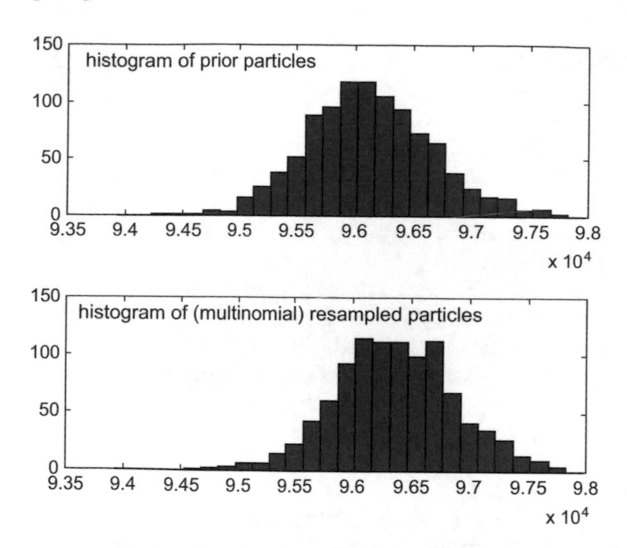

Fig. B.33 Histograms of prior and resampled particles (Fig. 1.51)

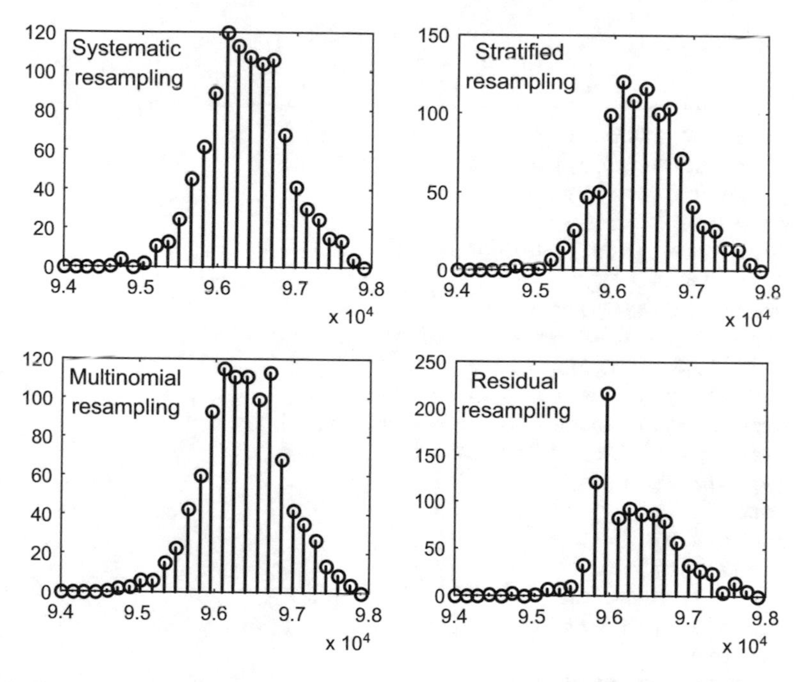

Fig. B.34 Histograms of systematic, stratified, multinomial, residual resampling example (Fig. 1.52)

Program B.10 Resampling methods

```
%Resampling methods
%Particle filter example
%Radar monitoring of falling body
randn('state',0); %initialization of random number generator
%------------------------------------------------------
%Prepare for the simulation of the falling body
T=0.4; %sampling period
g=-9.81;
rho0=1.225; %air density, sea level
k=6705.6; %density vs. altitude constant
L=100; %horizontal distance radar$<->$object
L2=L^2;
Nf=3; %maximum number of steps
tim=0:T:(Nf-1)*T; %time
%process noise
Sw=[10^5 0 0; 0 10^3 0; 0 0 10^2]; %cov
w11=sqrt(Sw(1,1)); w22=sqrt(Sw(2,2)); w33=sqrt(Sw(3,3));
w=[w11; w22; w33];
%observation noise
Sv=10^6; %cov.
v11=sqrt(Sv);
%------------------------------------------------------
%Prepare for filtering
x=[10^5; -5000; 400]; %initial state
xe=x; %initial estimation
%prepare particles
Np=1000; %number of particles
%reserve space
px=zeros(3,Np); %particles
apx=zeros(3,Np); %a priori particles
ny=zeros(1,Np); %particle measurements
vy=zeros(1,Np); %meas. dif.
pq=zeros(1,Np); %particle likelihoods
%reserve space for resampling posteriors
Spx=zeros(3,Np);
Mpx=zeros(3,Np);
Rpx=zeros(3,Np);
Fpx=zeros(3,Np);
%particle generation
wnp=randn(3,Np); %noise (initial particles)
for ip=1:Np,
  px(:,ip)=x+(w.*wnp(:,ip)); %initial particles
end;
%system noises
wx=randn(3,Nf); %process
wy=randn(1,Nf); %output
%------------------------------------------------------
%Behaviour of the system and the filter after initial state
```

```
nn=1;
while nn<Nf+1,
  %Simulation of the system
  %system
  rho=rho0*exp(-x(1)/k); %air density
  d=(rho*(x(2)^2))/(2*x(3)); %drag
  %next system state
  x(1)=x(1)+(x(2)*T);
  x(2)=x(2)+((g+d)*T);
  x(3)=x(3);
  x=x+(w.*wx(:,nn)); %additive noise
  %system output
  y=sqrt(L2+(x(1)^2))+(v11*wy(nn)); %additive noise
  ym=y; %measurement
  %Particle propagation
  wp=randn(3,Np); %noise (process)
  vm=randn(1,Np); %noise (measurement)
  for ip=1:Np,
    rho=rho0*exp(-px(1,ip)/k); %air density
    d=(rho*(px(2,ip)^2))/(2*px(3,ip)); %drag
    %next state
    apx(1,ip)=px(1,ip)+(px(2,ip)*T);
    apx(2,ip)=px(2,ip)+((g+d)*T);
    apx(3,ip)=px(3,ip);
    apx(:,ip)=apx(:,ip)+(w.*wp(:,ip)); %additive noise
    %measurement (for next state)
    ny(ip)=sqrt(L2+(apx(1,ip)^2))+(v11*vm(ip)); %additive noise
    vy(ip)=ym-ny(ip);
  end;
  %Likelihood
  %(vectorized part)
  %scaling
  vs=max(abs(vy))/4;
  ip=1:Np;
  pq(ip)=exp(-((vy(ip)/vs).^2));
  spq=sum(pq);
  %normalization
  pq(ip)=pq(ip)/spq;
  %Prepare for roughening
  A=(max(apx')-min(apx'))';
  sig=0.2*A*Np^(-1/3);
  rn=randn(3,Np); %random numbers
  %=============================================================
  %Resampling (systematic)
  acq=cumsum(pq);
  cmb=linspace(0,1-(1/Np),Np)+(rand(1)/Np); %the "comb"
  cmb(Np+1)=1;
  ip=1; mm=1;
  while(ip<=Np),
```

```
  if (cmb(ip)<acq(mm)),
    aux=apx(:,mm);
    Spx(:,ip)=aux+(sig.*rn(:,ip)); %roughening
    ip=ip+1;
  else
    mm=mm+1;
  end;
end;
%=============================================================
%Resampling (multinomial)
acq=cumsum(pq);
mm=1;
nr=sort(rand(1,Np)); %ordered random numbers (0, 1]
for ip=1:Np,
  while(acq(mm)<nr(ip)),
    mm=mm+1;
  end;
  aux=apx(:,mm);
  Mpx(:,ip)=aux+(sig.*rn(:,ip)); %roughening
end;
%=============================================================
%Resampling (residual)
acq=cumsum(pq);
mm=1;
%preparation
na=floor(Np*pq); %repetition counts
NR=sum(na); %total count
Npr=Np-NR; %number of non-repeated particles
rpq=((Np*pq)-na)/Npr; %modified weights
acq=cumsum(rpq); %for the monomial part
%deterministic part
mm=1;
for j=1:Np,
  for nn=1:na(j),
    Rpx(:,mm)=apx(:,j);
    mm=mm+1;
  end;
end;
%multinomial part:
nr=sort(rand(1,Npr)); %ordered random numbers (0, 1]
for j=1:Npr,
  while(acq(mm)<nr(j)),
    mm=mm+1;
  end;
  aux=apx(:,mm);
  Rpx(:,NR+j)=aux+(sig.*rn(:,j)); %roughening
end;
%=============================================================
%Resampling (stratified)
```

```
    acq=cumsum(pq);
    stf=zeros(1,Np);
    nr=rand(1,Np)/Np;
    j=1:Np;
    stf(j)=nr(j)+((j-1)/Np); %(vectorized code)
    stf(Np+1)=1;
    ip=1; mm=1;
    while(ip<=Np),
      if (stf(ip)<acq(mm)),
        aux=apx(:,mm);
        Fpx(:,ip)=aux+(sig.*rn(:,ip)); %roughening
        ip=ip+1;
      else
        mm=mm+1;
      end;
    end;
    %==========================================================
    px=Spx; %posterior (edit to select a resampling method)
    %Results
    %estimated state (the particle mean)
    xe=sum(px,2)/Np;
    nn=nn+1;
end;
%-----------------------------------------------------
%display
figure(1)
hist(pq,20);
title('histogram of weights');
figure(2)
plot(acq);
title('cumsum() of weights');
figure(3)
subplot(2,1,1)
bx=9.4e4:1.5e2:9.8e4;
hist(apx(1,:),bx);
title('histogram of prior particles')
subplot(2,1,2)
hist(Mpx(1,:),bx);
title('histogram of (multinomial) resampled particles');
figure(4)
Spt=hist(Spx(1,:),bx);
Mpt=hist(Mpx(1,:),bx);
Rpt=hist(Rpx(1,:),bx);
Fpt=hist(Fpx(1,:),bx);
subplot(2,2,1)
stem(bx,Spt,'k');
title('Systematic resampling')
subplot(2,2,2)
stem(bx,Fpt,'k');
```

```
title('Stratified resampling')
subplot(2,2,3)
stem(bx,Mpt,'k');
title('Multinomial resampling');
subplot(2,2,4)
stem(bx,Rpt,'k');
title('Residual resampling')
```

B.1.10 The Perspective of Numerical Integration (8.8.)

Next figure compares the results of the Laplace's method and the variational method for the approximation of the Student's T PDF (Fig. B.35).

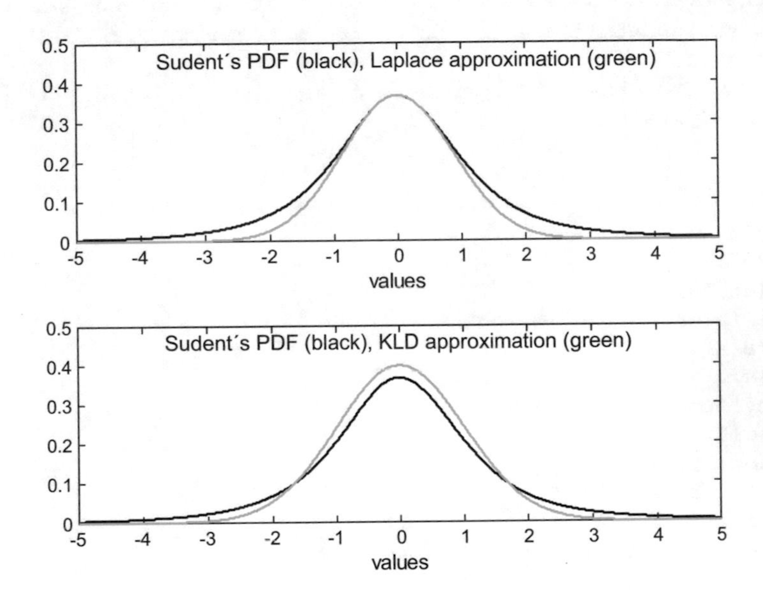

Fig. B.35 Approximations of Student's T PDF: (*top*) Laplace's method, (*bottom*) KLD minimization (Fig. 1.56)

Program B.11 Example of Laplace's and KLD approximations

```
%Example of Laplace's and KLD approximations
% Student's PDF
v=-5:0.01:5; %values set
N=length(v);
nu=3; %random variable parameter ("degrees of freedom")
ypdf=tpdf(v,nu); %chi-square PDF
py=max(ypdf); %the peak
%Laplace's (coincidence around the peak)
mE=1000; msig=0;
for sig=0.6:0.01:1,
  npdf=normpdf(v,0,sig); %normal PDF
  pn=max(npdf);
  Lpdf=(py/pn)*npdf;
  %error around the mode
  E=0; c=round(N/2);
  for nn=1:20,
    E=E+abs(ypdf(c+nn)-Lpdf(c+nn));
    E=E+abs(ypdf(c-nn)-Lpdf(c-nn));
  end;
  if E<mE, %take the minimum
    mE=E; msig=sig;
  end;
end;
npdf=normpdf(v,0,msig); %normal PDF
pn=max(npdf);
Lpdf=(py/pn)*npdf;
%search for min KLD
mKL=1000; msig=0;
for sig=0.6:0.01:1,
  xpdf=normpdf(v,0,sig);
  %KLD calculation
  KL=0;
  for nn=1:N,
    aux1=ypdf(nn); aux2=xpdf(nn);
    KL=KL+(aux2*log(aux2/aux1));
  end;
  if KL<mKL, %take the minimum
    mKL=KL; msig=sig;
  end;
end;
mpdf=normpdf(v,0,msig);
%display
figure(1)
subplot(2,1,1)
plot(v,ypdf,'k'); hold on; %plots figure
plot(v,Lpdf,'g'); %the approximation
axis([-5 5 0 0.5]);
xlabel('values');
```

```
title('Sudent's PDF (black), Laplace approximation (green)');
subplot(2,1,2)
plot(v,ypdf,'k'); hold on; %plots figure
plot(v,mpdf,'g'); %the approximation
axis([-5 5 0 0.5]);
xlabel('values');
title('Sudent's PDF (black), KLD approximation (green)');
mKL
```

B.1.11 Smoothing (8.10.)

Example of fixed-interval smoothing (Fig. B.36).

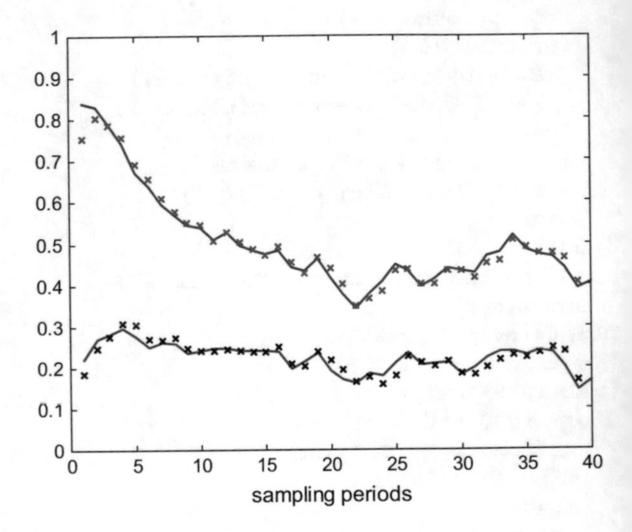

Fig. B.36 System estimated states (*cross marks*), and fixed-lagl smoothed states (*continuous*) (Fig. 1.59)

Program B.12 Fixed-lag smoothing example

```
%Fixed-lag smoothing example
L=5; %lag
%state space system model (2 tank system):
A1=1; A2=1; R1=0.5; R2=0.4;
cA=[-1/(R1*A1) 1/(R1*A1); 1/(R1*A2) -(1/A2)*((1/R1)+(1/R2))];
cB=[1/A1; 0]; cC=[1 0; 0 1]; cD=0;
Ts=0.1; %sampling period
%discrete-time model:
csys=ss(cA,cB,cC,cD); dsys=c2d(csys,Ts,'zoh');
[A,B,C,D]=ssdata(dsys); %retrieves discrete-time model matrices
%simulation horizon
Nf=40;
%process noise
Sw=[12e-4 0; 0 6e-4]; %cov
sn=zeros(2,Nf);
```

```
sn(1,:)=sqrt(Sw(1,1))*randn(1,Nf);
sn(2,:)=sqrt(Sw(2,2))*randn(1,Nf);
%observation noise
Sv=[6e-4 0; 0 15e-4]; %cov.
on=zeros(2,Nf);
on(1,:)=sqrt(Sv(1,1))*randn(1,Nf);
on(2,:)=sqrt(Sv(2,2))*randn(1,Nf);
% system simulation preparation
x=[1;0]; % state vector with initial tank levels
u=0.4; %constant input
% Kalman filter simulation preparation
%space for matrices
K=zeros(2,2); M=zeros(2,2); P=zeros(2,2);
xe=[0.5; 0.2]; % filter state vector with initial values
%space for recording xe(n),ym(n)
rxe=zeros(2,Nf-1);
rym=zeros(2,Nf);
rym(:,1)=C*x; %initial value
%behaviour of the system and the Kalman
% filter after initial state
% with constant input u
for nn=1:Nf-1,
  %system simulation
  xn=(A*x)+(B*u)+sn(nn); %next system state
  x=xn; %system state actualization
  ym=(C*x)+on(:,nn); %output measurement
  %Prediction
  xa=(A*xe)+(B*u); %a priori state
  M=(A*P*A')+ Sw;
  %Update
  K=(M*C')*inv((C*M*C')+Sv);
  P=M-(K*C*M);
  xe=xa+(K*(ym-(C*xa))); %estimated (a posteriori) state
  %recording xe(n),ym(n)
  rxe(:,nn)=xe;
  rym(:,nn+1)=ym;
end;
%Smoothing-----------------------------
% Smoothing preparation
N=zeros(2,2); P=zeros(2,2);
% augmented state vectors
axa=zeros(2*(L+1),1);
axp=zeros(2*(L+1),1);
% augmented input
bu=zeros(2*(L+1),1); bu(1:2,1)=B*u;
% augmented A matrix
aA=diag(ones(2*L,1),-2); aA(1:2,1:2)=A;
% augmented K
aK=zeros(2*(L+1),2);
```

```
% set of covariances
Pj=zeros(2,2,L);
%space for recording xs(n)
rxs=zeros(2,Nf-1);
jed=(2*L)+1; %pointer for last entries
%jed=1;
%action:
axa(1:2,1)=rxe(:,1); %initial values
for nn=1:Nf,
  M=(A*P*A')+Sw;
  N=(C*P*C')+Sv;
  ivN=inv(N);
  K=(A*P*C')*ivN;
  aK(1:2,:)=K;
  aK(3:4,:)=(P*C')*ivN;
  for jj=1:L,
    bg=1+(jj*2); ed=bg+1;
    aK(bg:ed,:)=(Pj(:,:,jj)*C')*ivN;
  end;
  aux=[A-K*C]';
  Pj(:,:,1)=P*aux;
  for jj=1:L-1,
    Pj(:,:,jj+1)=Pj(:,:,jj)*aux;
  end;
  axp=(aA*axa)+bu+aK*(rym(:,nn)-C*axa(1:2,1));
  P=M-(K*N*K');
  rxs(:,nn)=axp(jed:jed+1);
  axa=axp; %actualization (implies shifting)
end;
%-------------------------------------------------
% display of state evolution
figure(3)
plot(rxs(1,L:end),'r'); %plots xs1
hold on;
plot(rxs(2,L:end),'b'); %plots xs2
plot(rxe(1,:),'mx'); %plots xe1
plot(rxe(2,:),'kx'); %plots xe2
axis([0 Nf 0 1]);
xlabel('sampling periods');
title('Kalman filter states(x) and Smoothed states(-)');
```

Example of smoothing of state estimate at a fixed point (Figs. B.37 and B.38).

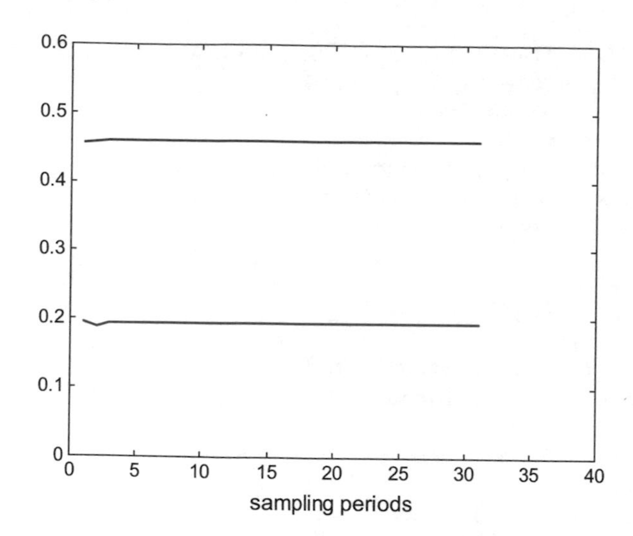

Fig. B.37 Smoothing of states with fixed-point smoothing (Fig. 1.60)

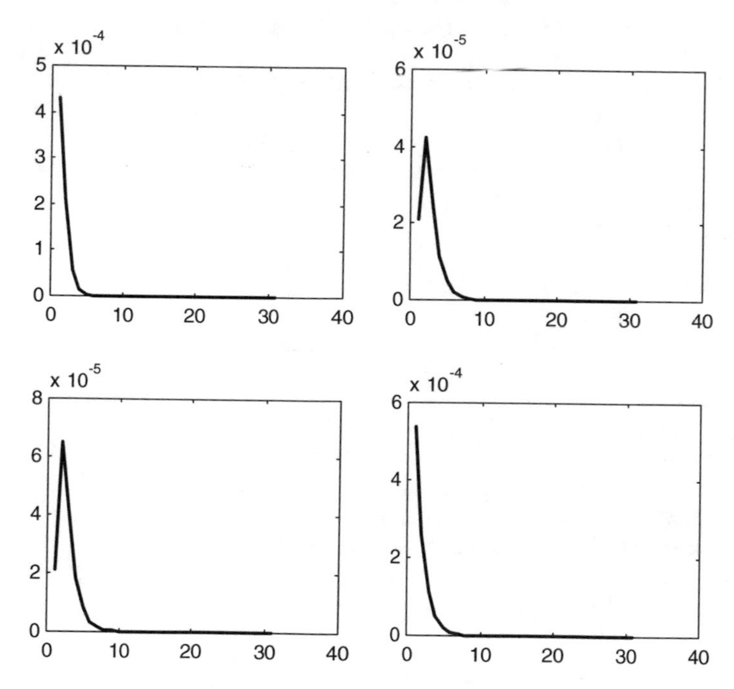

Fig. B.38 Evolution of covariances in the fixed-point smoothing example (Fig. 1.61)

Program B.13 Fixed-point smoothing example

```
%Fixed-point smoothing example
L=5; %lag
%state space system model (2 tank system):
A1=1; A2=1; R1=0.5; R2=0.4;
```

```
cA=[-1/(R1*A1) 1/(R1*A1); 1/(R1*A2) -(1/A2)*((1/R1)+(1/R2))];
cB=[1/A1; 0]; cC=[1 0; 0 1]; cD=0;
Ts=0.1; %sampling period
%discrete-time model:
csys=ss(cA,cB,cC,cD); dsys=c2d(csys,Ts,'zoh');
[A,B,C,D]=ssdata(dsys); %retrieves discrete-time model matrices
%simulation horizon
Nf=40;
%process noise
Sw=[12e-4 0; 0 6e-4]; %cov
sn=zeros(2,Nf);
sn(1,:)=sqrt(Sw(1,1))*randn(1,Nf);
sn(2,:)=sqrt(Sw(2,2))*randn(1,Nf);
%observation noise
Sv=[6e-4 0; 0 15e-4]; %cov.
on=zeros(2,Nf);
on(1,:)=sqrt(Sv(1,1))*randn(1,Nf);
on(2,:)=sqrt(Sv(2,2))*randn(1,Nf);
% system simulation preparation
x=[1;0]; % state vector with initial tank levels
u=0.4; %constant input
% Kalman filter simulation preparation
%space for matrices
K=zeros(2,2); M=zeros(2,2); P=zeros(2,2);
xe=[0.5; 0.2]; % filter state vector with initial values
%space for recording xe(n),ym(n),P
rxe=zeros(2,Nf-1);
rym=zeros(2,Nf);
rym(:,1)=C*x; %initial value
rP=zeros(2,2,Nf);
%behaviour of the system and the Kalman
% filter after initial state
% with constant input u
for nn=1:Nf-1,
  %system simulation
  xn=(A*x)+(B*u)+sn(nn); %next system state
  x=xn; %system state actualization
  ym=(C*x)+on(:,nn); %output measurement
  %Prediction
  xa=(A*xe)+(B*u); %a priori state
  M=(A*P*A')+ Sw;
  %Update
  K=(M*C')*inv((C*M*C')+Sv);
  P=M-(K*C*M);
  xe=xa+(K*(ym-(C*xa))); %estimated (a posteriori) state
  %recording xe(n),ym(n),P(n)
  rxe(:,nn)=xe;
  rym(:,nn+1)=ym;
  rP(:,:,nn+1)=P;
```

```
end;
%Smoothing-----------------------------
% Smoothing preparation
Nfix=10; %the fixed point
%space for matrices
N=zeros(2,2); P11=zeros(2,2); P21=zeros(2,2);
% augmented state vectors
axa=zeros(4,1);
axp=zeros(4,1);
% augmented input
bu=zeros(4,1); bu(1:2,1)=B*u;
% augmented A matrix
aA=diag(ones(4,1)); aA(1:2,1:2)=A;
% augmented K
aK=zeros(4,2);
%space for recording xs(Nfix), P11(n)
rxs=zeros(2,Nf);
rP11=zeros(2,2,Nf);
%action:
P11=rP(:,:,Nfix); P21=P11; %initial values
axa(1:2,1)=rxe(:,Nfix); %initial values
axa(3:4,1)=rxe(:,Nfix); %initial values
for nn=Nfix:Nf,
   M=(A*P11*A')+Sw;
   N=(C*P11*C')+Sv;
   ivN=inv(N);
   K=(A*P11*C')*ivN;
   Ka=(P21*C')*ivN;
   aK(1:2,:)=K; aK(3:4,:)=Ka;
   axp=(aA*axa)+bu+aK*(rym(:,nn)-C*axa(1:2,1));
   axa=axp; %actualization
   rP21(:,:,nn)=P21; %recording
   rxs(:,nn)=axp(3:4,1);
   P11=M-(K*N*K');
   P21=P21*(A-(K*C))';
end;
%-----------------------------------------------------
% display of smoothed state at Nfix
figure(3)
plot(rxs(1,Nfix:end),'r'); %plots xs1
hold on;
plot(rxs(2,Nfix:end),'b'); %plots xs2
axis([0 Nf 0 0.6]);
xlabel('sampling periods');
title('State smoothing at Nfix');
% display of Covariance evolution
figure(4)
subplot(2,2,1)
plot(squeeze(rP21(1,1,Nfix:end)),'k');
```

```
title('Evolution of covariance');
subplot(2,2,2)
plot(squeeze(rP21(1,2,Nfix:end)),'k');
subplot(2,2,3)
plot(squeeze(rP21(2,1,Nfix:end)),'k');
subplot(2,2,4)
plot(squeeze(rP21(2,2,Nfix:end)),'k');
```

B.2 Chapter 2: Sparse Representations

B.2.1 Patches. The K-SVD Method (9.4.2.)

Next three figures correspond to an example of image denoising using K-SVD (Figs. B.39, B.40 and B.41).

Fig. B.39 Original picture, and image with added Gaussian noise (Fig. 2.21)

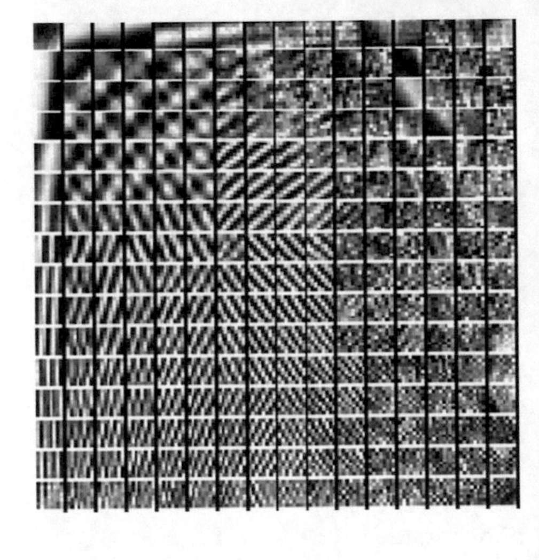

Fig. B.40 Patch dictionary obtained with K-SVD (Fig. 2.22)

Fig. B.41 Denoised image (Fig. 2.23)

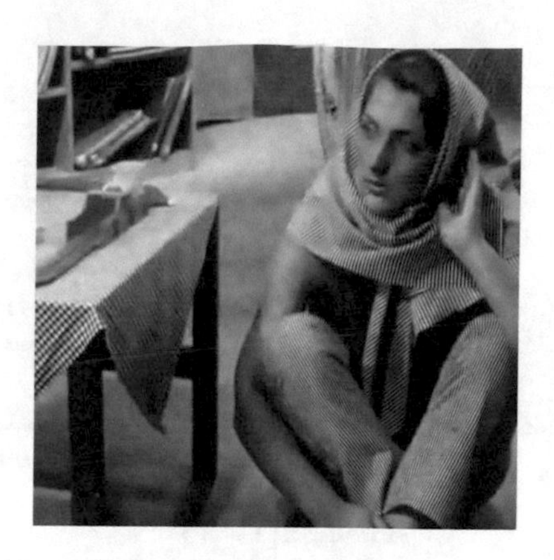

Program B.14 K-SVD denoising example

```
%K-SVD denoising example
% (using 8x8 patches)
p=imread('barbara.png');
[fc,fr]=size(p);
F0=im2double(p);
F0=F0*255;
sig=25; %noise var
etg=1.2*sig; %error target
Fn=F0+sig*randn(size(p)); %noisy image
K=256; %number of atoms in the final dictionary
%obtain from noisy image a collection of many patches
disp('obtaining many patches');
Npa=50000; %number of patches (you can add more)
PAM=zeros(64,Npa); %space for the matrix of patches
rPm=randperm(prod([fc,fr]-7)); %random permutation
spa=rPm(1:Npa); %selected patches
for nn=1:Npa,
  [r,c]=ind2sub([fc,fr]-7,spa(nn));
  aux=Fn(r:r+7,c:c+7); %a patch
  PAM(:,nn)=aux(:); %convert to column, insert into matrix
end;
%compute an initial dictionary
aux=ceil(sqrt(K));
Di=zeros(8,aux);
for nn=0:aux-1,
  S=cos([0:1:7]'*nn*pi/aux);
  if nn>0, S=S-mean(S); end;
  Di(:,nn+1)=S/norm(S);
end;
Di=kron(Di,Di); %obtain a 64xK matrix
%dictionary normalization
Di=Di*diag(1./sqrt(sum(Di.*Di)));
Di=Di.*repmat(sign(Di(1,:)),size(Di,1),1);
```

```
%Apply KSVD-----------------------------------------------
disp('apply K-SVD...please wait')
for kk=1:4, %4 KSVD iterations (you can add more)
  CFM=spc(Di,PAM,etg); %sparse coding
  %search for better dictionary elements (bDE)
  pr=randperm(K);
  for in=pr,
    %data indices that use jth dictionary elements
    rDI=find(CFM(in,:)); %relevant data indices
    if(length(rDI)<1),
      %when there are no such data indices
      aux=PAM-Di*CFM; aux2=sum(aux.^2);
      [d,i]=max(aux2);
      bDE=PAM(:,i);
      bDE=bDE./(sqrt(bDE'*bDE));
      bDE=bDE.*sign(bDE(1));
      CFM(in,:)=0;
    else
      %when there are such data indices
      Maux=CFM(:,rDI);
      Maux(in,:)=0; %elements to be improved
      ers=PAM(:,rDI)-Di*Maux; %vector of errors to minimize
      [bDE,SV,bV]=svds(ers,1); %the SVD
      CFM(in,rDI)=SV*bV'; %use sign of first element
    end;
    Di(:,in)=bDE; %insert better dictionary element;
  end;
  disp(['iteration: ',num2str(kk)]);
end;
%clean dictionary
B1=3; B2=0.99;
%removing of identical atoms
er=sum((PAM-Di*CFM).^2,1);
Maux=Di'*Di; Maux=Maux-diag(diag(Maux));
for i=1:K,
  aux=length(find(abs(CFM(i,:))>1e-7));
  if (max(Maux(i,:))>B2) || (aux<=B1),
    [v,ps]=max(er); er(ps(1))=0;
    Di(:,i)=PAM(:,ps(1))/norm(PAM(:,ps(1)));
    Maux=Di'*Di;Maux=Maux-diag(diag(Maux));
  end;
end;
disp('dictionary ready')
%Image denoising-------------------------------------------
Fout=zeros(fc,fr); %prepare space for denoised image
wgt=zeros(fc,fr); %weights
bks=im2col(Fn,[8,8],'sliding');
nub=size(bks,2); %number of blocks
```

```
ix=[1:nub];
disp('compute coefficients for denoising')
%Proceed with sets of 25000 coefficients
for nj=1:25000:nub,
  jsz=min(nj+25000-1,nub); %jump size
  cf= spc(Di,bks(:,nj:jsz),etg); %coefficients (sparse coding)
  bks(:,nj:jsz)=Di*cf;
  disp(['subset: ',num2str(nj),'-',num2str(jsz)]);
end;
disp('start denoising');
nn=1;
[r,c]=ind2sub([fc,fr]-7,ix);
for j=1:length(c),
  ic=c(j); ir=r(j);
  bk=reshape(bks(:,nn),[8,8]); %a block
  Fout(ir:ir+7,ic:ic+7)=Fout(ir:ir+7,ic:ic+7)+bk;
  wgt(ir:ir+7,ic:ic+7)=wgt(ir:ir+7,ic:ic+7)+ones(8);
  nn=nn+1;
end;
%combine with noisy image
Fd = (Fn+0.034*sig*Fout)./(1+0.034*sig*wgt);
%Result-----------------------------------------------------
disp('result display');
figure(1)
subplot(1,2,1)
imshow(F0,[]);
title('original picture')
subplot(1,2,2)
imshow(Fn,[]);
title('noisy image')
figure(2)
imshow(Fd,[])
title('denoised image')
figure(3)
cD=zeros(9,9,1,K); %collection of Dictionary patches
for np=1:K,
  ni=1;
  for j=1:8,
    for i=1:8,
      cD(i,j,1,np)=Di(ni,np);
      ni=ni+1;
    end;
    cD(9,j,1,np)=1; %white line
  end;
  %patch contrast augmentation
  xx=1:8;
  cD(xx,xx,1,np)=cD(xx,xx,1,np)-min(min(cD(xx,xx,1,np)));
  aux=max(max(cD(xx,xx,1,np)));
  if aux>0,
```

```
    cD(xx,xx,1,np)=cD(xx,xx,1,np)./aux;
  end;
end;
montage(cD);
title('patch dictionary');
```

Here is the function employed by the previous program.

Function B.15 spc

```
function [A]=spc(D,X,erg);
%sparse coding (OMP)
[n,P]=size(X); [n,K]=size(D);
e2 = erg^2*n;
mnc = n/2; %maximum number of coefficients
A = sparse(K,P); %sparse matrix
for mk=1:P,
  aux=[]; inx=[];
  x=X(:,mk); resd=x;
  rno2 = sum(resd.^2); %residual norm 2
  mj = 0;
  while rno2>e2 & mj < mnc,
    mj = mj+1;
    proj=D'*resd;
    ps=find(abs(proj)==max(abs(proj)));
    ps=ps(1); inx(mj)=ps;
    aux=pinv(D(:,inx(1:mj)))*x;
    resd=x-D(:,inx(1:mj))*aux;
    rno2 = sum(resd.^2);
  end;
  if (length(inx)>0), A(inx,mk)=aux; end;
end;
return;
```

B.2.2 Bregman Related Algorithms (9.5.2.)

Next figures shows an example of ROF-TV anisotropic denoising. The picture on the left hand side has salt and pepper noise. The denoised picture is shown on the right hand side (Fig. B.42).

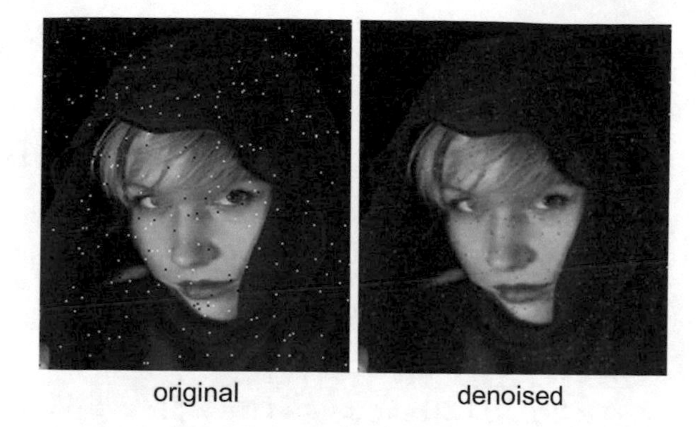

Fig. B.42 ROF total variation denoising using split Bregman (Fig. 2.37)

Program B.16 Example of picture denoising using Anisotropic TV

```
%Example of picture denoising using Anisotropic TV
% and Split Bregman
%
lambda=0.2; mu=0.1;
P=imread('face1.jpg'); %read image
A=imnoise(P,'salt & pepper',0.01); %add salt & pepper noise
u=double(A); uo=u; un=u;
[ly,lx]=size(u);
%initialization of variables
lx1=lx-1; ly1=ly-1;
dx=zeros(ly,lx); dy=zeros(ly,lx);
bx=zeros(ly,lx); by=zeros(ly,lx);
s=zeros(ly,lx); as=zeros(ly,lx);
g=zeros(ly,lx);
KL=lambda/(mu+(4*lambda)); KM=mu/(mu+(4*lambda));
%main loop
for niter=1:3,
  %solve first equation with Gauss-Seidel
  for i=2:ly1,
    j=2:lx1;
    un(i,j)=(KM*uo(i,j))+(KL*(u(i+1,j)+u(i-1,j)...
    + u(i,j+1)+u(i,j-1)...
    +dx(i-1,j)-dx(i,j)+dy(i,j-1)-dy(i,j)-...
    -bx(i-1,j)+bx(i,j)-by(i,j-1)+by(i,j)));
  end;
  u=un;
  %obtain the s -------------------
  for i=2:ly1,
    j=2:lx1;
    s(i,j)=sqrt(abs((u(i+1,j)-u(i-1,j))/2+bx(i,j)).^2...
    +abs((u(i,j+1)-u(i,j-1))/2+by(i,j)).^2);
  end;
  for i=1:ly1,
```

```
  j=1; s(i,j)=sqrt(abs((u(i+1,j)-u(i,j))+bx(i,j))^2...
  +abs((u(i,j+1)-u(i,j))+by(i,j))^2);
end;
for j=1:lx1,
  i=1; s(i,j)=sqrt(abs((u(i+1,j)-u(i,j))+bx(i,j))^2...
  +abs((u(i,j+1)-u(i,j))+by(i,j))^2);
end;
for i=2:ly,
  j=lx; s(i,j)=sqrt(abs((u(i,j)-u(i-1,j))+bx(i,j))^2...
  +abs((u(i,j)-u(i,j-1))+by(i,j))^2);
end;
for j=2:lx,
  i=ly; s(i,j)=sqrt(abs((u(i,j)-u(i-1,j))+bx(i,j))^2...
  +abs((u(i,j)-u(i,j-1))+by(i,j))^2);
end;
%obtain the dx, dy -------------------
ls=lambda*s; as=ls+1;
for i=2:ly1,
  for j=1:lx,
    dx(i,j)=(ls(i,j).*((u(i+1,j)-u(i-1,j))/2+bx(i,j)))...
    /as(i,j);
  end;
end;
for j=1:lx,
  i=1; dx(i,j)=(ls(i,j)*((u(i+1,j)-u(i,j))+bx(i,j)))/as(i,j);
end;
for j=1:lx,
  i=ly; dx(i,j)=(ls(i,j)*((u(i,j)-u(i-1,j))+bx(i,j)))...
  /as(i,j);
end;
for i=1:ly,
  for j=2:lx1,
    dy(i,j)=(ls(i,j).*((u(i,j+1)-u(i,j-1))/2+by(i,j)))...
    /as(i,j);
  end;
end;
for i=1:ly,
  j=1; dy(i,j)=(ls(i,j)*((u(i,j+1)-u(i,j))+by(i,j)))/as(i,j);
end;
for i=1:ly,
  j=lx; dy(i,j)=(ls(i,j)*((u(i,j)-u(i,j-1))+by(i,j)))...
  /as(i,j);
end;
%obtain the bx, by -------------------
for i=2:ly1,
  j=1:lx;
  bx(i,j)=bx(i,j)+((u(i+1,j)-u(i-1,j))/2-dx(i,j));
end;
for j=1:lx,
```

```
    i=1; bx(i,j)=bx(i,j)+((u(i+1,j)-u(i,j))-dx(i,j));
  end;
  for j=1:lx,
    i=ly; bx(i,j)=bx(i,j)+((u(i,j)-u(i-1,j))-dx(i,j));
  end;
  for i=1:ly,
    j=2:lx1;
    by(i,j)=by(i,j)+((u(i,j+1)-u(i,j-1))/2-dy(i,j));
  end;
  for i=1:ly,
    j=1; by(i,j)=by(i,j)+((u(i,j+1)-u(i,j))-dy(i,j));
  end;
  for i=1:ly,
    j=lx; by(i,j)=by(i,j)+((u(i,j)-u(i,j-1))-dy(i,j));
  end;
end;
%display
figure(1)
subplot(1,2,1)
imshow(A);
title('ROF-TV denoising using Split Bregman');
xlabel('original');
subplot(1,2,2)
imshow(uint8(un));
xlabel('denoised')
```

Index

J.M. Giron-Sierra, *Digital Signal Processing with Matlab Examples, Volume 3*,
Signals and Communication Technology, DOI 10.1007/978-981-10-2540-2

Printed in the United States
By Bookmasters